Metamaterial for Planar Antenna

Editor
Mohammad Tariqul Islam
Professor
Department of Electrical, Electronic and Systems Engineering
Universiti Kebangsaan Malaysia (UKM), Malaysia

CRC Press
Taylor & Francis Group
Boca Raton London New York

CRC Press is an imprint of the
Taylor & Francis Group, an **informa** business
A SCIENCE PUBLISHERS BOOK

First edition published 2025
by CRC Press
2385 NW Executive Center Drive, Suite 320, Boca Raton FL 33431

and by CRC Press
4 Park Square, Milton Park, Abingdon, Oxon, OX14 4RN

CRC Press is an imprint of Taylor & Francis Group, LLC

Library of Congress Cataloging-in-Publication Data (applied for)

ISBN: 978-1-032-83281-4 (hbk)
ISBN: 978-1-032-85132-7 (pbk)
ISBN: 978-1-003-51668-2 (ebk)

DOI: 10.1201/9781003516682

Typeset in Times New Roman
by Prime Publishing Services

Preface

Metamaterials have emerged as a transformative technology in the field of planar antenna design, enabling unprecedented control over electromagnetic waves and facilitating remarkable advancements in communication systems. This book, *Metamaterial for Planar Antenna*, is dedicated to exploring the design and integration of metamaterial structures with planar antennas, offering innovative solutions for the challenges faced in modern wireless communication technologies, including 5G and beyond. With a focus on theoretical foundations, practical designs, and experimental validations, this book provides a comprehensive roadmap for researchers and engineers working at the intersection of antenna design and metamaterials. The book is organized into nine key chapters, addressing a crucial aspect of metamaterial design and applications in planar antennas.

The **first chapter** introduces metamaterial-based frequency-selective surfaces (FSS) specifically designed for 5G applications. These structures are known for their versatility. The chapter details the challenges of designing miniaturized FSS, ensuring incident angle stability, and achieving conformality. By covering a wide frequency spectrum and providing robust stopband and passband performance, these FSS designs cater to high-performance EMI shielding and antenna gain enhancement, paving the way for next-generation 5G and millimeter-wave (mmWave) applications.

In the **second chapter**, polarization-insensitive dual-band meta-surfaces are discussed in depth. These structures overcome significant challenges in antenna and meta-surface design by maintaining consistent performance regardless of the polarization of incident electromagnetic waves. The chapter delves into the design methodologies, parametric analysis, and experimental validations of meta-surfaces capable of operating efficiently across multiple frequency bands. Such innovations are particularly crucial for applications requiring robust and reliable performance in diverse environments, including wireless communication systems, and EMI shielding.

Metamaterial structures loaded with one or more splits can cause short wire- like even or odd modes of resonances at low frequency regions. Unlike the half-wave length resonators, they can give resonance with the size of one tenth of the operating wavelength. These resonators can be utilized to form compact antennas with a proper feeding network. One drawback is that the impedance bandwidth of the resonances is very poor. A promising solution to this problem is to defect the metallic ground plane. Defected ground structure (DGS) can cause additional even or odd modes of resonances. In the **third chapter,** an optimization of the desired dimensions can place these resonances in such a way that the bandwidth becomes the maximum. Overall, these structures can effectively nurture the next generation compact antenna applications by adopting some bandwidth enhancement techniques without additional circuitry like DGS.

The **fourth chapter** focuses on high-refractive-index metamaterials designed to enhance radiation aggregation and antenna gain. A detailed exploration of metamaterial unit cell designs, broadband dipole antenna integration, and metamaterial incorporation provides readers with a solid foundation in designing compact and efficient antennas. This chapter highlights the potential of metamaterials to improve antenna performance in terms of bandwidth, gain, and directivity while addressing fabrication challenges and experimental validations. These findings are particularly significant for applications requiring high-performance antennas in compact and space-constrained environments.

The demand for compact, efficient, and high-performance antennas has grown exponentially with advancements in modern wireless communication systems. Among various antenna types, multiband patch antennas have gained prominence for their simplicity, low profile, and capability to operate across multiple frequency bands. However, their inherent limitations in gain and bandwidth often necessitate innovative solutions to meet evolving communication standards. The **fifth chapter** explores the integration of near-zero index metamaterials (NZIMs) as a transformative approach to overcoming these challenges. NZIMs, with their unique electromagnetic properties, enable enhanced wave manipulation, resulting in improved antenna gain and directivity. Through comprehensive theoretical analysis, simulation, and experimental validation, the chapter underscores the potential of NZIMs in revolutionizing multiband patch antenna designs. By addressing key performance bottlenecks, this work aims to contribute to the development of more efficient and adaptable wireless communication systems.

5G communication is an unprecedented breakthrough in wireless communications with faster information rate, shorter latency, higher reliability, and channel capacity. 5G sub-6 GHz spectrums are the most popular and easy to implement with existing technology, which serves an excellent balanced connectivity with high coverage and capacity. The **sixth chapter** explores Multiple-input multiple-output (MIMO) antennas are crucial enablers of 5G communications that enhance data transmission rate, spectrum utilization, channel quality, and reliability. However, mutual coupling among compact MIMO antennas is a crucial limitation, which significantly deteriorates the overall antenna performance as well as channel capacity and reliability. Thus, a highly isolated compact MIMO antenna is profoundly desirable for 5G communications. Metamaterials introduced a new paradigm for mitigating mutual coupling in MIMO antennas. The innovative T-structured metamaterial shield can enhance isolation significantly between two-port compact 5G MIMO antennas. Moreover, this tailored metamaterial shield provides a very low envelope correlation coefficient.

The **seventh chapter** explores the fast-growing demand for compact, frequency-flexible, and high information rate 5G smart systems; designing and integrating multiple antennas in a compact MIMO system is very challenging. Moreover, miniature and simple antenna geometry is highly desirable for enhancing MIMO antenna elements within the limited space. The inclusion of parasitic metamaterial into the antenna substrate leads to miniaturization and improves antenna characteristics. Also, the metamaterial superstrate can enhance the antenna operating spectrum, gain, isolation, and efficiency. MIMO antenna elements enhancement flexibility strategy has the potential for facilitating massive MIMO applications, leading the way in 6G communications. By bridging the parasitic and superstrate metamaterial in MIMO antenna technology, this study has the prospect of shaping the future of MIMO antenna technology.

The **eighth chapter** is dedicated to metamaterial-enabled beam-tilting mechanisms for mmWave antenna applications. The chapter presents a thorough analysis of polarization-sensitive triple-band metamaterials and their integration with beamforming antennas. By addressing challenges such as incident angle sensitivity, gain enhancement, and directional beam control, this chapter offers novel designs for advanced beam-tilting applications. These innovations are particularly important for 5G and mmWave systems, where precise beam alignment is essential for achieving optimal communication performance in dense urban environments.

The exploration of metamaterials has opened new frontiers in manipulating electromagnetic properties, enabling groundbreaking advancements in stealth, sensing, and communication technologies. In the **ninth chapter**, by leveraging the unique capabilities of tailored structural designs, such as dual T-shaped resonators, metamaterials provide exceptional control over absorption characteristics, particularly in the terahertz frequency range. This work emphasizes the integration of advanced computational techniques, notably Random Forest regression, to streamline the design process, significantly reducing simulation time and computational resources. Through precise hyperparameter optimization and robust validation, this approach highlights the potential of machine learning in revolutionizing material design workflows. Furthermore, the practical implications extend beyond absorber performance, as integrating these innovations with antenna optimization techniques enhances the overall functionality of terahertz systems. By bridging theoretical and experimental frameworks, this study demonstrates how the synergy between metamaterials and computational methodologies can shape the future of high-performance terahertz technologies.

Through these chapters, the book emphasizes practical and scalable solutions to address real-world challenges in planar antenna design. The integration of metamaterials enables the development of compact, lightweight, and highly efficient antennas suitable for a wide range of applications, from 5G communication to satellite systems, wearable devices, and IoT-enabled platforms. By bridging the gap between theoretical insights and practical implementations, the book provides a robust foundation for advancing the field of metamaterial-loaded antennas.

This work would not have been possible without the dedicated efforts of my collaborators, researchers, and students, whose contributions have enriched the content of this book. I am deeply grateful to the institutions, funding agencies, and organizations that supported this research, enabling the realization of novel designs and experiments. I also extend my sincere appreciation to the reviewers and publishers for their valuable feedback and guidance throughout the preparation of this manuscript.

It is my hope that this book serves as a valuable resource for researchers, engineers, and students working in the fields of electromagnetic design and wireless communication. I am confident that the insights shared in this work will inspire further advancements in metamaterials and planar antenna technology, shaping the future of next-generation communication systems.

Contents

List of Contributors

Md. Mhedi Hasan

Department of Electrical, Electronic and Systems Engineering, Faculty of Engineering and Built Environment, Universiti Kebangsaan Malaysia (UKM), 43600 UKM Bangi, Selangor, Malaysia.

Md. Samsuzzaman

Department of Computer and Communication Engineering, Patuakhali Science and Technology University, Patuakhali, Bangladesh.

Mohammad Lutful Hakim

Department of Electrical, Electronic and Systems Engineering, Universiti Kebangsaan Malaysia, Bangi, Selangor, Malaysia.

Mohammad Tariqul Islam

Department of Electrical, Electronic and Systems Engineering, Universiti Kebangsaan Malaysia, Bangi, Selangor, Malaysia.

Prince Jain

Department of Mechatronics Engineering, Parul University, Vadodara, Gujarat, India.

Touhidul Alam

Pusat Sains Angkasa (ANGKASA), Institut Perubahan Iklim, Universiti Kebangsaan Malaysia, 43600 UKM Bangi, Selangor, Malaysia.

CHAPTER 1

Metamaterial Frequency Selective Surface for 5G Applications

Mohammad Tariqul Islam[1*], *Mohammad Lutful Hakim*[2] and *Touhidul Alam*[3]

1. Introduction

Over the past two decades, composite materials have drawn significant attention from researchers. These composite structures are also known as **metamaterials**. Engineers and scientists find it interesting to create structures that allow them to control their electromagnetic response freely due to their widespread applications, such as multiband reflectors in curved secondary mirrors, multiband feed systems, and spatial filters employed in side-emission radar and planar metamaterial or frequency selective surface **(FSS)**. The FSS are arrays of conducting components frequently printed on dielectric substrates with one or more support layers, allowing or reflecting electromagnetic (EM) waves at the frequency of the array resonance and can serve as filters. An array of metallic elements illuminated by an EM wave produces an electric current in the conductors, whose amplitude depends on how strongly the wave and the elements are coupled. The coupling is strongest for linear elements like dipoles at the resonance frequency when element length $l = \lambda/2$ [1, 2]. The coupling is affected by the distance between the elements, their polarisation, and the incident angle of the illuminating wave with regard to the element surface.

This chapter studies the basic concepts of metamaterials and summarises the relevant theoretical basics for FSS designs. It also provides a revision of the research areas of this work and aims to: furnish the current trend of research on the applications of FSS on single-band and multi-band; summarise all the basic required knowledge

[1,2] Department of Electrical, Electronic and Systems Engineering, Universiti Kebangsaan Malaysia, Bangi, Selangor, Malaysia.

[3] Pusat Sains Angkasa (ANGKASA), Institut Perubahan Iklim, Universiti Kebangsaan Malaysia, 43600 UKM Bangi, Selangor, Malaysia.

* Corresponding author: tariqul@ukm.edu.my, lutful@ukm.edu.my

for the design of FSS; and the challenge of wireless communication at 5G mmWave band. It also highlights the scope of utilisation of FSS at this frequency to improve the system performance; the applications of FSS for EMI shielding and antenna performance enhancement; and review the subject to understand this work's current challenges and contributions and build a solid base for the analysis. Moreover, works on tuneable metamaterials are also reviewed and a summary is made of some of those works. Based on the findings of this chapter, new designs will be initiated targeting to overcome some limitations of the existing works along with new features.

1.1 Millimetre Wave 5G Frequency Communications

The mmWave fifth-generation (5G) technology is exponentially growing to fulfil the demands of high data rates with low latency and inadequate bandwidth in the **5G** microwave frequency (sub-6 GHz) [3]. The advanced communication system will be implemented in millimetres wave frequency region [4]. Thanks to key 5G innovations like the flexible framework, self-contained time division duplex subframe, massive **multiple-input multiple-output (MIMO)**, mobilizing millimetre wave (mmWave), network slicing, and more, 5G new radio (NR) is planned to provide significantly more network capacity, higher user throughput, more uniform experience, lower latency, better reliability, and increased network flexibility [5]. Based on Q2, 2021 analysis by Ookla of Speedtest Intelligence, average 5G mmWave download speeds in the US are more than 38x faster than those of 4G LTE.

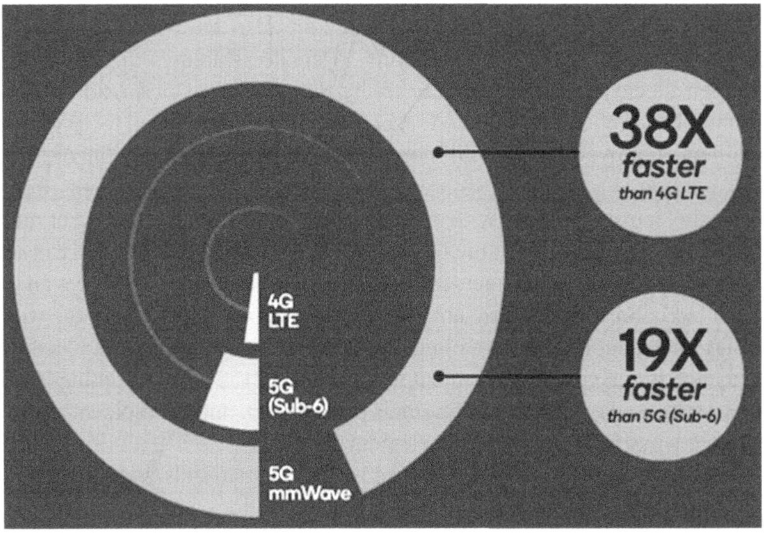

Fig. 1 Advantage of 5G mmWave
Source: Qualcomm, 2021.

It also enables average US mmWave 5G download speeds that are even 19x faster than those of sub-6 GHz 5G. The advantages of the 5G **mmWave** communication are shown in Figure 1[6]. The standardisation organisation, telecom companies, and government authorities are working on deploying and standardising mmWave 5G communications systems. The 26 GHz, 28 GHz, 37 GHz, and 38 GHz frequencies are allocated by different countries for 5G NR [7–9]. The Malaysian Communications and Multimedia Commission (MCMC) have been allocated 26 and 28 GHz for mmWave **5G** communications [10]. The frequency allocation by the different countries for mmWave 5G communications is presented in Figure 2, and the different allocated frequency band with uplink/downlink is shown in Figure 3. The 5G mmWave communications are expected to revolutionise daily life through its excellent connectivity with modern technology like smart factories [11], automotive cars [12], smart homes [13], telemedicine, virtual reality [14], etc. [15–17]. Figure 4 shows a few examples of **mmWave** 5G application fields. As a result, **antenna** development in mmWave frequency has gained significant importance in recent years [18].

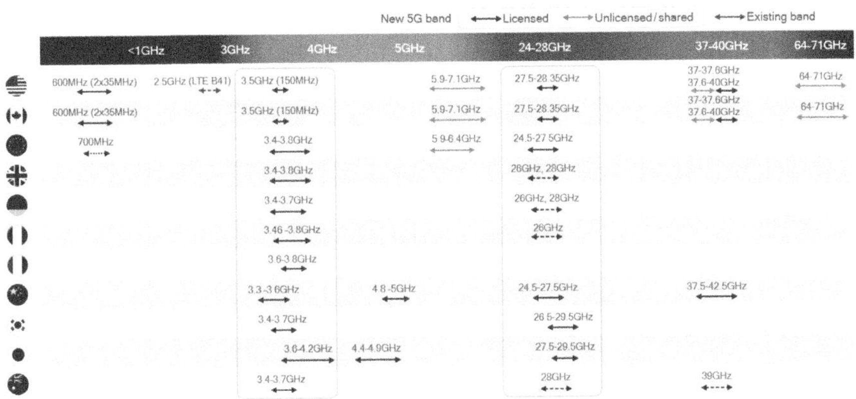

Malaysia 5G frequency band by MCMC

700 (MHz), 2300 MHz, 2600 MHz, 3.5 (GHz) and 26/28 GHz

Fig. 2 Global snapshot of allocated mmWave 5G spectrum.
Source: Qualcomm Technologies, 2017.

Band	Frequency (GHz)	Uplink/downlink (GHz)
n257	28	26.50 – 29.50
n258	26	24.25 – 27.50
n259	41	39.50 – 43.50
n260	39	37.00 – 40.00
n261	28	27.50 – 28.35

Fig. 3 5G NR mmWave band.
Source: 3GPP, 2023.

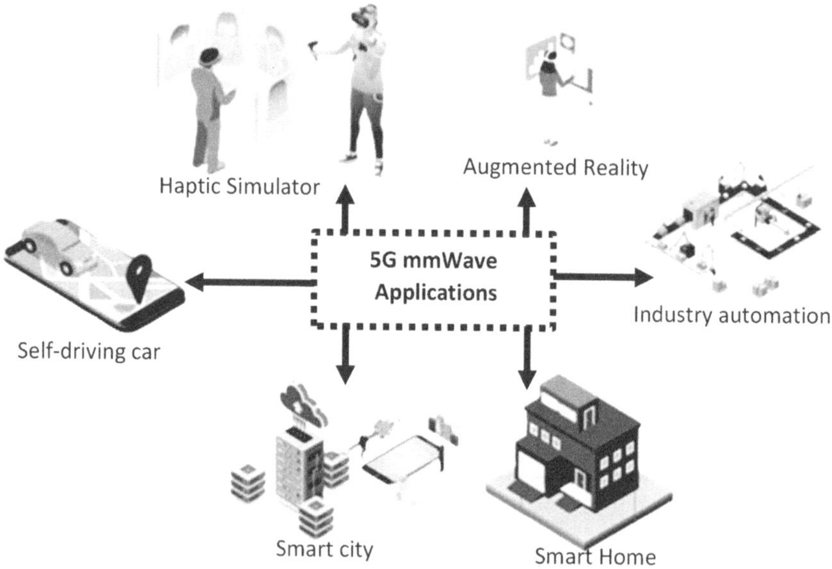

Fig. 4 Example of 5G application fields
Source: Zhang et al., 2022.

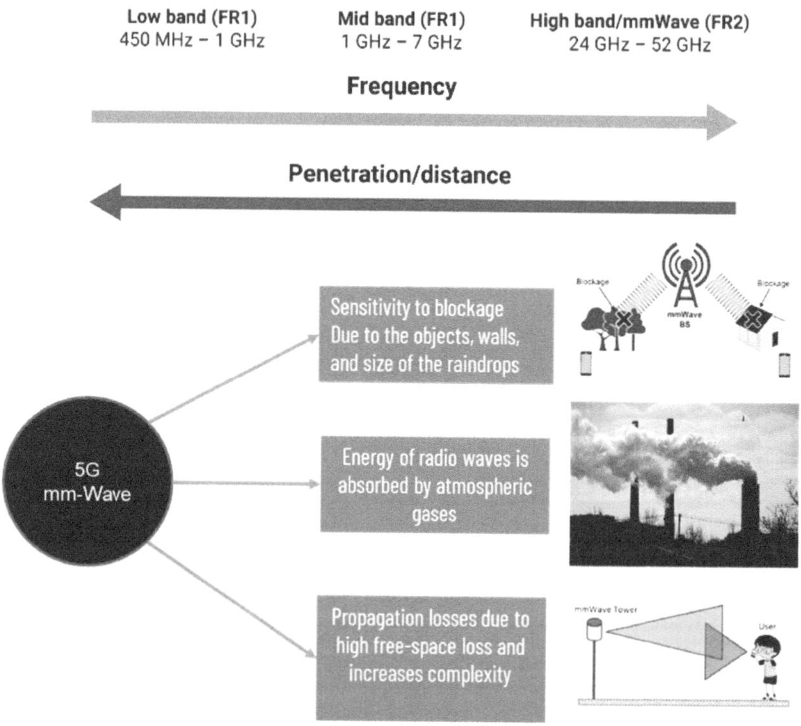

Fig. 5 Challenges in mmWave 5G spectrum.

Figure 5 shows challenges in **mmWave** frequency. This indicates that mmWave signals have a limited range and are easily attenuated by atmospheric absorption, rain, and obstacles like buildings and trees. This limits their ability to cover large areas and penetrate obstacles [19]. Therefore, this frequency requires a high gain **antenna** to overcome higher path loss and high atmospheric absorption compared with lower frequencies and sustain the link budget at this frequency [20].

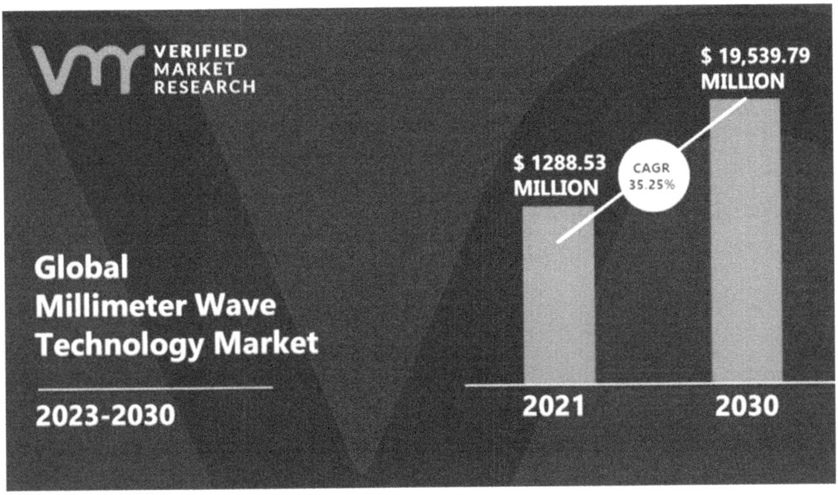

Fig. 6 Global mmWave technology market size.
Source: Verified market research 2023.

The mmWave technology market size was valued at USD 1288.53 Million in 2021 and is projected to reach USD 19,539.79 Million by 2030, growing at a CAGR of 35.25% from 2023–2030 [21]. Figure 6 shows the Global mmWave technology market size. Designing the FSS at mmWave 5G spectrum can contribute to improving the **5G** technology [22]. Figure 7 shows the technology improvement of 5G by Shannon Theorem, where the mmWave FSS can be utilised with the mmWave antenna to increase the gain, beamforming beam tilting, stopband **FSS** to interference cancellation by **EMI shielding**, and MIMO antenna isolation improvement.

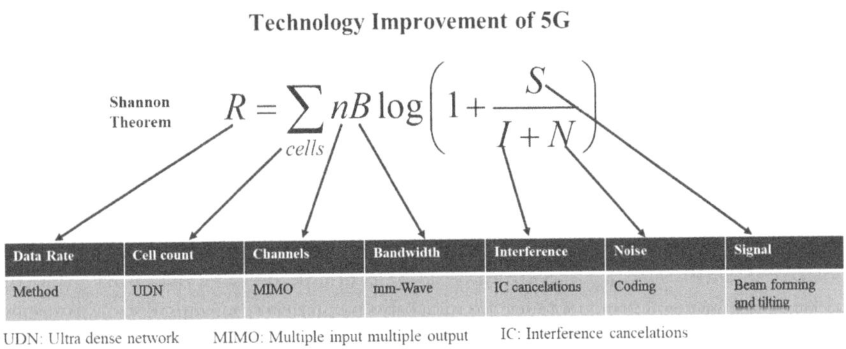

Fig. 7 Technology improvement of 5G.

1.2 Metamaterials and Characteristics

Metamaterials' characteristics can be described using macroscopic electric permittivity (ε) and magnetic permeability (μ) metrics. The categorisation separates the macroscopic characteristics of each type of material to describe wave-matter interaction performance accurately [23]. The capacity of a material to store electrical energy when an electric field is applied is known as its electric permittivity. Similarly, the magnetic permeability of the material is its ability to get magnetized in response to an applied magnetic field. The permittivity and the permeability of free space or air are denoted by ε_o and μ_o, respectively, where $\varepsilon_o = 8.858 \times 10^{-12}$ F/m and $\mu_o = 4\pi \times 10^{-7}$ H/m. All other materials' permittivity and permeability are determined relative to the free space permittivity and permeability. The relative permittivity (ε_r) and the relative permeability (μ_r) are defined as $\varepsilon_r = \varepsilon/\varepsilon_o$ and $\mu_r = \mu/\mu_o$, respectively. Additionally, a metamaterial's refractive index is determined by $\eta = \sqrt{\varepsilon_r . \mu_r}$ [24].

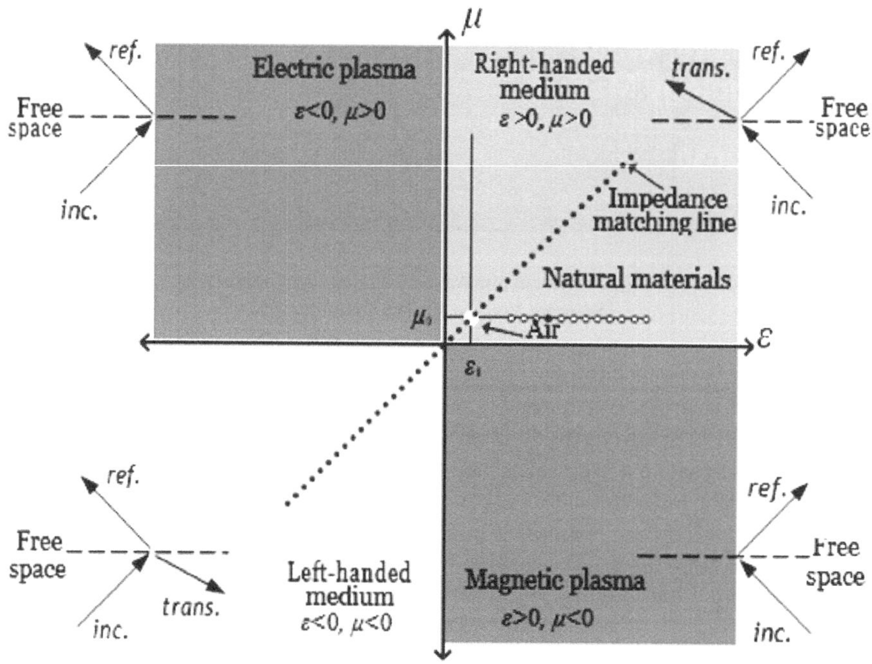

Fig. 8 Classification of metamaterials based on parameters ε and μ. Incident, reflected and transmitted waves are represented by inc., ref., and trans., respectively.

The classification of **metamaterials** according to their permittivity and permeability values is shown in Figure 8. Additionally, it displays how an EM wave behaves when its incident faces an air-material interface. The materials that have both positive quantities (ε_r, μ_r) and are referred to as double positive materials are located in the first quadrant; most of them are naturally occurring materials. Epsilon-negative substances with positive permeability and negative permittivity are shown in the second quadrant, known as epsilon-negative metamaterial. Surface waves may be supported by these materials, which have a strong **EM** field reaction and are frequently observed at optical frequencies (in the infrared/visible frequency range). The third quadrant represents double-negative metamaterials with negative permittivity and permeability. The characteristic of such materials is that although the energy is delivered in the forward direction, the wave propagates backwards. Additionally, it has been noted that these materials follow Snell's law in reverse. As a result, these materials are also known as Veselago's material (LH medium) or negative index media (NIM). The fourth quadrant represents materials having negative permeability and positive permittivity. They are known as mu-negative metamaterials. Gyrotropic materials are found to have this characteristic. The effect of Faraday rotation might also be seen in these materials. The majority of the natural materials are located on the line $\mu = \mu_o$ with $\varepsilon \geq \varepsilon_o$. The diagonal line in Figure 8 shows the impedance matching scenario with free space. As a result, for normal incidence (no reflection), materials with identical permittivity and permeability ($\mu_r = \varepsilon_r$) produce perfect matching at the material interface because their intrinsic impedance is $Z = \sqrt{\frac{\mu}{\varepsilon}} = \sqrt{\frac{\mu_o \mu_r}{\varepsilon_o \varepsilon_r}} = \sqrt{\frac{\mu_o}{\varepsilon_o}} =$ 376.6Ω which is the free space impedance [25]. It's interesting to note that the medium is referred to as nihility when the permittivity and permeability are both zero, which can produce a perfect tunnelling effect.

1.3 Scattering Matrix

Equivalent circuit models can depict **metamaterial** structures, allowing one to address microwave and **mmWave** problems using circuit analysis principles rather than Maxwell's equations. At low frequencies, there is very little phase shift along the circuit between any two sites because the circuit dimensions are relatively small in relation to the wavelength. However, at high frequencies, this is not the case. Different pots along the transmission line's voltages and currents may not be the same at high frequencies. As a result, the transmission line network has scattered elements used in network analysis. Using Maxwell's equations, network analysis is more straightforward than field analysis and can be simply changed to combine several distributed parts. However, the circuit analysis solution may be oversimplified in some circumstances, producing inaccurate findings. To find electric or **magnetic fields** at random points in space, for example, one must apply field analysis utilising Maxwell's equations in such circumstances [26].

The S-parameter analysis can be described as a network that deals with incident, reflected, and transmitted waves more effectively. The scattering matrix explains the relationships between power waves that are incident or moving from one port to another and those that are reflected or moving away from the ports. The scattering matrix of a two-port network is presented by Equation (2.1) [26].

$$S_{ij} = \left. \frac{V_i^-}{V_j^+} \right|_{V_k^+ = 0,\ f \text{ or } k \neq j} \qquad ...(1)$$

Where (subscripts denote the port number and superscripts denote the wave direction), V_j and V_i stand for the voltage of the wave incident on port j and the voltage of the wave reflected off port i, respectively. Additionally, Equation (1) [26] shows that when S-parameters are identified, all ports are frequently matched. An extensive **scattering matrix** of a two-port network is given by Equation (2) [26]

$$\begin{bmatrix} V_1^- \\ V_2^- \end{bmatrix} = \begin{bmatrix} S_{11} & S_{12} \\ S_{21} & S_{22} \end{bmatrix} = \begin{bmatrix} V_1^+ \\ V_2^+ \end{bmatrix} \quad ...(2)$$

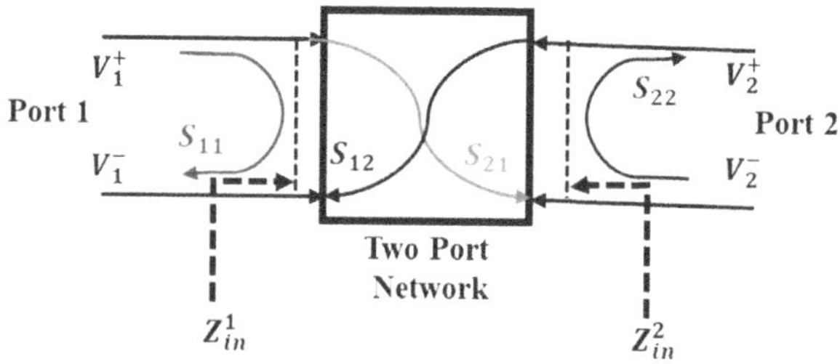

Fig. 9 S-paraments and two-port network.

When all other ports are matched and not excited, S_{ii} represents the reflection coefficient for port i. As all other ports are matched and not excited, S_{ij} represents the transmission coefficient from port j to port i. Figure 9 depicts a two-port network diagram with both transmitted and **reflected waves**. The dashed line shows the terminal plane, which is the reference plane for current and voltage phasors. Z_{in}^1 stands for the input impedance when viewed from port 1 and looking into the network. The input impedance has the formula $Z_{in} = R + jX$, with the real impedance drawn to power accepted and the imaginary impedance related to network energy storage. The impedance's real part equals zero for a lossless network. The network accepts all of the input power and dissipates zero on average. For a capacitive load, the imaginary portion of the impedance is negative, but for an inductive load, it is positive. S_{11} and Z_{in} are also related by Equation (3), [26] where Z_o is the output port impedance.

$$Z_{in}^i = \frac{1+S_{ii}}{1-S_{ii}} Z_o \quad ...(3)$$

1.4 Metamaterial Inspired FSS

FSSs are spatial filters that provide bandpass, band stop, and absorption behaviour in response to an incident EM wave. On a dielectric substrate, metallic components, including patches, slots, wire grids, and resonators, are spaced out to create FSSs. Strong electric fields are created in the gaps, and strong **magnetic fields** are generated around the electric currents in metallic strips when an **EM** wave is impacted on a surface with resonant elements.

Furthermore, there is coupling between the resonant components. The general behaviour of the FSS is a result of these two kinds of interactions. Unit cell element geometry and element spacing are thus the two determining criteria. Incident angle

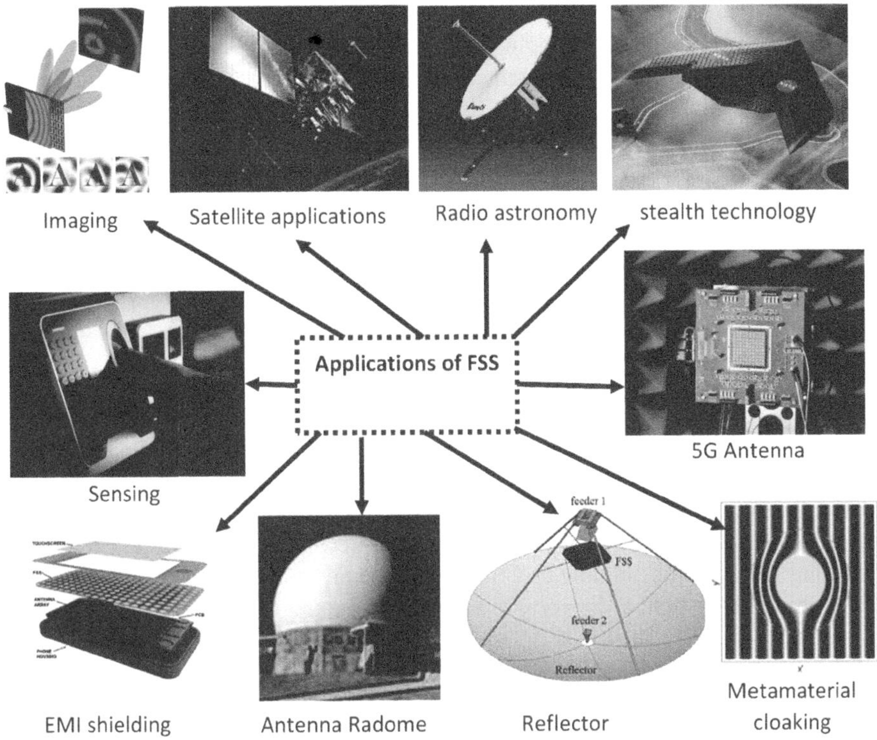

Fig. 10 Applications of FSS in different fields.

stability and polarization angle stability present the most significant design problems for FSSs. If an FSS is positioned following an **antenna**, a relatively tiny region will be lit from various angles, which can cause grating lobes to form unless a mitigating strategy is used. Furthermore, a uniform phase front is necessary, which is not possible to achieve by a large FSS. Unit cells, which are tiny compared to the free space wavelength and compact to allow for dense arranging, are the solution to these challenges. Hence, metamaterials are the source of inspiration for this method of creating FSSs. Figure 10 shows the applications of FSS in different fields.

1.5 A Brief History of FSS

The ability of an FSS to operate as a surface with bandpass or bandstop filtering qualities to incident waves are its distinguishing characteristic. A periodic arrangement of conductive elements that capacitively and inductively connect achieves this when **EM** waves are incident like a plane wave. One of the earliest types of an FSS was a parabolic reflector grid that used a collection of resonant dipoles, designed and patented by Marconi and Franklin in 1919 [27]. Before the 1960s and 1970s, very little research had been done on what are now known as FSSs. During this time, the Air Force Avionics Laboratory was interested in developing FSS for radar and stealth applications [27, 28]. This study used novel FSS element designs, such as crossed dipoles, which were more adaptable than previously studied single **resonant dipoles**. FSSs were useful for some applications, such as stealth radomes and multi-band Cassegrain reflector dishes in antenna systems [27, 29], because they offered better performance, including insensitivity to an angle of incidence (the angle between the direction normal to the plane of the FSS and a plane wave's direction of propagation). A new trend in related research started a few decades ago. It concentrated on renewed approaches to developing FSS for widespread usage. The use of numerical solvers, made possible by more recent advancements in computing technology, has given rise to various FSS designs and applications, including active FSSs [30] and three-dimensional FSS structures [31].

1.6 Practical Design Problems of FSS

The incident angle of an impinging plane wave can generate the resonance frequency. The FSS's curvature can significantly change the frequency response. The FSS's inter-element spacing must be adjusted to less than $\lambda/2$ to avoid grating lobes. On the other hand, the frequency response at OIA also changes due to the phase change of incoming electromagnetic waves. **Miniaturisation** is the technique to avoid this frequency shifting. To comprehend how an FSS would operate in a real system, assessing the FSS structure's practical concerns is necessary.

1.7 Wave Incident Angle

Waves that are incident normally will excite an FSS. The OIA will impact the FSS's resonant frequency and bandwidth. These effects are primarily caused by the fact that the inductance and capacitance values of an FSS depend on the angle of incidence, the polarisation type (TE or TM), and other factors. The key factors affecting an FSS's resonant frequency are the levels of reactance (capacitance and inductance), which alter depending on the OIA (θ) and whether the **polarisation** is TE or TM. The changes in wave impedance at OIA that affect the loaded quality factor of a resonator in an FSS structure can be blamed for the fluctuations in the structure's bandwidth. The wave impedance for the TE mode shifts to $Z_o/\cos\theta$ [32], where θ is the incident angle. As a result, for parallel resonators, the bandwidth drops due to an increase in the loaded equality factor (Q), but for series resonators, the bandwidth increases when there is a large OIA (θ). As the OIA varies in TM mode, the wave impedance shifts to $Z_o \cos \theta$ [32]. Because of this, the bandwidth increases for the parallel resonator and drops for the series resonator for large OIA, where Q decreases. To operate under various OIA, an FSS might need to be designed for a practical demonstration. Three design strategies can usually be used to lessen the impact of OIA on the frequency response of an FSS. First, the impacts can be mitigated when the **FSS** is embedded in dielectric materials [33]. Second, by making the FSS element sizes smaller, the system can have an adequate number of components, which can help to lessen the effects of variations in OIA [34]. Reducing the space between elements is the third technique [27]. Doing this makes it possible to minimize the distance between the elements while also lessening the impact of the OIA-induced phase difference between the parts.

1.8 Curved FSS

The possibility of demanding a curved FSS structure presents another practical challenge for its deployment. Stealth radome structures and sub-reflector **antennas** are two uses for a curved FSS [35]. Depending on the type of curvature, an FSS's frequency response may alter as it is curved. Consideration should be given to the single-curved FSS and the doubly-curved FSS, two distinct types of curvature [36]. While the double-curved FSS adheres to a conically rounded or spherical shape, the single-curved FSS has a cylinder shape. In actuality, the single curvature FSS has less impact on frequency response than the double curvature. A curved FSS's resonant response will differ from an equal planar's FSS. This is mostly due to the possibility of changes in **coupling** between FSS parts caused by curvature. As a result, each FSS element in the structure has a variable incident angle due to the curvature, which results in variations in each element's magnitude and phase [36]. Furthermore, the curvature may affect the phase difference over the FSS and the coupling of nonadjacent elements.

1.9 Grating Lobe

A "grating lobe" refers to an undesirable phenomenon occurring when the FSS structure is designed to operate at specific angles or frequencies. Grating lobes are secondary radiation beams that occur at angles away from the main beam direction in an antenna or phased array system. They are a result of the periodic nature of the FSS structure or the array elements. In the context of FSS, when the dimensions of the FSS elements are on the order of the wavelength of the incident EM wave, the FSS can act as a diffraction grating, causing energy to radiate at angles other than the desired direction. Grating lobes are typically unwanted because they divert energy away from the main beam, potentially reducing the gain and directivity of the **antenna** or array. They can lead to poor coverage of the intended target area and increase the risk of interference with other systems or objects nearby. Grating lobes become more prominent when the spacing between FSS elements or array elements is larger than the operating wavelength [27]. Several strategies can be employed to mitigate the effects of grating lobes in FSS-based antenna systems. Decreasing the spacing between FSS elements can help suppress the occurrence of grating lobes. However, this might lead to increased complexity in design and fabrication [37]. Designing the FSS with higher-order modes can also help in suppressing grating lobes. These modes can have narrower angular operation ranges and reduced grating lobe effects [38]. Gradually varying the size or spacing of the **FSS** elements across the structure can mitigate grating lobe effects by reducing the abrupt transition between different modes of operation. Implementing curved surfaces in the FSS design can help mitigate grating lobes by introducing curvature-induced phase shifts. Combining FSS with other structures, such as reflectors or lenses, can help reshape the radiation pattern and reduce grating lobes [27].

1.10 Types of FSS Based on Transmission Attributes

FSSs are classified as an **absorber**, bandpass, and bandstop FSS to control the transmission and reflection of specific frequency ranges, as shown in Figure 11. Each

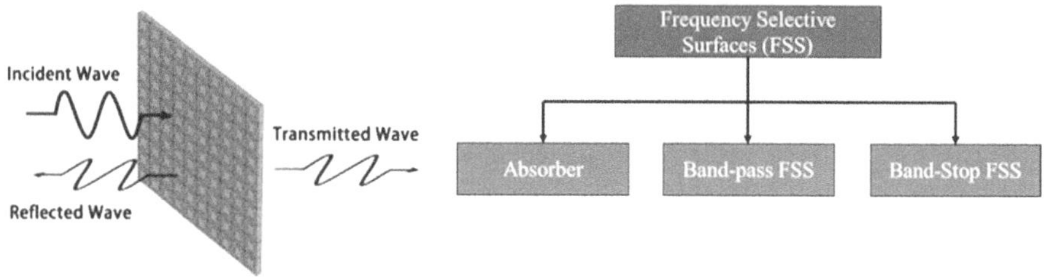

Fig. 11 Types of FSS based on transmission attributes.
Source: Dey & Dey, 2022.

type serves different purposes and applications based on their frequency response characteristics.

1.11 Absorber

The absorber is a kind of FSS designed to absorb **EM** energy within a specific frequency range, effectively attenuating and dissipating the incident waves. Instead of reflecting or transmitting the energy, an absorber FSS converts it into heat. An absorber is particularly useful for applications that require reducing reflections, echo, or unwanted signals. It's often used in EM compatibility testing and stealth technology to minimise radar cross-section and reduce detectability. A splits ring resonator-based polarisation-sensitive **metamaterial** absorber is presented in [24] for sub-6 GHz applications. The asymmetrical design achieved polarisation sensitivity. Figure 12 shows an asymmetrical design of the metamaterial absorber and measurement setup. A modified ring-shaped metamaterial absorber was presented in [39] for X and Ku band applications where resonant frequency appeared at 8.23 GHz, 10.33 GHz, 14.32 GHz, and 15.58 GHz with 98.87%, 95.11%, 93.27%, and 99.99% **absorption**, respectively.

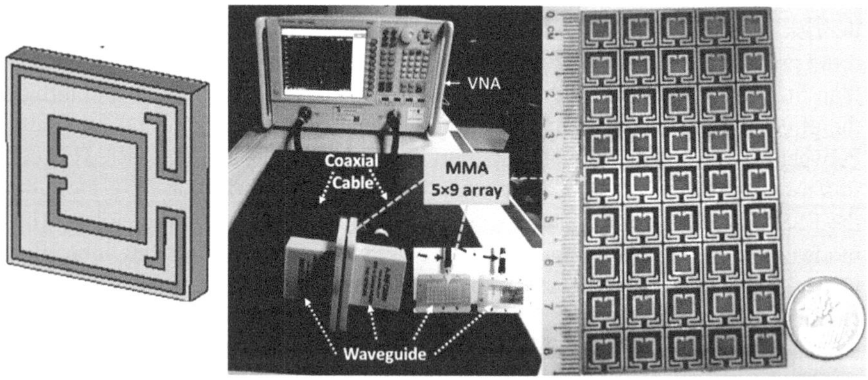

Fig. 12 Polarisation-sensitive metamaterial absorber for sub-6 GHz applications and measurement setup.
Source: Hakim, Islam et al., 2023

The absorber for the Ka/5G spectrum is designed by [40], which shows polarisation sensitivity for its asymmetrical patch. The FR-4 (1.6 mm thickness) substrate dielectric constant of 4.4 and a loss tangent of 0.02 are used in the design. In [41], glass and nematic liquid crystal substrate-based multilayer **microstrip line** metamaterial absorber is designed for ka/5G triband applications at 25.4 GHz, 28.7 GHz, and 32.8 GHz frequency with around 90% peak absorption. Multiple fractal splits ring-based multiband metamaterial absorber from microwave to **mmWave** presented in [25]. The peak absorption was achieved at 6.75 GHz, 19.60 GHz, 24.26 GHz, 26.62 GHz, and 30.05 GHz frequencies. Figure 13 shows a symmetrical design of the metamaterial absorber and measurement setup.

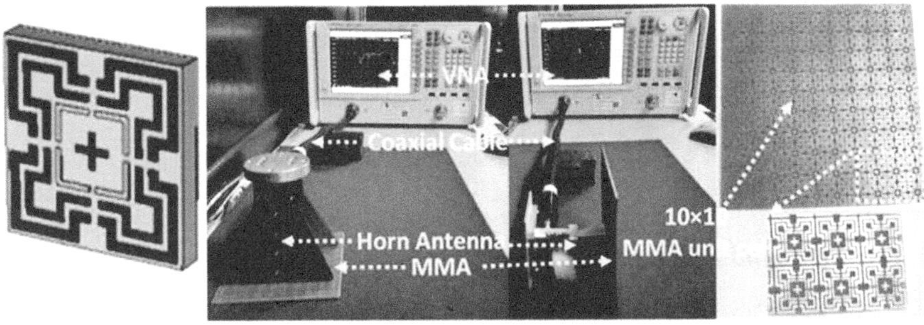

Fig. 13 Polarisation-insensitive metamaterial absorber for microwave to mmWave applications and measurement setup.
Source: Hakim, Alam, Islam, et al., 2022

1.12 Bandpass FSS

The bandpass FSS allows a specific frequency range to pass through while attenuating frequencies outside that range. It acts like a filter, transmitting signals within a specific bandwidth and reflecting or attenuating signals outside that bandwidth. Bandpass FSS is often used when there is a requirement to allow specific frequencies to pass through for communication, sensing, or other applications. It finds applications in **wireless communication**, radar systems, and frequency-specific sensing, etc. 3D multiple-layer passband FSS [38] shows highly selective behaviour at the 3.5 GHz 5G frequency band. Implementing five metallic layers of FSS forms two layers of parallel LC resonators with an inductive impedance inverter layer in between, providing a higher-order filtering response. Due to the multiple layers and rigid substrate, the **FSS** is unsuitable for conformal applications. A 2.5-D via-based complex FSS achieved the passband from 26.90–29.02 GHz; Figure 14 shows the absorber's unit sell design and measurement setup [34]. A meandered line resonator completes the miniaturisation of the FSS, but due to the complexity and rigid substrate, it is not applicable for conformal applications. Based on cyclic olefin copolymer (COC), a transparent and low-loss material FSS on a glass window is presented for 5G millimetre-wave communication signal selectivity [42]. The suggested FSS based on dual-loop units is constructed on both sides of the COC, which aids in realising the band-pass filtering effect in the desired frequency range of 27.5–29.5 GHz under normal and oblique-incident conditions

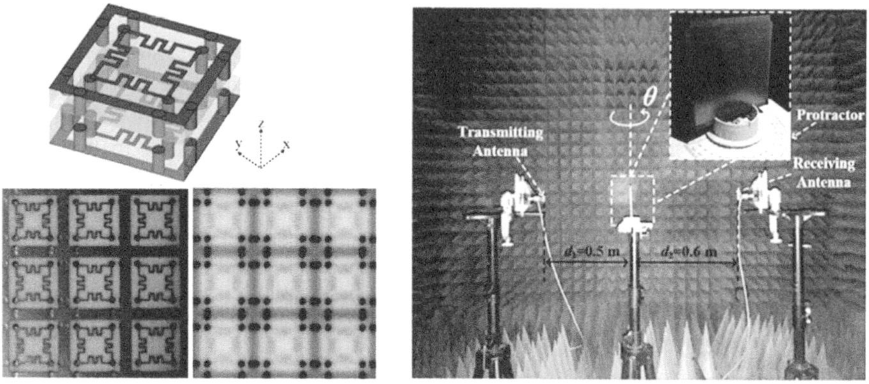

Fig. 14 2.5-D via-based passband FSS and measurement setup. *Source:* Li et al., 2017.

1.13 Stopband FSS

The stopband FSS, or notch FSS, is designed to block a specific frequency range while allowing other frequencies to pass through. It's the opposite of a bandpass FSS, serving as a notch filter that prevents interference from specific frequencies. The stopband FSS suppresses unwanted signals and reduces **EM** interference (EMI) from specific sources. It's commonly applied in scenarios where there is a need to mitigate interference or protect sensitive equipment from specific frequencies. Different stopband FSSs have recently been developed for various microwave-to-**mmWave** frequency range applications. The miniaturised microwave FSSs are presented with single or multiple stopband applications [43]. A thin dielectric substrate-based stopband FSS [37] has stable responses to polarisation and incidence angle changes. A unit cell size of 0.091 λ, where λ is the wavelength at the lowest resonance, is used in this design to maximise spatial efficiency. The conformal stopband FSS has three transmission zeros at 4, 6.95, and 11.3 GHz, where an FR-4 substrate with 0.25 mm thickness is used to design the FSS. The stopband bandwidth is applicable for shielding satellite communication's C/X band frequencies. Figure 15 shows the unit cell design geometry, fabricated prototype, and measurement setup [37]. The different **mmWave** frequency bands are selected for possible utilisations in 5G communications. However, only a few FSSs are designed in the mmWave region [32, 44, 45]. Researchers are exploring novel miniaturised conformal **FSS** with polarisation and high OIA stability for extensive use in mmWave 5G applications. A miniaturised non-conformal stopband FSS is reported in [32] for only n261 band applications, where a cross-loop resonator with a wire grid achieves miniaturisation. However, it shows an OIA stability of only 35°. A conformal bandstop FSS is presented in [44], where meshed resonators achieve n257, and a more comprehensive range of stopband (5–40 GHz) and passband (3.12–4.64 GHz) are realised by the multilayer approach [45]. Nevertheless, this broad stopband microwave to mmWave FSS is incompatible with conformal applications.

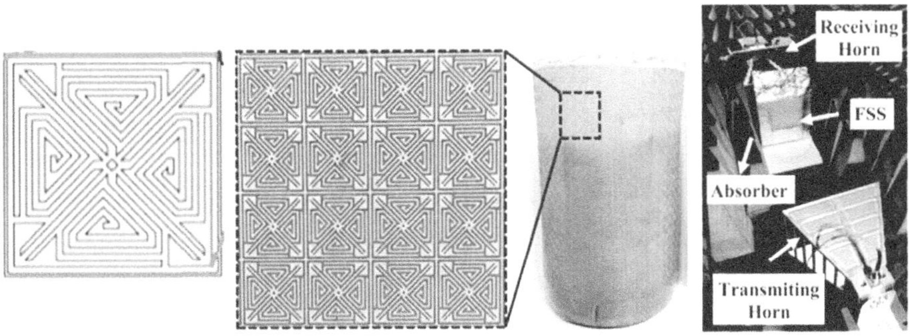

Fig. 15 Miniaturised stopband FSS, fabricated prototype, and measurement setup. ***Source:*** Dey & Dey, 2022.

1.14 Polarisation Sensitivity and Insensitivity of the FSS

FSS is considered polarisation sensitive when its response to incident **EM** waves varies based on the polarisation direction of those waves. Conversely, an FSS is polarisation insensitive when its performance remains consistent regardless of the polarisation orientation of the incident waves [25, 37]. In more detail, a polarisation-sensitive stopband FSS exhibits different stopband characteristics (e.g., reflecting or attenuating specific frequencies) for different polarisation orientations of the incident waves. For instance, the FSS might effectively block a particular frequency for waves polarised in one direction but not for waves polarised in a different direction. This can occur due to the **asymmetric geometry** and arrangement of the FSS elements, which may interact differently with different polarisation orientations [37]. The asymmetric arrangement of carbon fibres into the epoxy resin shows polarisation-sensitive stopband characteristics. [46], Also, mechanical flexibility makes it suitable for conformal applications. The symmetric and two asymmetric resonator arrangement exhibits polarisation-sensitive stopband behaviour. Figure 16 shows a unit cell design with an FR-4 substrate with height h = 1.6 mm, a fabricated prototype, and a measurement setup [47]. The TM **polarisation** mode attained a stopband from 12.4–15.8 GHz, and the TE polarisation mode attained 8.4–12.4 GHz. Another via-based asymmetric FSS was implemented for the passband for TE polarisation mode from 31.7 to 36.3GHz, but the structural complexity limits it for conformal applications. Table 1 compares existing polarisation-sensitive FSS regarding their unit cell design geometry, unit cell size, filtering types, operating frequency, BW%, covered **5G** mmWave band, OIA stability, and filter types.

A polarisation-insensitive stopband FSS maintains consistent stopband characteristics regardless of the polarisation of the incident waves. In other words, the FSS behaves similarly for all polarisation orientations. Figure 17 shows in a rotational symmetrical structure, the wave propagation direction k, and the e-filed and h-field vector direction are presented for TE and TM polarization modes where φ is the field vetro polarisation angle [25]. Polarisation insensitivity is desirable in many practical applications as it ensures that the FSS's performance remains reliable and predictable across a range of operating conditions. For applications where polarisation sensitivity

Fig. 16 Polaristion-sensitive asymmetrical FSS unit cell, fabricated prototype, and measurement setup.
Source: Abbasi et al., 2018.

is not desirable, efforts are often made to design FSS structures that minimise this sensitivity. Polarisation insensitivity can be achieved through careful design and arrangement of rotational symmetric FSS elements to ensure that they interact with different polarisations in a balanced and uniform manner.

A polygon-shaped symmetrical resonator-based FSS is designed [43] to filter x band frequency. The symmetrical resonator achieved polarisation-insensitive behaviour. Another Jerusalem cross resonator-based stopband FSS shows polarisation-insensitive behaviour due to its **symmetrical geometry** [51]. Similarly, there are different types of symmetrical resonators, like fractal resonators [34], meshed resonators [44], and cross-loop resonators with a wire grid [32] are implemented in mmWave frequency, which achieves polarisation insensitivity. Figure 18 shows the symmetrical geometrical structure of the meshed resonator for polarisation-insensitive transparent stopband FSS at **mmWave** 5G spectrum and measurement setup. The meshed resonator achieved optical transparency [44]. Table 2 compares existing polarisation-insensitive **FSSs** regarding their unit cell design geometry, unit cell size, filtering types, operating frequency, BW%, covered 5G **mmWave** band, **OIA stability**, and filter types.

Table 1 Comparison of polarisation-sensitive FSSs.

Ref	*Asymmetrical Resonator Geometry.*	*Unit Cell Size (mm^2)*	*Filter Type*	*Frequency (GHz)*	*BW (%)*	*Covers 5G Band*	*Conformal*	*OIA Stability*
[46]	Conductive cylindrical fiber	9 × 9	Stop band	10.2	23.4	–	Yes	60°
[47]	Split ring resonator	7.6 × 7.6	Stop band	10.4 14.1	38.4, 24.1	–	No	60°
[48]	Via-based slotted patch	5.5 × 5.5	Pass band	34	13.5	–	No	60°
[49]	Modified asymmetric resonator	6 × 5	Stop band	40	25	n259, n260,	N/A	N/A
[50]	Slotted patch	2.87 × 2.87	Stop band	29	28.2-30	Partial n257	No	40

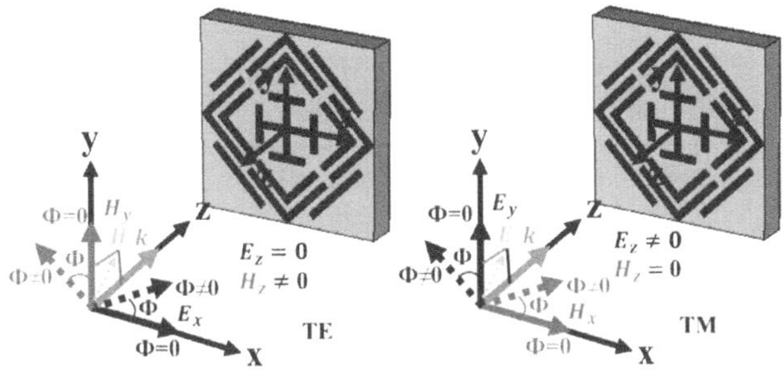

Fig. 17 Vector direction of the TE and TM polarisation mode.
Source: Hakim, Alam, Islam et al., 2022.

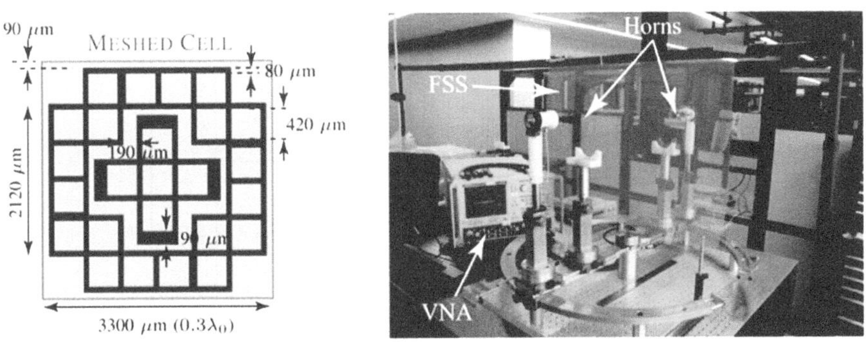

Fig. 18 Symmetrical geometrical structure of meshed resonator for polarisation insensitive transparent stopband FSS and measurement setup.
Source: King et al., 2023.

1.15 Miniaturisation of FSS to Achieve OIA Stability

An FSS responds to waves with different OIA primarily due to the size and form of the array element. The **miniaturisation** of the FSS unit cell can provide high OIA stable filter behaviour by reducing the phase shifting at OIA compared to normal incidents of EM wave [37]. Developing an FSS with an array of elements with a precise response and small size is challenging since it necessitates excellence to get the desired result [52]. Numerous methods have been suggested to reduce the size of the FSS array elements, like increasing the corresponding **lumped-element** values for resonant structures and lengthening each array element's electrical length, which is measured in terms of wavelength can increase the electrical length [53]. Capacitor and inductor lumped components can be used in an array element to increase electrical length [54]. A complex structure is another way to accomplish it [34].

A convoluted geometry is implemented [37] to miniaturise the unit cell, which shows the periodicity of the unit cell is only 0.091λ, where λ is the wavelength at the lowest resonance. The **FSS** shows resonance frequency at 4, 6.95, and 11.3 GHz. The stable stopband filtering properties achieved up to 60° OIA. A cross-loop resonator with a wire grid achieved another miniaturized stopband FSS at mmWave

Table 2 Comparison of polarisation-insensitive FSSs.

Ref	*Symmetrical Resonator Geometry.*	*Unit Cell Size (mm^2)*	*Filter type*	*Frequency (GHz)*	*BW (%)*	*Covered mm5G Band*	*Conformal*	*OIA Stability*
[38]	Fractal cross resonator	10 × 10	Pass band	3.34	3.7	--	No	60°
[43]	Polygon shape	8.3 × 8.3	Stop band	10	40	--	Yes	60°
[51]	Jerusalem cross resonator	9.25 × 9.25	Stop band	10	51		Yes	60°
[32]	Cross loop resonator with a wire grid.	4.4 × 4.4	Stop band	27.2, 33.1	8.3, 13.8	n261	No	35°
[34]	Vertical via based 2.5 D mandarin line	2.87 × 2.87	Stop band	27.75	9.6	n261	No	60°
[44]	Transparent Meshed resonators	4 × 4	Stop band	28	29.3	n257	Yes	60°
[42]	Dual loop resonator	2.4 × 2.4	Pass band	28	7.017	n261	No	50°

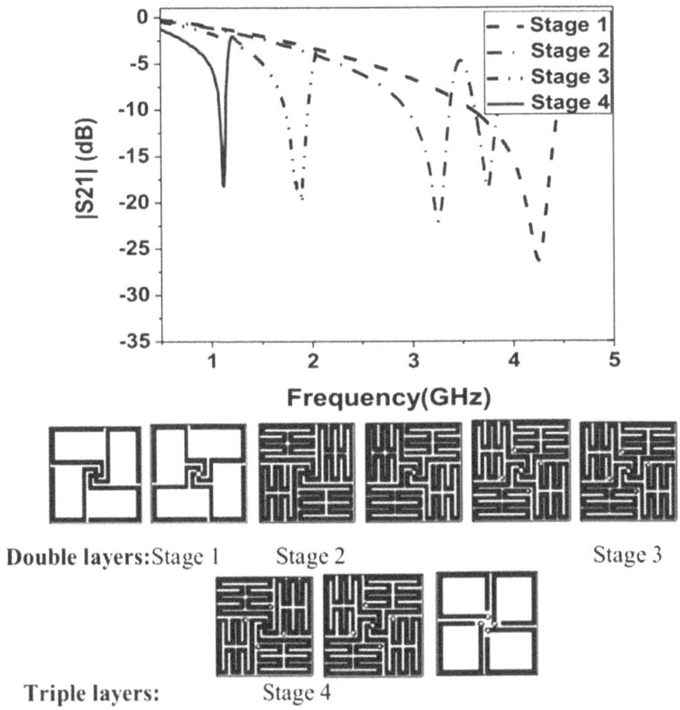

Fig. 19 Multiple layers with convoluted arms, metallic via-based miniaturisations, and corresponding S_{21}.
Source: Ashvanth et al., 2021.

5G frequency band, which shows OIA stability up to 35° [32]. Multiple layers with convoluted arms and metallic via-based structures achieve ultra-miniaturisation [53]. Figure 19 shows the corresponding S_{21} gradual shifts towards the lower frequency by **convoluted geometry** (stage 2), metallic via (stage 3), and multilayers (stage 4) techniques.

1.16 Frequency Tuning of FSS

Frequency tuning of stopband FSS is essential for various applications where the frequency-selective behaviour needs to be adjusted to match specific operational requirements. This can include creating multiband filters, adapting to changes in operating frequencies, or achieving compatibility with different wireless communication standards. The FSS [55] comprises repeating patterns of circular loops coupled to one another by tuneable operation of varactor diodes. The design has a broad adjusting frequency range, where the band stop response exhibits a tuneable range from 0.54–2.50 GHz under the reverse bias voltage. The varactor diodes in the structure's prototype are biased using a cutting-edge biasing method. But, implementing this approach in highly **miniaturised** structures is complex. Another varactor-tuneable FSS [56] achieved high high-transmittance. When the bias voltage applied to the varactors is tuned from 5–20 V, the transmission rate of the varactor-tuneable FSS is tuned from 4.84–5.9 GHz (S_{21} = 1.09 dB). But, implementing this approach in highly miniaturised structures is complex. A passive frequency shifting or sensing metamaterial absorber is presented by [57, 58] by placing a dielectric substrate material which is less complex.

1.17 Applications of FSS for EMI Shielding

FSS plays a significant role in EMI shielding, which plays a vital role in improving system performance. They offer tailored frequency responses that can be utilised to selectively block or allow specific frequency ranges, enhance antenna capabilities, and address various EM challenges in a wide range of applications. The transmission-blocking feature by FSS is also attractive in the **EMI shielding** application for filtering unwanted frequencies. The transceiver of the radiofrequency device suffers due to unwanted EMI. Traditionally, sensitive devices are protected from EMI by applying a metal shield, but this approach blocks all EM waves [59, 60]. The conductive ink-coated EMI shielding is used for flexible applications but has limited durability and does not show the filtering capability of specific narrow-band frequencies [61]. The FSS is an attractive solution for blocking a specific frequency and allowing all other frequencies to propagate. In [43], a polygon-shaped resonator-based transmission blocking FSS is presented for 8-12 GHz **EMI shielding**. The unit cell size of the FSS occupies 8 × 8 mm^2. The EMI shielding is completed by a twisted spiral triangle cross dipole resonator and Jerusalem cross resonator [37] and [51], respectively. The EMI shielding by conventional metallic box and FSS is shown in Figure 20. EMI shielding metallic boxes are commercially available [62], which can protect PCB from all incident EM waves, but in FSS-methods specific frequencies are blocked, and others are transmissible.

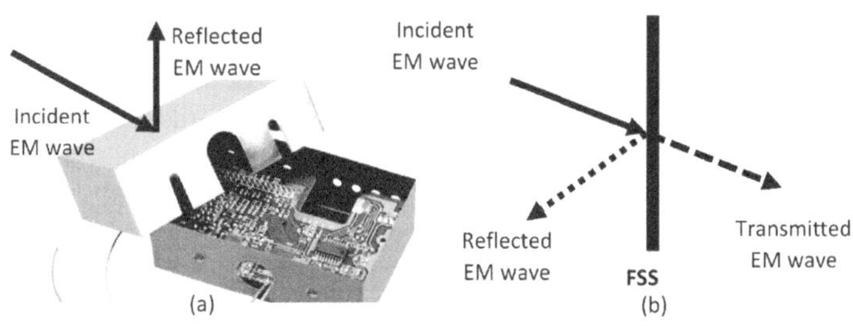

Fig. 20 EMI shielding by: (a) conventional metal box, and (b) FSS.

The FSS is a periodic array of artificial structures that selectively filter or control the transmission and reflection of **EM** waves at the required frequency [2, 63-65]. Structural geometry, spacing, and dimensions are designed to interact with incident EM waves at specific frequencies. FSSs find applications in a wide range of fields, including **wireless communication** systems [37, 66], EM shielding [34], radar systems antenna design [67], reflectors [68], absorbers [69-71], and EM wave manipulation [72]. They are particularly useful in controlling and managing EMI in electronic devices and systems [37, 73]. The advantages of conformal planar FSS are their versatility, lightweight nature, conformal integration capabilities, customisable performance, and ease of manufacturing [43, 44]. These qualities make them a valuable choice for conformal integration applications. Various telecommunication authorities allocated n257 or 28 GHz frequency (26.50–29.50 GHz) for 5G band applications [74, 75]. FSS for 28 GHz 5G EMI shielding is essential for ensuring EM compatibility, reducing interference, and maintaining the reliable operation of 5G devices and systems in increasingly crowded frequency spectrums [44]. Depending on the miniaturisation method, the FSS can be classified by applied techniques such as fractal geometry [76], convoluted patterns [37], multilayer [38, 77], and lumped elements [78, 79]. In the existing literature, various multiband and single-band conformal FSSs are investigated to accomplish the requirements at the microwave frequency range [37, 43, 46]. However, only a few FSS are designed in mmWave 5G frequency spectrum with conformal features, and different miniaturisation techniques were applied to achieve the FSS structure to maintain its response and performance characteristics under a different angular incident angle (AIA) of the incoming **EM** waves [32, 34, 42, 44, 80]. Among them, the vertical via-based 2.5-D mandarin line resonator [34] and dual loop resonator [42] are designed for passband applications at the n261 band with 60° and 50° AIA stability. Moreover, both FSSs are based on rigid substrates, which do not show conformal features. The stopband non-conformal mmWave n261 band FSS is presented in [32]. However, it shows low AIA stability of 35° and high peak transmission zero (TZ) frequency shifting of <4.1% due to the larger periodicity of the unit cell. On the other hand, conformal FSS based on back-to-back resonator in [80] shows dual stopband at n257 and n260 band but has the disadvantage of high peak TZ frequency shifting of <6.1% at 60° at AIA. The peak TZ frequency shifting is improved to <2.25% by screen-printed transparent meshed resonators in [44], which shows stopband at n257 band and ultra-thin structure facilitate conformal applications. Therefore, designing a conformal miniaturised stopband FSS for the 5G n257 band is highly desirable with high stopband stability and low shifting of peak TZ frequency at high AIA. Moreover, polarisation insensitivity will provide effective EMI shielding at the desired n257 frequency band.

This chapter presents a thin flexible stopband metamaterial for 5G n257 band. The specific target of designing metamaterial is to explore a conformal miniaturised stopband metamaterial for the 5G n257 band with low shifting of peak TZ frequency at high OIA and polarisation insensitivity, which can provide effective EMI shielding at the desired n257 frequency band. The novelty of **metamaterial** includes both aspects of design geometry and results performances. The unit cell of the metamaterial consists of a fractal square ring resonator and an interconnecting fractal cross resonator. The size of the unit cell is 1.9×1.9 mm^2 only, or periodicity is $0.154\lambda_o$, where λ_o is the lower stopband frequency. The reformed geometry gives a high selective stopband filtering response that reflects the signal at 27.84 GHz with an $S_{21} < -10$ dB stopband **bandwidth** (BW) percentage of 21.49%. The designed metamaterial rotational symmetrical geometry demonstrates unique stopband features for TM and TE polarisation modes. The ultra-miniaturised unit cell is completed by fractal technology and leads to a high stability of stopband at an OIA up to 60°. Moreover, the low periodicity significantly reduces the peak TZ frequency shifting at high OIA, which is only 0.21% and the peak TZ is tuneable by adding an additional substrate layer.

2. Design Analysis

2.1 Unit Cell Design

The design configuration of the FSS is shown in Figure 21. The CST studio suite software is utilized to design and simulate the FSS. The frequency domain solver and default tetrahedral mesh are applied for simulation. The **EM wave** is excited along the z-axis via the floquet port, and the unit cell boundary is applied along the x and y-axis. The top resonator layer consists of a miniaturised fractal cross resonator and a fractal square ring resonator. The thin Rogers RT5880 substrate material is employed to design the FSS, whose thickness and dielectric constant are 0.0245 mm and 2.2, respectively. This thin substrate material facilitates conformal application. At OIA, the FSS array experiences a phase delay during excitation. As illustrated in Figure 22, for the $N \times N$ FSS array, the highest phase delay (φ) can be estimated by Equation (4) [32] when the angular incident angle is θ.

$$\varphi = \frac{2\pi f}{\upsilon_c}(N \times P_c \sin\theta) \qquad ...(4)$$

Where υ_c is the free space wave velocity, f is the centre frequency, and P_c is the periodicity of the unit cell. The peak TZ frequency of the FSS array shifts at OIA due to the phase delay (φ), which can be reduced by **miniaturised** unit cell size. In various applications, like radome or curved surfaced, flexible FSS is required, which strengthens its useability. The transmission coefficient of miniaturized of FSS shows stable characteristics while installed over a conformal surface. The thin, rotation-symmetric FSS is miniaturised, with a unit cell periodicity (Pc) of only 1.9 mm or $0.154\lambda_o$, where λ_o is the lowest stopband frequency. The other design parameters shown in Figure 21 are $L1$ = 1.75 mm, $L2$ = 0.7 mm, $L3$ = 0.36 nm, $G1$ = 0.08 nm, $W1$ = 0.53 nm, $W2$ = 0.09 nm, W = 0.05. The ultra-miniaturised feature achieved high OIA stable stopband filter response for normal and conformal EMI shielding at n257 band applications. In addition, rotation symmetric geometry achieved **polarisation** insensitivity.

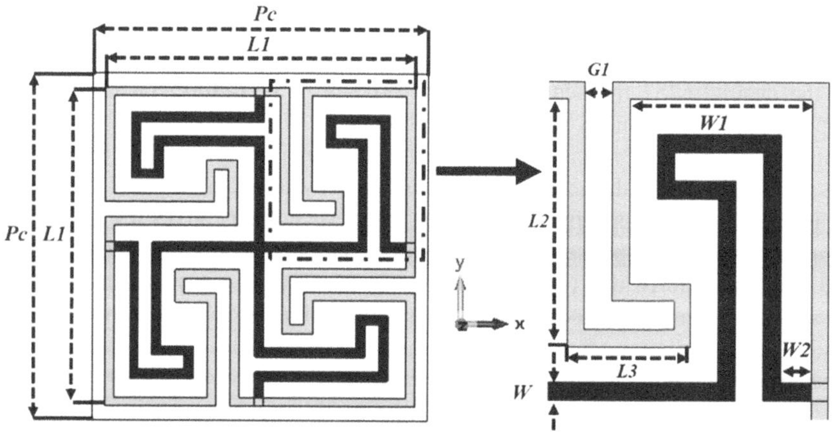

Fig. 21 Front view of the developed FSS.

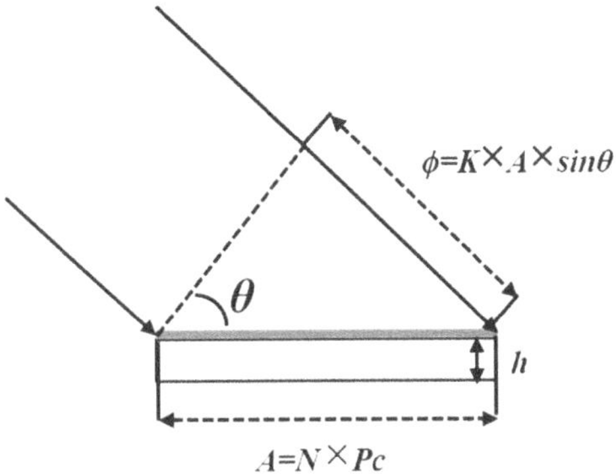

Fig. 22 Phase difference between parallel rays at OIA on the FSS array.

2.2 Design Evolution

The transmission coefficient (S_{21}) of the intermediate and developed FSS is illustrated in Figure 23 for an easy understanding of **miniaturisation**. The reference design is indicated as Design-I, consisting of a miniaturised fractal cross-resonator with a strip line width of *W*. At normal incident, ($\theta = 0°$) and OIA ($\theta = 60°$) of EM wave peak TZ frequency is obtained at 34.53 GHz and 34.25 GHz frequency. The peak TZ frequency is shifted 0.81% at 60° compared with 0°. The design reduces the peak TZ frequency shifting and improves OIA stability without altering the unit cell dimension by incorporating a miniaturised fractal **square ring resonator**. The peak TZ frequency is obtained at 27.84 GHz (at $\theta = 0°$) which is 6.69 GHz shifted towards a lower frequency than Design-I. Moreover, the final design shows at OIA ($\theta = 60°$), the peak TZ frequency shifting is only 0.21%.

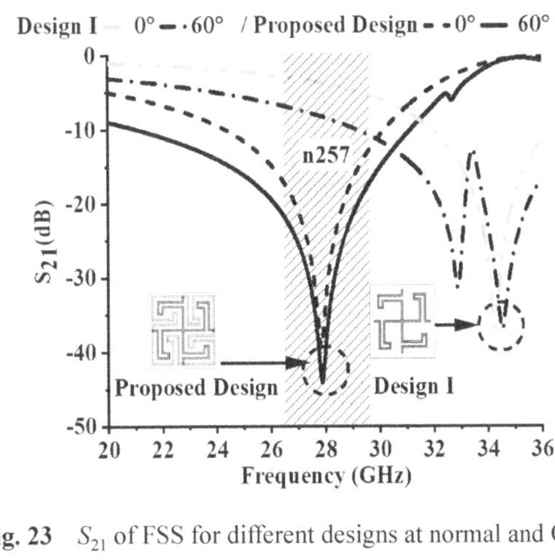

Fig. 23 S_{21} of FSS for different designs at normal and OIA.

2.3 Parametric Analysis

A parametric investigation on strapline width (*W*) is also presented in Figure 24 to show the best physical parameters of the **FSS**. The increment of *W* increases the inductivity of the FSS; therefore, peak TZ shifts towards a lower frequency. Almost similar $S_{21} < -10$ dB bandwidth is achieved for $W = 0.04$ mm and $W = 0.05$ mm, but when $W = 0.06$ mm, the $S_{21} < -10$ dB is 2.34% shifted towards lower frequency due to the increased inductivity. Finally, $W = 0.05$ mm, is chosen as the final value.

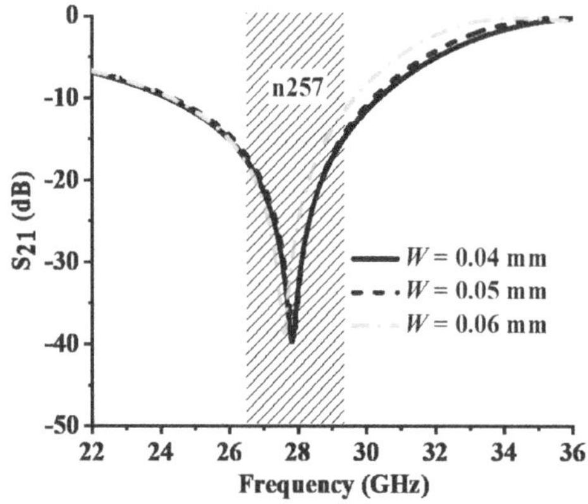

Fig. 24 S_{21} of FSS for different transmission line widths (W).

3. Results Analysis

3.1 Equivalent Circuit

The **equivalent circuit model** (ECM) of the FSS is illustrated in Figure 25(a). The ECM is modelled based on [42, 43, 81]. The inductance L_1 is considered due to the metallic part of the unit cell, and the gap ($G1$) between the unit cell causes capacitance C_1. The equivalent capacitance $C2$ is considered due to the space $L1$. The substrate is presented by transmission line Z_{TL}, which is calculated by $Z_{TL} = \sqrt{\frac{\mu_o \mu_r}{\varepsilon_o \varepsilon_r}} = Z_o \sqrt{\frac{\mu_r}{\varepsilon_r}}$. Where, Z_o is the free space impedance, μ_r, and ε_r are the substrate's permeability and permittivity, respectively. The impedance ($Z_{METAMATERIAL}$) of the stopband **metamaterial** is calculated by Equation (5) [82, 83].

Admittance,
$$Y_{FSS} = Y_1 + Y_2 = \left(j\omega L - \frac{j}{\omega C_1} \right) + j\omega C_2$$

$$= \left\{ -\frac{j\omega L_1}{\omega^2} LC_1 - 1 \right\} + j\omega C_2$$

$$= \frac{j\omega(\omega^2 L_1 C_1 - C_1 - C_2)}{(\omega^2 L_1 C_1 - 1)}$$

$$Z_{FSS} = \frac{1}{Y_{FSS}} = \frac{(\omega^2 L_1 C_1 - 1)}{j\omega(\omega^2 L_1 C_1 - C_1 - C_2)} \quad \text{...(5)}$$

The inductance L_1 is placed iteratively, and the C_1 and C_2 are derived by Equation (6) [82], where ω_Z and ω_{p2} are the zero and second poles, respectively. At the same time, the first pole (ω_{p1}) is located at zero.

$$C_1 = \frac{}{\omega_Z^2 L_1}, \text{ and } \qquad \frac{}{\omega_{P2}^2 L_1 C_1} \qquad \ldots(6)$$

The designed ECM is simulated by utilising Advance Design Systems (**ADS**), and the stopband frequency is adjusted by tuning the parameters. The simulated transmission coefficient of the **ECM** and CST agree well shown in Figure 25(b). Alternatively, the transmission parameters (S_{21}) of the metamaterial can be estimated by transmission line theory [43, 82, 84]. The S_{21} is expressed by Equation (7) [82].

$$S_{21} = \frac{2Z_o}{AZ_o + B + CZ_o^2 + DZ_o} \qquad \ldots(7)$$

where, the coefficient A, B, C, and D refer $ABCD$ matrix of the systems having n number of dielectric substrate layers.

$$\begin{bmatrix} A & B \\ C & D \end{bmatrix} = [M_{FSS}][M_1] \ldots\ldots [M_n] \qquad \ldots(8)$$

where, M_n is the scattering matrix of the nth dielectric substrate layer. In the designed metamaterial, only one dielectric substrate layer is used, so Equation (8) [82] becomes

$$\begin{bmatrix} A & B \\ C & D \end{bmatrix} = [M_{FSS}]\,[M_1]$$

$$[M_{FSS}] = \begin{bmatrix} 1 & 0 \\ \dfrac{1}{Z_{FSS}} & 1 \end{bmatrix}$$

$$[M_1] = \begin{bmatrix} \cos(k_{z1}h) & jZ_{TL}\sin(k_{z1}h) \\ j\dfrac{\sin(k_{z1}h)}{Z_1} & \cos(k_{z1}h) \end{bmatrix}$$

where, Z_1 is the characteristics impedance, Kz_1 is the wave number, and h is the thickness of substrate layer 1.

3.2 Metamaterial Property Analysis of FSS

The stopband feature is also related to the negative relative permittivity (ε_r) or epsilon negative (ENG) and negative relative permeability (μ_r) or mu negative MNG) characteristics. The single negative (SNG) behaviour of the **metamaterial** (either ENG or MNG) at the desired frequency generated a stopband filter response. Different approaches can be implemented to extract these parameters. One method is to employ the S-Parameters template built in the post-processing module of CST. Another method for obtaining the effective parameters using MATLAB code is based on **Nicolson-Ross-Weir** (NRW) method presented in the literature [85]. The scattering parameter values are taken from CST and used in MATLAB code for extracting relative **permittivity** (ε_r), permeability (μ_r), relative refractive index (η), and effective impedance (z) by Equations (9-11) [25].

$$\varepsilon_r = \frac{2}{jK_o d} \times \frac{1-(S_{11}+S_{21})}{1+(S_{11}+S_{21})} \quad ...(9)$$

$$\mu_r = \frac{2}{jK_o d} \times \frac{1-(S_{21}+S_{11})}{1+(S_{21}+S_{11})} \quad ...(10)$$

$$\eta = \sqrt{\varepsilon_r \, \mu_r} \quad ...(11)$$

where, S_{11} = reflection coefficient, S_{21} = transmission coefficient $K_o = 2\pi f/c$, c is the velocity of light, and d is the thickness of the substrate.

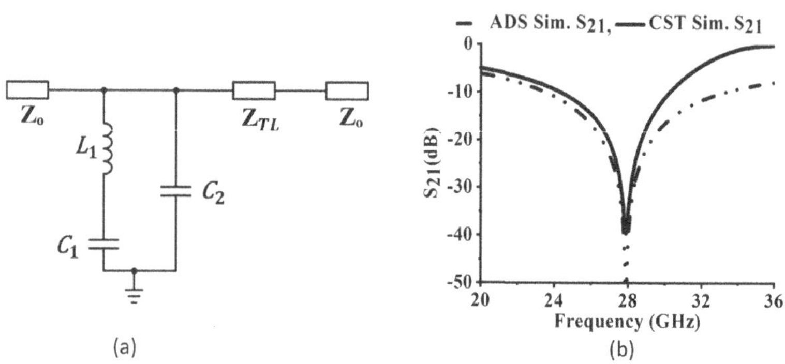

Fig. 25 (a) ECM of FSS (L_1 = 0.604 nH, C_1 = 0.054 pF, C_2 = 0.048 pF, $Z_o = 377\Omega$, and $Z_{TL} = Z_o/\sqrt{\varepsilon}$), and (b) simulated S_{21} from ECM and CST design.

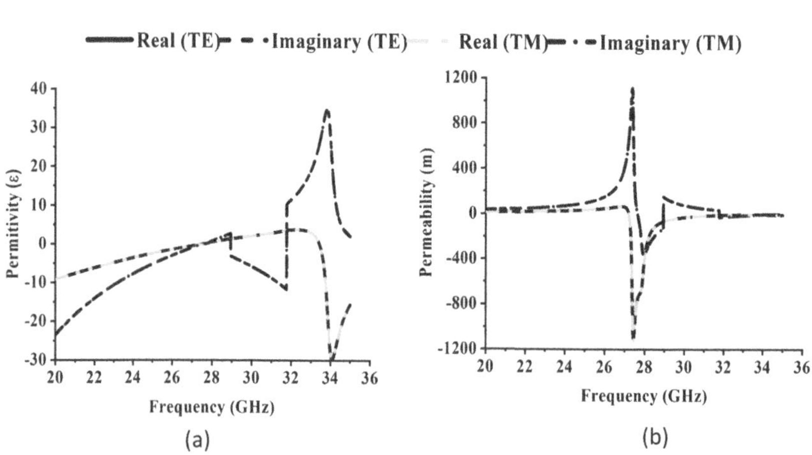

Fig. 26 (a) permittivity, and (b) permeability of the FSS at TE and TM polarisation modes.

3.3 Polarisation Insensitivity and Incident Angle Stability

The simulated reflection coefficient (S_{11}) and transmission coefficient (S_{21}) of the developed FSS at the normal incident ($\theta = 0°$) of the EM wave are presented in Figure 27. The rotational symmetrical design achieved polarisation insensitivity at both transverse electric (TE) and transverse magnetic (TM) polarised waves. The peak TZ of the TE and TM modes are achieved at 27.85 GHz frequency. The stopband ($S_{21} < -10$ dB) of the developed FSS is realized from 24.33–30.19 GHz frequency with BW% of 21.49%, and S_{11} of the FSS at **stopband frequency** is between −0.09 to −0.61 dB. Moreover, the desired n257 frequency band achieved $S_{21} < -13.5$ dB and $S_{11} > -0.3$ dB.

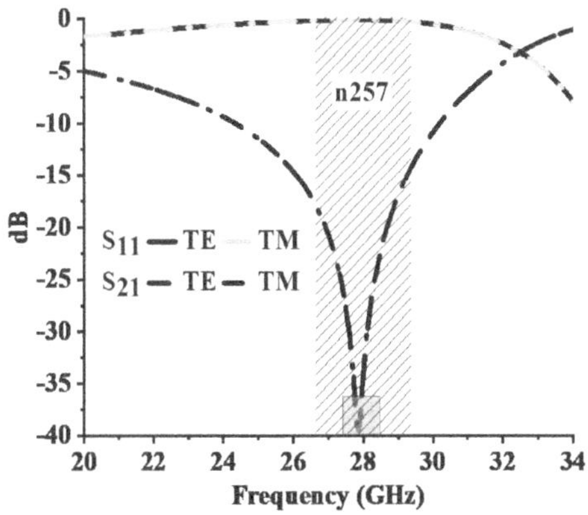

Fig. 27 S_{11} and S_{21} of the FSS at the normal incident of TE and TM polarisation modes.

The S_{21} of the developed FSS at OIA ($\theta \neq 0°$) is illustrated in Figure 28. High OIA stable $S_{21} < -10$ dB is realised by ultra miniaturised design for both polarisation modes shown in Figure 28 (a, b). The peak TZ frequency shows an almost stable position for all possible OIA due to the lowest phased delay at high OIA, which is achieved by the lowest periodicity of the unit cell. The proportional relationship between the periodicity and phased delay was discussed earlier in Equation (4) [32]. A constant BW% of stopband at TE and TM polarisation modes is achieved up to 20°, but with the increment of the OIA, the BW% increases at TE polarisation mode and decreases at TM polarisation mode. Finally, $S_{21} < -10$ dB is achieved for the n257 band up to 60° at TM polarisation mode, while for further increments of the OIA, the $S_{21} < -10$ dB becomes narrow and does not cover the n257 band. This characteristic arises because of the impedance variation of the **metamaterial FSS** ($Z^{METAMATERIAL}_{TE/TM}$) at OIA for both TE and TM polarization, which can be expressed by Equation (12) [37].

$$Z^{FSS}_{TE} = \frac{Z_o}{\cos\theta}, \text{ and } Z^{FSS}_{TM} = Z_o \cos\theta \qquad ...(12)$$

The peak TZ frequency and BW% of the developed FSS at different OIA for both modes are illustrated in Figure 28 (c). The shifting of the peak TZ at OIA is calculated by Equation (13) [37], where The TZ frequency shows only <0.21% of shifting at 60° for both **TE** and **TM** polarized waves.

$$\text{Shifting (\%)} = \frac{f_{\text{shift}} - f_{\text{Inital}}}{f_{\text{Inital}}} \times 100\% \qquad ...(13)$$

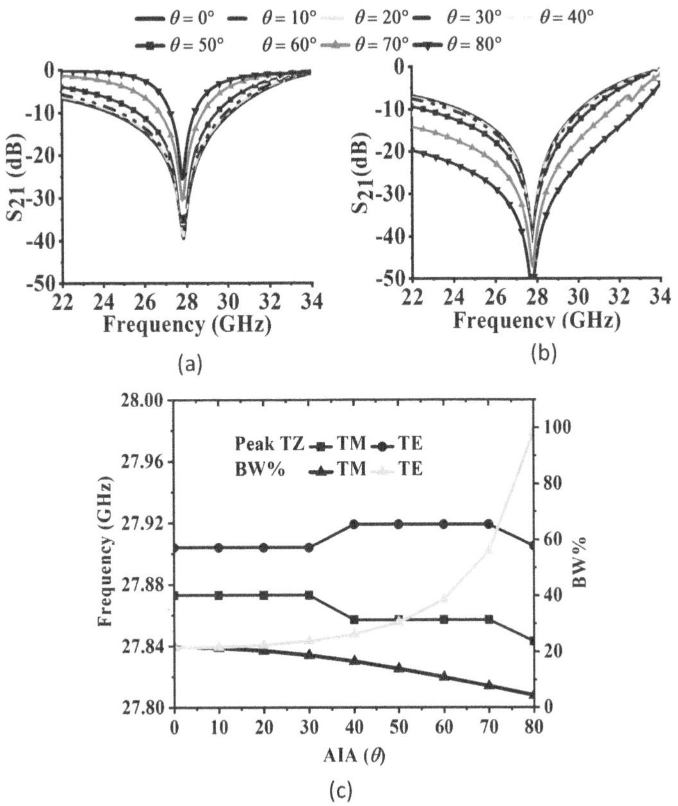

Fig. 28 S_{21} of metamaterial FSS for different OIA at: (a) TM, and (b) TE polarised wave, and (c) peak TZ frequency and BW% at TE and TM polarisation mode.

At the peak TZ, the FSS behaves as a series *LC* resonance, where the quality factor is given by Equation (14) [37]. The increment of OIA is responsible for the increase and decrease of Z_{TE}^{FSS} and Z_{TM}^{FSS} in Equation (5.6). (12) [37]. This feature leads to a reduced Q-factor in TE polarisation mode and an increased **Q-factor** in TM polarisation mode, according to Equation (14) [37]. Particularly, BW% decreases for TM and increases for **TE polarisation** mode.

$$Q = \frac{1}{Z_{TE/TM}^{FSS}} \sqrt{\frac{L}{C}} \qquad ...(14)$$

The stopband stability can be analysed by a figure of merit (**FoM**) shown in Equation (15) [37] where, $FBW(\theta_{0^\circ})$ denotes the fractional BW at 0° OIA and $FBW(\theta_{max})$ expresses the fractional BW at high OIA (θ_{max} = 60°). The FSS shows FoM of 0.58 and 0.64 at TE and TM polarised wave. This value is smaller than the miniaturised FSS presented in

$$FoM_{TE/TM} = \frac{2\,|\,FBW(\theta_{\max}) - FBW(\theta_{\theta^\circ})\,|}{|\,FBW(\theta_{\max}) + FBW(\theta_{\theta^\circ})\,|} \qquad ...(15)$$

3.4 Surface Current Electric Field and Magnetic Field Analysis

The surface current distribution of the **FSS** at peak TZ frequency (27.85 GHz) and two lower TZ frequencies (24.33 GHz and 30.19 GHz) for TE and TM polarisation mode are presented in Figure 29. To analyse the stopband filter behaviour. For TE mode, At 24.33 GHz frequency, the current traverses a longer path in the vertical arm of the fractal cross resonator. The current is high at the upper-left and lower-right portions of the FSS and less at the opposite side, which is similar to a half-wave dipole. A similar current distribution also appeared at 30.19 GHz stopband frequency; however, the intensity is slightly higher due to the high frequency. At 27.85 GHz frequency, the half-wave dipole behaviour is similar to other frequencies, but the intensity is low, indicating the highest **transmission block**. As discussed earlier, the current has a uniform magnitude due to the fractal square ring loading, achieving a stable filter behaviour at high OIA. A similar surface current distribution appeared for TM mode with a 90° rotational form.

Fig. 29 Surface current distributions of the FSS at TE polarisation mode: (a) 24.33 GHz, (b) 27.85 GHz, (c) 30.19 GHz; and TM polarisation mode: (d) 24.33 GHz, (e) 27.85 GHz, (f) 30.19 GHz.

4. Frequency Tuning

The transmission coefficient (S_{21}) of the FSS is directly related to the permittivity and thickness of the substrate, which is earlier expressed in Equation (8) [82]. Therefore, the peak **TZ** frequency and stopband can be tuneable by varying the substrate permittivity and thickness. The investigation of **frequency tuning** is completed by adding an additional substrate layer with the FSS, as shown in Figure 30(a). Due to the two-substrate layer, Equation (8) [82] becomes:

$$\begin{bmatrix} A & B \\ C & D \end{bmatrix} = [M_{FSS}][M_1][M_2] \qquad ...(16)$$

Figure 30(b) shows the S_{21} of the FSS, where the permittivity of substrate layer-2 is varied from 1–10, keeping the loss tangent = 0 and thickness = 0.25 mm. The peak TZ frequency is shifting towards the lower frequency with the increment of **permittivity** and BW% slightly reduced. When permittivity is 10, the peak TZ is shifted to 26.8 GHz, and the TZ frequency shifting with an additional substrate layer is 3.73%. The peak TZ frequency for different permittivity values is presented in Figure 30(c). The district symbol denotes the simulation data, and the continuous line is the polynomial fit data, which is estimated by the parabolic Equation of (17) [83] where, f_{TZ}^{FSS} is the shifted TZ frequency, εr is the permittivity of substrate layer-2, $X1$, $X2$, $X3$, and $X4$ are the polynomial constants. The above fourth-order terms are ignored due to their very small value.

$$f_{TZ}^{FSS} = X_1 + X_2\, \varepsilon_{r,\,MUT} + X_3\, \varepsilon_{r,\,MUT}^2 + X_4\, \varepsilon_{r,\,MUT}^3 \qquad ...(17)$$

Figure 14(c) shows the continuous line begins from $\varepsilon_r = 1$, which indicates air or no substrate layer 2 on the back side of the **FSS**. As the presence of substrate layer 2 and increment of permittivity value and the Equation (17) [83] can be expended with respect to ($\varepsilon_{r,\,sub2} - 1$), thereon peak TZ is X_1 when $\varepsilon_{r,\,sub2} = 1$,

$$f_{TZ}^{FSS} = X_1 + X_2\,(\varepsilon_{r,\,sub2} - 1) + X_3(\varepsilon_{r,\,sub2} - 1)^2 + X_4(\varepsilon_{r,\,sub2} - 1)^3 \qquad ...(18)$$

From the fitting curve presented in Figure 30(c), the polynomial constants of Equation (18) [83] are derived as follows.

$$f_{TZ}^{FSS} = 27.84 - 0.20416(\varepsilon_{r,\,sub2} - 1) + 0.01095(\varepsilon_{r,\,sub2} - 1)^2 + 0.0039(\varepsilon_{r,\,sub2} - 1)^3 ...(19)$$

Equation (19) [83] is validated by utilising Rogers RO4350B (thickness 0.254 mm) material in substrate layer-2 with a permittivity of 3.66. Figure 31 shows the S_{21} of the single-layer and dual-layer FSS. The peak **TZ frequency** is shifted to 27.3 GHz from 27.84 GHz frequency. By placing $\varepsilon_{r,\,sub2} = 3.66$ in Equation (19) [83] the peak TZ is calculated as 27.3 GHz, similar to simulations.

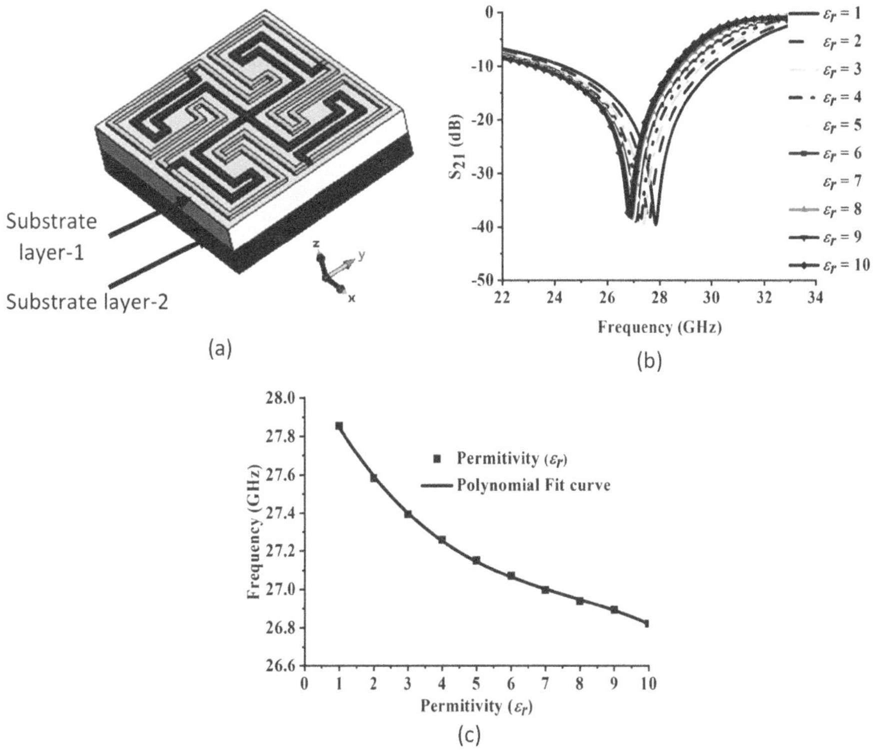

Fig. 30 (a) passive frequency tuneable FSS, (b) S_{21} of FSS for different permittivity values of substrate layer-2, (c) peak TZ frequency at different permittivity values of substrate layer-2 and fit curve.

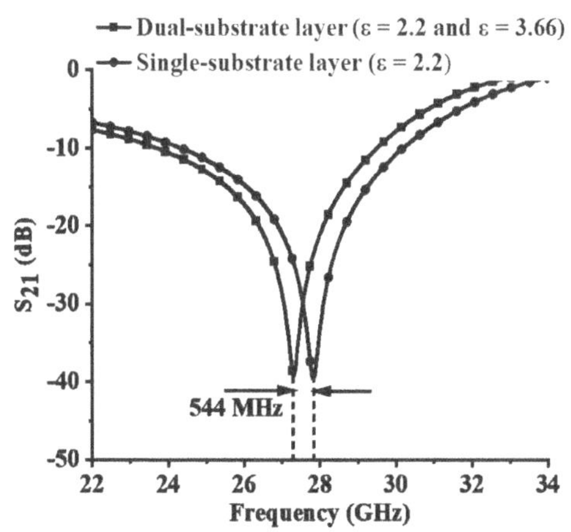

Fig. 31 S_{21} of the FSS for single and dual layer substrate.

5. Conformal METAMATERIAL FSS

The flexible and thin profile of the developed FSS enables it for conformal applications. To evaluate the stopband filter response at the conformal surface, the **FSS** is bent shown in Figure 32.

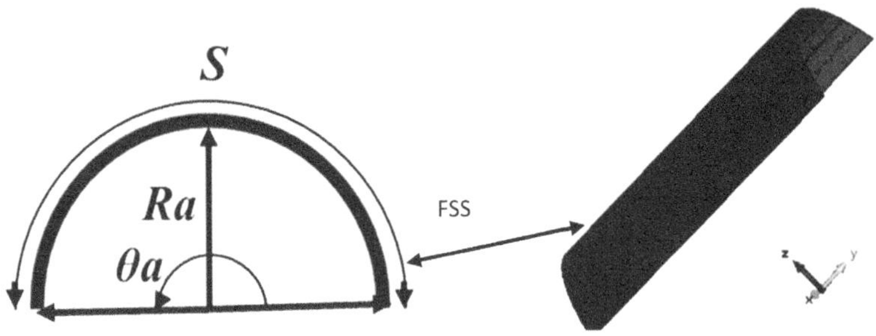

Fig. 32 Curved conditions of developed FSS.

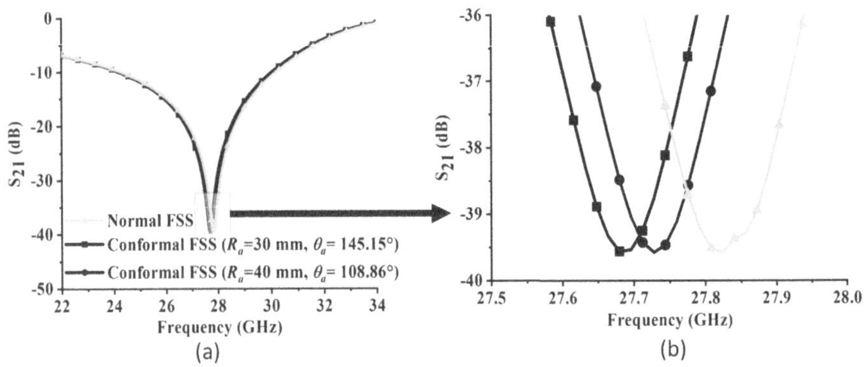

Fig. 33 (a) Simulated S_{21} of conformal FSS at different conformal angles for TE polarised wave at the normal incident of EM wave, (b) enlarged view at peak TZ frequency.

The FSS is bent as a semicircle whose radius is R_a, and the subtends angle of θa by curved FSS at the centre. The arc length S is $40P_c$ (P_c is the unit **cell periodicity**), and the radius of the curved FSS is calculated by Equation (20) [83].

$$R_a = 180 \times \frac{40\ P_c}{\pi\theta_a} \qquad \ldots(20)$$

where conformal angle = θa (degree) and arc length = $40P_c$.

The FSS is simulated for two different values of θc, as shown in Figure 33. In contrast, maintaining the array size constant, the SE is similar to the normal plain FSS surface.

6. Experimental Results

To validate the designed FSS, a 40 × 40 arrays prototype is fabricated with a lateral dimension of 76 × 76 mm^2. The **transmission coefficient** (S_{21}) is measured using the free space measurement method, as explained in [32]. Figure 34(a, b) shows the fabricated prototype of **FSS** and the experimental arrangement, respectively.

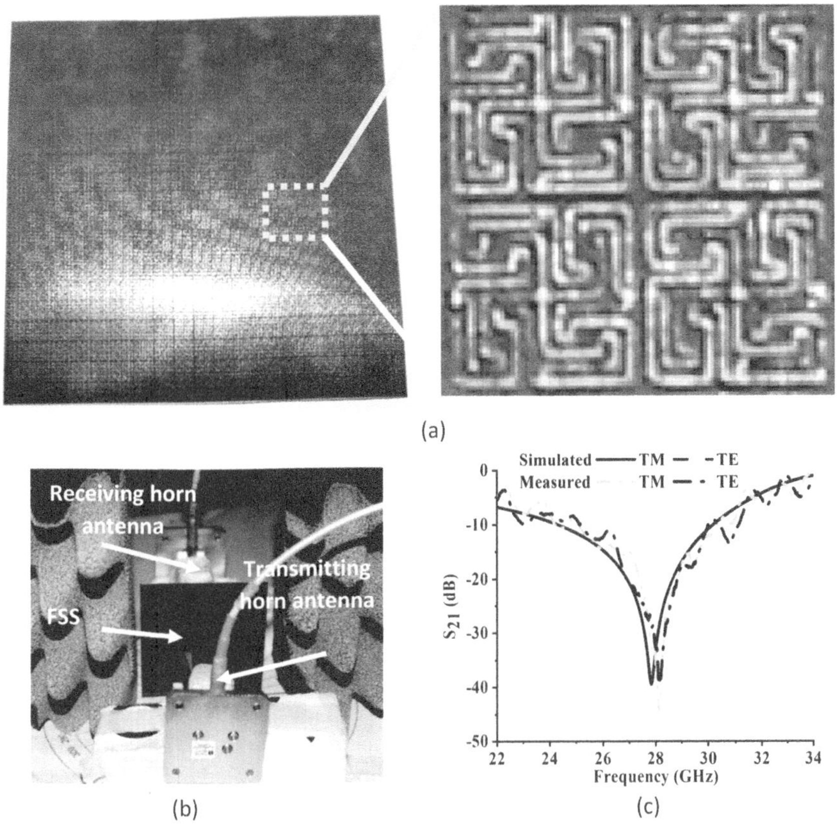

Fig. 34 (a) fabricated 40 × 40 unit cell FSS prototype, (b) measurement setup, and (c) measured S_{21} at the normal incident of EM wave for TE and TM polarisation mode.

Two dual-ridged broadband horns are positioned on both sides of the FSS at a distance of $2D^2/\lambda$, where the dimension of the **FSS** is D. The horn antenna operates from 18–40 GHz, a nominal gain of 12–17 dBi, and 3dB beamwidth of 20–43°. At the normal incidence of the EM wave, the simulated and measured S_{21} of the FSS for both TE and TM polarisation is presented in Figure 34(c). The measured resonance frequency of the FSS obtained at 28.11 GHz frequency and $S_{21} < -10$ dB is achieved from 25.15–30.20 GHz with BW% of 24.58%. The measured resonance frequency and filter response slightly shift to a higher frequency compared to the simulated filter response. This mismatch is attributed to the **fabrication tolerances**, measurement error, and edge diffraction because of the finite array dimension.

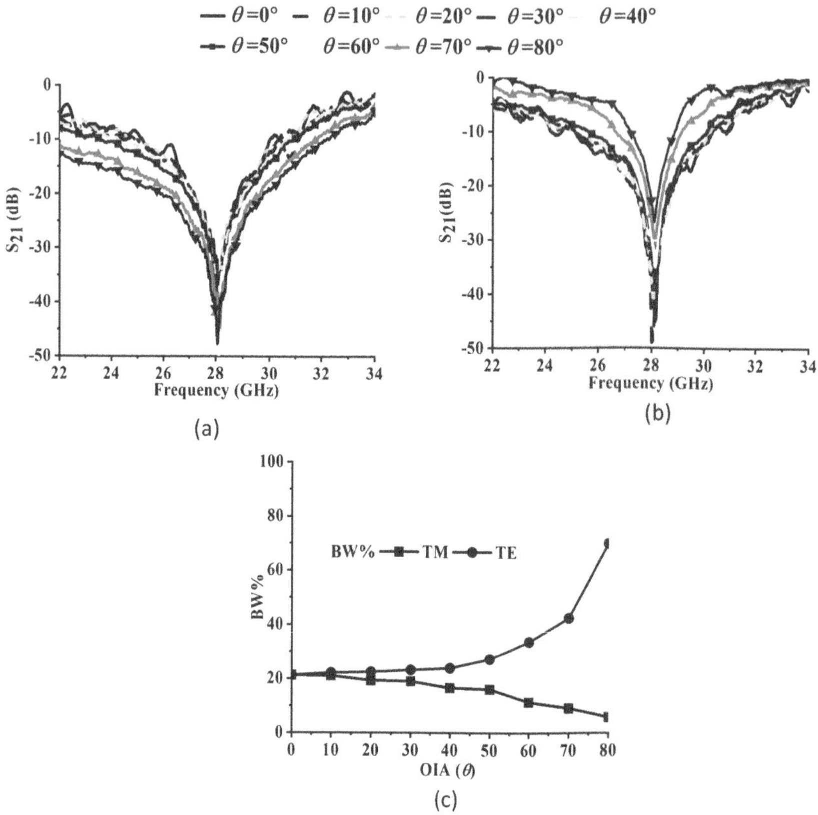

Fig. 35 Measured S_{21} of the developed metamaterial FSS: (a) TE, (b) TM polarisation mode, and (c) BW% at OIA for TE and TM polarisation mode.

The measured results show a stable filter response of the developed FSS at different **OIA** for both TE and TM polarisations, as illustrated in Figure 35(a, b). At high OIA, the measured results show an almost stable position of peak TZ frequency for both TE and TM polarisation. On the other hand, the stopband BW% increases for TE polarised wave and decreases for TM polarised wave with the increment of OIA, like simulation results. Figure 35(c) shows BW% at OIA for TE and TM polarisation mode. Figure 36(a) exhibits the measurement setup of the conformal **FSS**, where the centre of the receiving and transmitting horn **antenna** aperture coincides with the centre of the conformal FSS. The multiple reflections between the receiving and transmitting horn antenna are reduced by the microwave absorber at the edge of conformal FSS. Figure 36(b) shows the S_{21} of the FSS has stable performance at different conformal conditions, as indicated in Equation (20) [83]. When the conformal condition is $Ra = 40$ mm, $\theta a = 108.86°$ the measured peak TZ is achieved at 28.6 GHz and this peak TZ is slightly lower shifted to 28.04 GHz when the conformal condition is $Ra = 30$ mm, $\theta a = 145.15°$ which agrees well with the simulation results. The peak TZ frequency tuning is validated by measurement. Figure 37(a) shows the fabricated dual substrate layer FSS, where Rogers RO4350B (thickness 0.254 mm, ε = 3.66) is used as layer 2. The peak **TZ frequency** is shifted to a lower frequency like simulation. The measured peak TZ is achieved at 27.57 GHz by 540 MHz (Figure 37(b)).

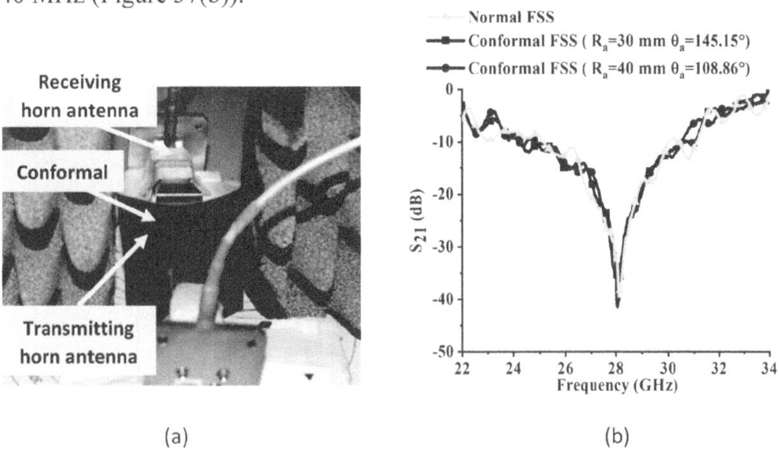

Fig. 36 (a) conformal FSS with measurement setup, (b) S_{21} of measured normal and conformal FSS at the TE polarised normal incident EM wave.

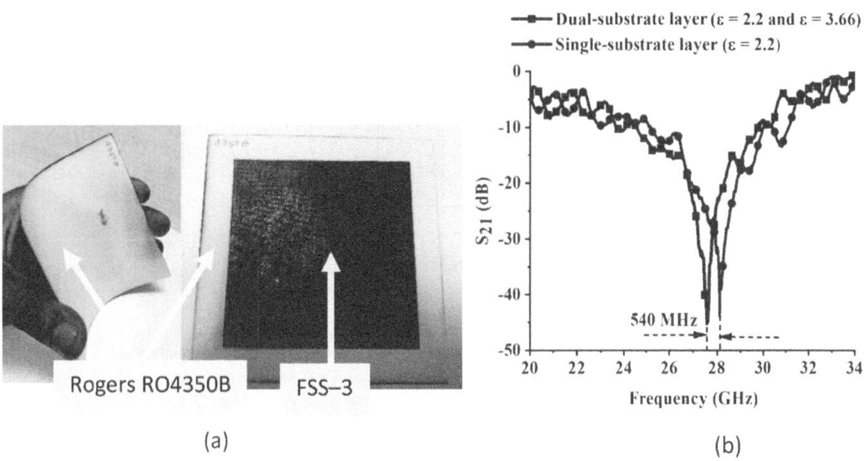

Fig. 37 (a) fabricated dual layer FSS, and (b) measured S_{21} of single and dual substrate layer FSS.

7. Performance Comparison with Existing METAMATERIAL FSS

The comparison of developed FSS and related works presented in different literature is shown in Table 3. The stopband FSS in [43, 46] shield microwave frequency and mmWave 5G band **shielding** FSSs are presented in [32, 44]. On the other hand, the passband mmWave FSS is designed in [34, 42]. The rotational symmetrical design of all listed FSS achieves polarisation insensitivity. Hence, the resonance frequency varies with the OIA and depends on the periodicity of the FSSs. The developed miniaturised FSS has the lowest unit cell size of only 1.9 × 1.9 mm^2 compared to other

Table 3 Comparison with others polarisation insensitive FSSs.

Ref	*Miniaturization Technique*	*Filter Type*	*Periodicity (λ_o)*	*Size (mm^2)/ Periodicity (λ_o)*	*Frequency (GHz)*	*BW (%)*	*Covers 5G Band*	*Conformal*	*OIA Stability*	*Peak TZ Shifting at 60°*
[43]	Polygon shape	Stop band	0.26 λ_o	8.3 × 8.3	10	40	--	Yes	60°	<3%
[32]	Cross loop resonator with a wire grid.	Stop band	0.4 λ_o	4.4 × 4.4	27.2, 33.1	8.3, 13.8	n261	No	35°	<4.1%
[34]	Vertical via based 2.5 D mandarin line	Pass band	0.26 λ_o	2.87 × 2.87	27.75	9.6	n261	No	60°	–
[44]	Meshed resonators	Stop band	0.3 λ_o	4 × 4	28	29.3	n257	Yes	60°	<2.25%
[42]	Dual loop resonator	Pass band	0.23 λ_o	2.4 × 2.4	28	7.017	n261	No	50°	–
Developed FSS	Fractal resonator	Stop band	0.154 λ_o	1.9 × 1.9	27.84	21.49	n257	Yes	60°	<0.21%

listed FSSs in Table 3. Also, the periodicity is only 0.154 λ_o, which is smaller than other's work. Therefore, the FSS exhibits improved miniaturisation by fractal cross and fractal square resonator. The developed FSS offers stable stopband filter response at n257 band up to 60° OIA for both TE and TM polarised wave, which is higher than [32, 42]. Most importantly, due to the extensive **miniaturisation,** the peak TZ shifting is only 0.21%, whereas other FSSs exhibit a very high percentage of shifting. The peak TZ of [43, 32, 44] shift <3%, <4.1%, and <2.25%, respectively, at 60° OIA,. Another important feature of the developed FSS is conformal applications which are accomplished by flexible and thin substrate materials. Where FSSs in [32, 34, 42] are not suitable for conformal applications because of the rigid substrate material, the stopband of the developed can be easily **tuneable** towards lower frequencies up to 540 MHz by adding an additional substrate layer (ε = 3.66, thickness = 0.254 mm). Finally, compared to the others listed FSSs, the developed FSS has the advantages of polarisation insensitivity, high OIA stability, low peak TZ frequency shifting, conformal applications, and frequency tunability.

8. 5G EMI Shielding Application of FSS

The developed stopband **FSS** has potential applications for EMI shielding at n257 band (mmWave) 5G frequency [44]. The amount of shielding EM waves is calculated by an **FoM**, expressed as shielding effectiveness (SE). The SE is calculated from the ratio of the transmitted (*Et*) to incident (*Ei*) wave presented in Equation (21) [83].

$$\text{SE(dB)} = -20\ \log|E_t/E_i| = -S_{21}\ \text{(dB)} \qquad ...(21)$$

Figure 38 shows the SE of the FSS for both TE and TM **polarised waves** at OIA, which shows stable responses up to 60° at the n257 band. The peak SE is 35 dB at 60° for TM mode and increases to 50 dB at 60° for TE mode.

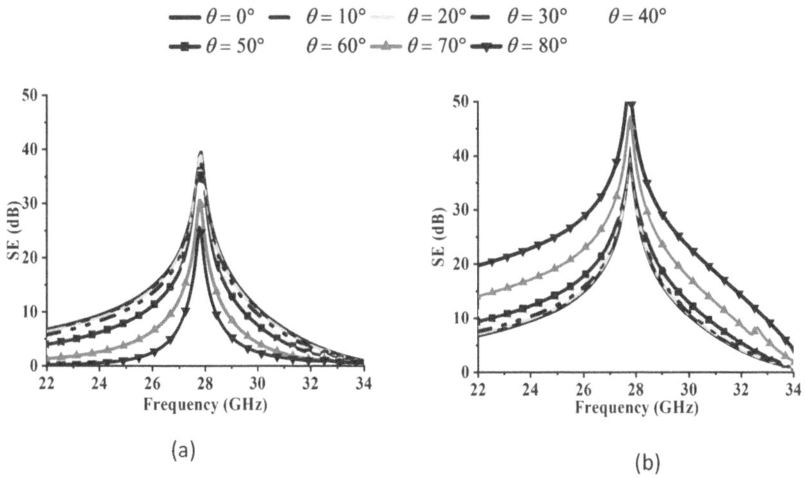

Fig. 38 The SE of FSS at: (a) TM, and (b) TE polarisation modes.

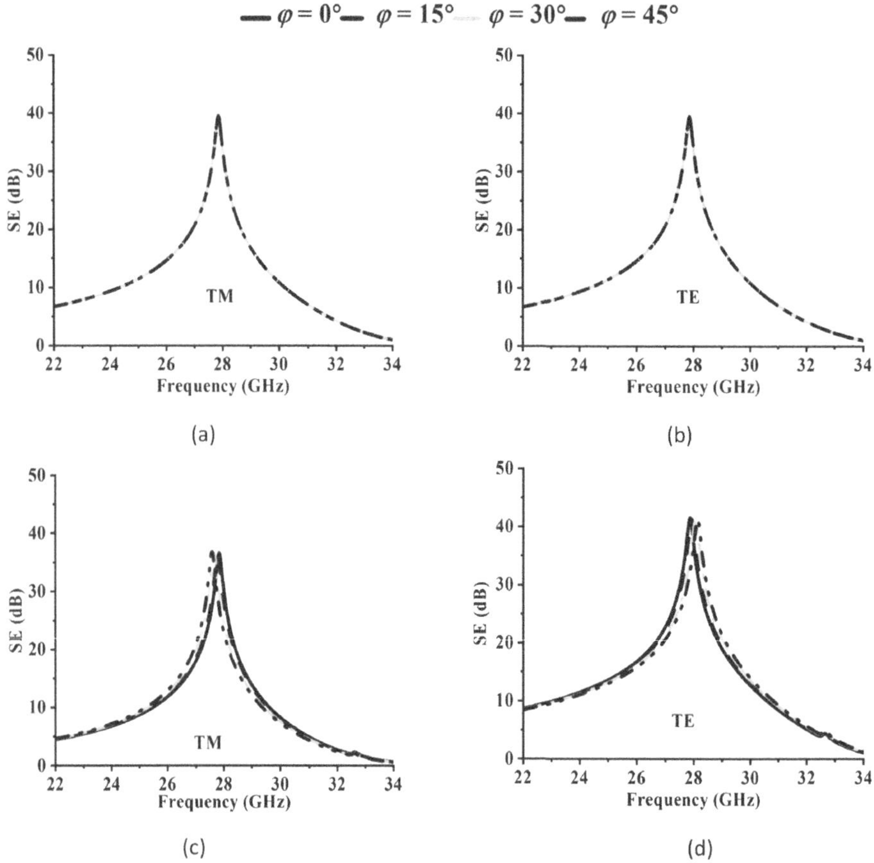

Fig. 39 The SE of FSS at $\theta = 0°$ (a) TM, (b) TE polarisation modes; and at $\theta = 45°$ (c) TM, (d) TE polarisation modes.

The FSS demonstrates **polarisation** insensitive stable filter response for different azimuthal angles (φ) due to the 90° rotational symmetric design. Figure 39 shows the SE of the designed FSS at different φ-angles when OIA is $\theta = 0°$ and $\theta = 45°$. At the normal incidence of **EM** wave, FSS shows peak **TZ frequency** at a constant position of 27.85 GHz for different φ-angles. The stopband ($S_{21} < -10$ dB) of the FSS is realised from 24.33–30.19 GHz frequency, illustrated in Figure 39(a, b). On the other hand, Figure 39(c) shows at OIA ($\theta = 45°$) peak TZ shifting to a lower frequency for TM polarised wave with the increment of φ-angles although the shifting percentage is only 1.14 % at $\varphi = 45°$ compared to $\varphi = 0°$. The stopband for TM mode at $\theta = 45°$ is reduced as discussed in Equation (12) [37] and $S_{21} < -10$ dB achieved from 25.45–29.56 GHz. The peak TZ frequency is shifted towards a higher frequency at OIA for TE mode, shown in Figure 39(d). The peak TZ frequency shifts by 0.91% for $\varphi = 45°$ compared to $\varphi = 0°$. In addition, the stopband is increased for TM mode at $\theta = 45°$. The measured SE of developed **FSS** is presented in Figure 40, which agrees well with the simulated results.

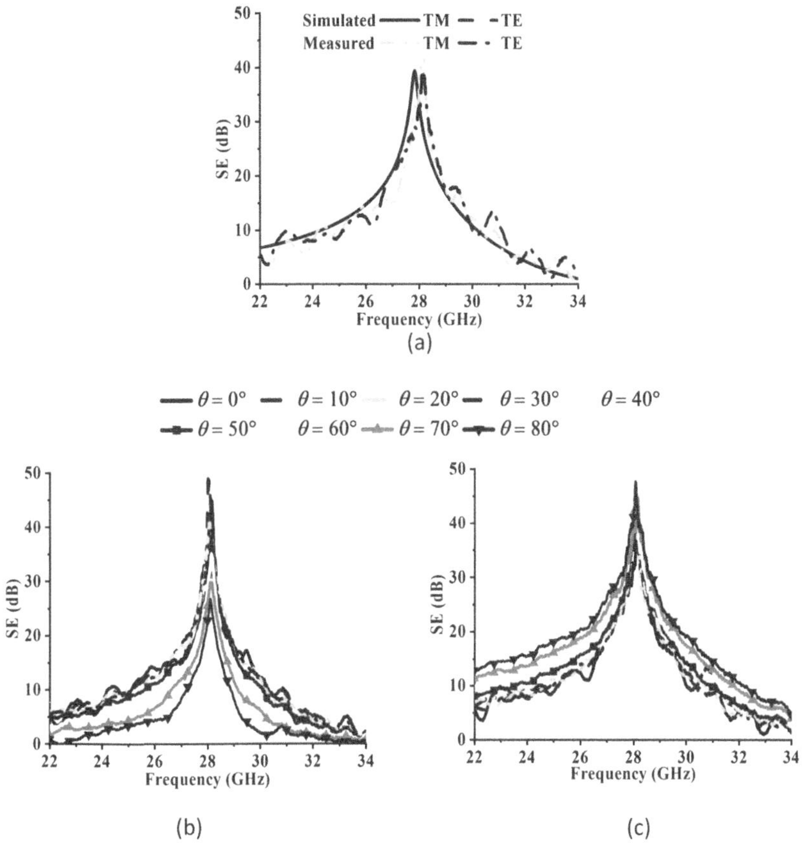

Fig. 40 (a) simulated measured SE of FSS at the normal incident of EM wave for TE and TM polarisation mode; SE at OIA for (b) TM, and (c) TE polarisation mode.

9. Summary

This chapter explores the design, characteristics, and applications of frequency selective surfaces (FSS) using metamaterials for 5G millimeter-wave (mmWave) communication systems. The chapter begins by introducing metamaterials, emphasizing their potential in controlling electromagnetic (EM) waves due to their unique structure, which allows them to exhibit specific filtering properties. The chapter details the design and performance of FSS, which are planar arrays of conducting elements that can selectively filter EM waves based on frequency and polarization. It highlights the challenges posed by mmWave 5G communications, such as high attenuation and limited range, and how FSS can mitigate these issues by improving antenna performance and providing electromagnetic interference (EMI) shielding.

The chapter goes on to explain the design fundamentals of FSS, including unit cell geometry, polarization sensitivity, and the impact of the angle of incidence (OIA) on the resonant frequency. It further discusses various strategies for overcoming the challenges of miniaturization, polarization insensitivity, and incident angle stability for mmWave FSS designs, particularly for 5G applications. Moreover, it delves into the materials and substrates used, such as Rogers RT5880, and highlights the importance of achieving high OIA stability and low peak frequency shifting in FSS designs.

Additionally, the chapter presents various types of FSS, including absorbers, bandpass, and stopband filters, each suited to different 5G applications like EMI shielding and gain enhancement. The polarization-insensitive nature of the designs is particularly useful for ensuring reliable performance across varying incident angles and polarization modes. Simulation and measurement results are provided, showcasing the high performance of the miniaturized FSS, which operates efficiently within the n257 band (28 GHz), with performance improvements in incident angle stability and polarization insensitivity. The chapter concludes by emphasizing the potential of these metamaterial-based FSS to significantly improve the performance of 5G systems, particularly in EMI shielding and antenna gain enhancement.

References

[1] T. Wu. (Ed). *Frequency Selective Surface and Grid Array* (A Wiley Interscience Publication), 1995.

[2] B.A. Munk. *Frequency Selective Surfaces: Theory and Design*. John Wiley & Sons, 2005.

[3] L. Li et al. "The path to 5G: mmWave aspects," *Journal of Communications and Information Networks*, vol. 1, no. 2, pp. 1–18, 2016.

[4] FCC. "America's 5G Future." FCC. https://www.fcc.gov/5G (accessed 1 March, 2023).

[5] A. Ghosh, A. Maeder, M. Baker, and D. Chandramouli. "5G evolution: A view on 5G cellular technology beyond 3GPP release 15," *IEEE Access*, vol. 7, pp. 127639–127651, 2019.

[6] Qualcomm. "Qualcomm 5G—Redefining Speed." https://www.qualcomm.com/products/technology/5g/5g-speeds#:~:text=Powered%20by%20the%20Snapdragon%20Modem%2DRF%20system%2C%20mmWave%205G%20can,5G%20mmWave%20in%20Q2%202021. (accessed.

[7] N. Hussain, M.-J. Jeong, A. Abbas and N. Kim. "Metasurface-based single-layer wideband circularly polarized MIMO antenna for 5G millimeter-wave systems," *IEEE Access*, vol. 8, pp. 130293–130304, 2020.

[8] 3GPP. *NR; User Equipment (UE) Radio Transmission and Reception; Part 2: Range 2 Standalone*, Ed: Sophia Antipolis Valbonne, France, 2023.

[9] N. Qualcomm Technologies. *Making 5G a reality: Addressing the Strong Mobile Broadband Demand in 2019 & Beyond*, 2017. [Online]. Available: https://www.qualcomm.com/content/dam/qcomm-martech/dm-assets/documents/whitepaper_-_making_5g_a_reality_-_addressing_the_strong_mobile_broadban_1.pdf

[10] MCMC. *Final Report on Allocation of Spectrum Bands for Mobile Broadband Service in Malaysia*. MCMC. https://www.mcmc.gov.my/en/media/press-releases/final-report-on-allocation-of-spectrum-bands-for-m (accessed 1 August, 2023).

[11] S.K. Rao and R. Prasad. "Impact of 5G technologies on industry 4.0," *Wireless Personal Communications*, vol. 100, pp. 145–159, 2018.

[12] F. Giust et al. "Multi-access edge computing: The driver behind the wheel of 5G-connected cars," *IEEE Communications Standards Magazine*, vol. 2, no. 3, pp. 66–73, 2018.

[13] K.E. Skouby and P. Lynggaard. "Smart home and smart city solutions enabled by 5G, IoT, AAI and CoT services," in *2014 International Conference on Contemporary Computing and Informatics (IC3I)*, 2014: IEEE, pp. 874–878.

[14] Z. Zhang, F. Wen, Z. Sun, X. Guo, T. He and C. Lee. "Artificial intelligence-enabled sensing technologies in the 5G/internet of things era: From virtual reality/augmented reality to the digital twin," *Advanced Intelligent Systems*, vol. 4, no. 7, p. 2100228, 2022.

[15] N.K. Mallat, M. Ishtiaq, A. Ur Rehman and A. Iqbal. "Millimeter-wave in the face of 5G communication potential applications," *IETE Journal of Research,* pp. 1–9, 2020.

[16] S. Wang, X. Fu, R. Ruby and Z. Li. "Pilot spoofing detection for massive MIMO mmWave communication systems with a cooperative relay," *Computer Communications,* vol. 202, pp. 33–41, 2023.

[17] I. Baştürk. "Green communication for OFDMA cellular networks with multiple antennas," *Computer Communications,* vol. 168, pp. 93–101, 2021.

[18] K. Sultan, M. Ikram and N. Nguyen-Trong. "A Multiband Multibeam Antenna for Sub-6 GHz and mm-Wave 5G Applications," *IEEE Antennas and Wireless Propagation Letters,* vol. 21, no. 6, pp. 1278–1282, 2022.

[19] Y. Niu, Y. Li, D. Jin, L. Su and A.V. Vasilakos. "A survey of millimeter wave communications (mmWave) for 5G: opportunities and challenges," *Wireless Networks,* vol. 21, pp. 2657–2676, 2015.

[20] X. Wang et al. "Millimeter wave communication: A comprehensive survey," *IEEE Communications Surveys & Tutorials,* vol. 20, no. 3, pp. 1616–1653, 2018.

[21] V.M. Research. "Millimeter Wave Technology Market Size and Forecast." https://www.verifiedmarketresearch.com/product/global-millimeter-wave-technology-market-size-and-forecast-to-2025/ (accessed 1 August 2023).

[22] M. Parker. *Digital Signal Processing 101: Everything You Need to Know to Get Started.* Newnes, 2017.

[23] T.J. Cui, D.R. Smith and R. Liu. *Metamaterials.* Springer, 2010.

[24] M.L. Hakim et al. "Triple-Band Square Split-Ring Resonator Metamaterial Absorber Design with High Effective Medium Ratio for 5G Sub-6 GHz Applications," *Nanomaterials,* vol. 13, no. 2, p. 222, 2023.

[25] M.L. Hakim, T. Alam, M.T. Islam, M.H. Baharuddin, A. Alzamil and M.S. Islam. "Quad-band polarization-insensitive square split-ring resonator (SSRR) with an inner Jerusalem cross metamaterial absorber for Ku-and K-Band sensing applications," *Sensors,* vol. 22, no. 12, p. 4489, 2022.

[26] D.M. Pozar. *Microwave Engineering.* John Wiley & Sons, 2011.

[27] B.A. Munk. *Frequency Selective Surfaces: Theory and Design.* New York, NJ John Wiley & Sons, Ed: Inc, 2000.

[28] B. Munk, R. Kouyoumjian and L. Peters. "Reflection properties of periodic surfaces of loaded dipoles," *IEEE Transactions on Antennas and Propagation,* vol. 19, no. 5, pp. 612–617, 1971.

[29] E.A. Parker. "The gentleman's guide to frequency selective surfaces," 1991.

[30] T.K. Chang, R.J. Langley and E.A. Parker. "Active frequency-selective surfaces," *IEE Proceedings-Microwaves, Antennas, and Propagation,* vol. 143, no. 1, pp. 62–66, 1996.

[31] A.K. Rashid, B. Li and Z. Shen., "An overview of three-dimensional frequency-selective structures," *IEEE Antennas and Propagation Magazine,* vol. 56, no. 3, pp. 43–67, 2014.

[32] S. Dey, S. Dey and S.K. Koul. "Second-order, single-band, and dual-band bandstop frequency selective surfaces at millimeter wave regime," *IEEE Transactions on Antennas and Propagation,* vol. 70, no. 8, pp. 7282–7287, 2022.

[33] P. Callaghan, E.A. Parker and R.J. Langley. "Influence of supporting dielectric layers on the transmission properties of frequency selective surfaces," in *IEE Proceedings H (Microwaves, Antennas, and Propagation)*, 1991, vol. 138, no. 5: IET, pp. 448–454.

[34] D. Li, T.-W. Li, E.-P. Li and Y.-J. Zhang. "A 2.5-D angularly stable frequency selective surface using via-based structure for 5G EMI shielding," *IEEE Transactions on Electromagnetic Compatibility,* vol. 60, no. 3, pp. 768–775, 2017.

[35] Z. Sipus and M. Bosiljevac. "Efficient analysis of curved frequency selective surfaces," in *2009 3rd European Conference on Antennas and Propagation*, 2009: IEEE, pp. 2726–2730.

[36] S. Savia, E.A. Parker and B. Philips. "Finite planar-and curved-ring-element frequency-selective surfaces," *IEE Proceedings-Microwaves, Antennas, and Propagation,* vol. 146, no. 6, pp. 401–406, 1999.

[37] S. Dey and S. Dey. "Conformal miniaturized angular stable triband frequency selective surface for EMI shielding," *IEEE Transactions on Electromagnetic Compatibility,* vol. 64, no. 4, pp. 1031–1041, 2022.

[38] W. Afzal, A. Ebrahimi, M.R. Robel and W.S. Rowe. "Low-Profile Higher-Order Narrowband Bandpass Miniaturized-Element Frequency-Selective Surface," *IEEE Transactions on Antennas and Propagation,* vol. 71, no. 4, pp. 3736–3740, 2023.

[39] T. Wu, Y.-M. Ma, J. Chen and L.-L. Wang. "A low profile quadruple-band polarization insensitive metamaterial absorber," *Progress In Electromagnetics Research M,* vol. 90, pp. 69–79, 2020.

[40] S.A. Naqvi, M.A. Baqir, G. Gourley, A. Iftikhar, M. Saeed Khan and D.E. Anagnostou. "A Novel Meander Line Metamaterial Absorber Operating at 24 GHz and 28 GHz for the 5G Applications," *Sensors,* vol. 22, no. 10, p. 3764, 2022.

[41] Q. Wang, Y.K. Liu, G.C. Wang, W.X. Jiang and T.J. Cui. "Tunable triple-band millimeter-wave absorbing metasurface based on nematic liquid crystal," *AIP Advances,* vol. 12, no. 1, p. 015127, 2022.

[42] H. Chen, H. Chen, X. Xiu, Q. Xue, and W. Che. "Transparent FSS on Glass Window for Signal Selection of 5G Millimeter-Wave Communication," *IEEE Antennas and Wireless Propagation Letters,* vol. 20, no. 12, pp. 2319–2323, 2021.

[43] M. Bilal, R. Saleem, Q.H. Abbasi, B. Kasi and M.F. Shafique. "Miniaturized and flexible FSS-based EM shields for conformal applications," *IEEE Transactions on Electromagnetic Compatibility,* vol. 62, no. 5, pp. 1703–1710, 2020.

[44] D.J. King, K. Hettak, M.R. Chaharmir and S. Gupta. "Flexible Ink-Minimized Screen-Printed Frequency Selective Surfaces With Increased Optical Transparency for 5G Electromagnetic Interference Mitigation," *IEEE Transactions on Components, Packaging, and Manufacturing Technology,* vol. 13, no. 1, pp. 110–119, 2023.

[45] T. Hong, S. Guo, W. Jiang and S. Gong. "Highly selective frequency selective surface with ultrawideband rejection," *IEEE Transactions on Antennas and Propagation,* vol. 70, no. 5, pp. 3459–3468, 2021.

[46] M.H. Nisanci and F. de Paulis "Easy-to-design-and-manufacture frequency selective surfaces for conformal applications," *IEEE Antennas and Wireless Propagation Letters,* vol. 20, no. 5, pp. 753–757, 2021.

[47] S. Abbasi, J. Nourinia, C. Ghobadi, M. Karamirad and B. Mohammadi. "A sub-wavelength polarization sensitive band-stop FSS with wide angular response for X-and Ku-Bands," *AEU-International Journal of Electronics and Communications,* vol. 89, pp. 85–91, 2018.

[48] X.-C. Zhu, P.-P. Zhang, Y.-X. Zhang, J.-X. Ge and Z.-H. Gao. "A high-gain filtering antenna based on folded reflectarray antenna and polarization-sensitive frequency selective surface," *IEEE Antennas and Wireless Propagation Letters,* vol. 19, no. 8, pp. 1462–1465, 2020.

[49] F. Khajeh-Khalili, M.A. Honarvar, M. Naser-Moghadasi and M. Dolatshahi. "Gain enhancement and mutual coupling reduction of multiple-intput multiple-output antenna for millimeter-wave applications using two types of novel metamaterial structures," *International Journal of RF and Microwave Computer-Aided Engineering,* vol. 30, no. 1, p. e22006, 2020.

[50] M. Mantash, A. Kesavan and T.A. Denidni. "Beam-tilting endfire antenna using a single-layer FSS for 5G communication networks," *IEEE Antennas and Wireless Propagation Letters,* vol. 17, no. 1, pp. 29–33, 2018.

[51] V. Krushna Kanth and S. Raghavan. "Design and optimization of complementary frequency selective surface using equivalent circuit model for wideband EMI shielding," *Journal of Electromagnetic Waves and Applications,* vol. 34, no. 1, pp. 51–69, 2020.

[52] Y. Zhang, X. Guo and J. Xu. "A novel dual-band miniaturized frequency selective surface with high selectivity," *Microwave and Optical Technology Letters,* vol. 65, no. 2, pp. 548–558, 2023.

[53] B. Ashvanth, B. Partibane and M.G.N. Alsath. "An ultraminiaturized frequency selective surface with angular and polarization stability," *IEEE Antennas and Wireless Propagation Letters,* vol. 21, no. 1, pp. 114–118, 2021.

[54] N. Liu, X. Sheng, C. Zhang, J. Fan and D. Guo. "A design method for synthesizing miniaturized FSS using lumped reactive components," *IEEE Transactions on Electromagnetic Compatibility,* vol. 60, no. 2, pp. 536–539, 2017.

[55] S. Ghosh and K.V. Srivastava. "Broadband polarization-insensitive tunable frequency selective surface for wideband shielding," *IEEE Transactions on Electromagnetic Compatibility,* vol. 60, no. 1, pp. 166–172, 2017.

[56] M. Guo et al. "Analysis and design of a high-transmittance performance for varactor-tunable frequency-selective surface," *IEEE Transactions on Antennas and Propagation,* vol. 69, no. 8, pp. 4623–4632, 2020.

[57] M.L. Hakim, T. Alam, A.F. Almutairi, M.F. Mansor and M.T. Islam. "Polarization insensitivity characterization of dual-band perfect metamaterial absorber for K band sensing applications," *Scientific Reports,* vol. 11, no. 1, p. 17829, 2021.

[58] W. Zhang, J.-Y. Li and J. Xie. "High sensitivity refractive index sensor based on metamaterial absorber," *Progress in Electromagnetics Research M,* vol. 71, pp. 107–115, 2018.

[59] S.K. Dhawan, A.P. Singh and A. Ohlan. *Smart Materials Design for Electromagnetic Interference Shielding Applications*, 2022. Bentham Books.

[60] A. Pandey, R. Kumar, D.P. Mondal, P. Kumar and S. Singh. "Metal and Metal Matrix 2D Nanomaterial Composites: Attractive Alternatives for EMI Shielding Applications," in *Adapting 2D Nanomaterials for Advanced Applications*: ACS Publications, 2020, pp. 347–373.

[61] S. Geetha, K. Satheesh Kumar, C.R. Rao, M. Vijayan and D. Trivedi. "EMI shielding: Methods and materials: A review," *Journal of Applied Polymer Science,* vol. 112, no. 4, pp. 2073–2086, 2009.

[62] Hollandshielding. "EMI-shielding housings/enclosures." https://hollandshielding.com/EMI-shielding-housings-enclosures (accessed 1 August 2023).

[63] T. Hong, S. Guo, W. Jiang and S. Gong. "Highly Selective Frequency Selective Surface With Ultrawideband Rejection," *IEEE Transactions on Antennas and Propagation,* vol. 70, no. 5, pp. 3459–3468, 2022.

[64] C.L. Valle, G.T. Carranza and R.C. Rumpf. "Conformal Frequency Selective Surfaces for Arbitrary Curvature," *IEEE Transactions on Antennas and Propagation,* vol. 71, no. 1, pp. 612–620, 2022.

[65] M.L. Hakim, M.T. Islam and T. Alam. "Incident Angle Stable Broadband Conformal mmWave FSS for 5G (n257, n258, n260, and n261) band EMI Shielding Application," *IEEE Antennas and Wireless Propagation Letters,* vol. 23, no. 2, pp. 488–492, 2024.

[66] M.A. Shukoor and S. Dey. "Novel dual-mode polarization insensitive wide angular stable circular ring based deca-band absorber for RCS and EMI shielding applications," *IEEE Transactions on Electromagnetic Compatibility,* vol. 64, no. 5, pp. 1337–1345, 2022.

[67] W.-J. Liao, W.-Y. Zhang, Y.-C. Hou, S.-T. Chen, C. Y. Kuo and M. Chou. "An FSS-integrated low-RCS radome design," *IEEE Antennas and Wireless Propagation Letters,* vol. 18, no. 10, pp. 2076–2080, 2019.

[68] M. Pazokian, N. Komjani and M. Karimipour "Broadband RCS reduction of microstrip antenna using coding frequency selective surface," *IEEE Antennas and Wireless Propagation Letters,* vol. 17, no. 8, pp. 1382–1385, 2018.

[69] P.M. Sainadh, A. Sharma and S. Ghosh. "Polarization-Insensitive Absorptive/Transmissive Reconfigurable Frequency Selective Surface with Embedded Biasing," *IEEE Antennas and Wireless Propagation Letters,* vol. 22, no. 1, pp. 164–168, 2022.

[70] Y. Li, P. Ren, R. Chen, B. Xu, Z. Xiang and M. Wang, "Design of Miniaturized Frequency Selective Rasorber Based on Complementary Bandpass FSS for 5G Applications," *IEEE Transactions on Electromagnetic Compatibility,* 2024.

[71] L.T. Amudhu, R.V. Samsingh, S.E. Florence and B.S. Abirami. "Low-Profile Polymer Composite Radar Absorber Embedded with Frequency Selective Surface," *IEEE Transactions on Components, Packaging, and Manufacturing Technology,* 2023.

[72] F.A. Dicandia and S. Genovesi. "Linear-to-Circular Polarization Transmission Converter Exploiting Meandered Metallic Slots," *IEEE Antennas and Wireless Propagation Letters,* vol. 21, no. 11, pp. 2191–2195, 2022.

[73] Y. Han, T. Liu, Q. Li, W. Che and Q. Xue. "A Bandpass Frequency Selective Surface Based on Aramid-paper Honeycomb Sandwich Structure," *IEEE Transactions on Antennas and Propagation*, 2023.
[74] FCC. "America's 5G Future." fcc.gov. https://www.fcc.gov/5G (accessed 16 December 2022).
[75] D. Brenner, "Global 5G spectrum update," in *Spectrum Strategy & Technology Poilcy*: SVP, 2020.
[76] S. Dey and S. Dey. "Miniaturized Near All Angle Stable 2.5 D FSS For Dual Band Shielding in Transmission Mode and Multiband Absorption in Reflection Mode," *IEEE Transactions on Antennas and Propagation*, vol. 71, no. 9, pp. 7323–332, 2023.
[77] Z. Sun, L. Yan, X. Zhao and R.X.-K. Gao. "An Ultra-Wideband Frequency Selective Surface Absorber with High Polarization-Independent Angular Stability," *IEEE Antennas and Wireless Propagation Letters*, 2022.
[78] H. Jiang, S. Liao and Q. Xue. "Hybrid Capacitive and Inductive Coupling Paths for Implementing Quasi-Elliptical High-Order Bandpass FSS With High Out-of-Band Suppression," *IEEE Transactions on Antennas and Propagation*, vol. 71, no. 4, pp. 3294–3303, 2023.
[79] Y. Fan et al. "Ultrawideband dual-polarized frequency-selective absorber with tunable reflective notch," *IEEE Transactions on Antennas and Propagation*, vol. 71, no. 3, pp. 2855–2860, 2023.
[80] M.L. Hakim, T. Alam and M.T. Islam. "Polarization Insensitive and Oblique Incident Angle Stable Miniaturized Conformal FSS for 28/38 GHz mm-Wave band 5G EMI Shielding Applications," *IEEE Antennas and Wireless Propagation Letters*, vol. 22, no. 11, pp. 2644–2648, 2023.
[81] T.-W. Li et al. "A novel miniaturized multiband strong coupled-FSS structure insensitive to almost all angles and all polarizations," *IEEE Transactions on Antennas and Propagation*, vol. 69, no. 12, pp. 8470–8478, 2021.
[82] V. Chaudhary and R. Panwar. "Machine Learning Empowered Magnetic Substrate Coupled Broadband and Miniaturized Frequency Selective Surface," *IEEE Transactions on Electromagnetic Compatibility*, vol. 65, no. 2, pp. 406–413, 2023.
[83] M.L. Hakim, M.T. Islam and T. Alam. "Ultra-Miniaturized Conformal Polarization Insensitive and Incident Angle Stable FSS for n257 Band 5G EMI Shielding Applications," *IEEE Transactions on Antennas and Propagation*, 2024.
[84] U. Farooq, M.F. Shafique and M.J. Mughal. "Polarization insensitive dual band frequency selective surface for RF shielding through glass windows," *IEEE Transactions on Electromagnetic Compatibility*, vol. 62, no. 1, pp. 93–100, 2019.
[85] A. Nicolson and G. Ross, "Measurement of the intrinsic properties of materials by time-domain techniques," *IEEE Transactions on Instrumentation and Measurement*, vol. 19, no. 4, pp. 377–382, 1970.

Chapter 2

Polarization Insensitive Dual Band Planar Metamaterial Surface Design for Antenna Applications

Mohammad Tariqul Islam[1], *Mohammad Lutful Hakim*[2] and *Touhidul Alam*[3]

1. Introduction

Planar **metamaterial** surfaces (MS) are periodic structures known for their ability to selectively filter electromagnetic (EM) waves based on frequency and polarization properties [1]. Acting as EM filters, MS can exhibit either bandstop or bandpass characteristics, making them highly versatile for various applications. These include electromagnetic interference (EMI) reduction [2], reflectors [3], **antenna** radomes [4], and polarization converters [5]. EMI shielding is particularly important for wireless communications and IoT systems operating at 5G millimeter-wave (mm-wave) frequencies [2, 6]. In electronic devices, EMI shielding prevents interference from external EM signals and contains generated signals to avoid disruptions to nearby systems. To address these needs, specific frequency bands such as n257 (26.50–29.50 GHz) and n260 (37.00–40.00 GHz) have been allocated globally for 5G mm-wave applications [7, 8]. Traditional metal shields, although effective, block all **EM waves** indiscriminately. In contrast, MS structures selectively block specific frequencies while allowing others to propagate, offering a more targeted EMI reduction solution. Furthermore, their low-profile design and ease of fabrication enhance their suitability for practical applications [9]. MS miniaturization is achieved through techniques like convoluted patterns, lumped elements, multilayer configurations, and fractal

[1,2] Department of Electrical, Electronic and Systems Engineering, Universiti Kebangsaan Malaysia, Bangi, Selangor, Malaysia.

[3] usat Sains Angkasa (ANGKASA), Institut Perubahan Iklim, Universiti Kebangsaan Malaysia, 43600 UKM Bangi, Selangor, Malaysia.

* Corresponding author: tariqul@ukm.edu.my, lutful@ukm.edu.my

geometries. Moreover, multifunctional requirements can be addressed by designing MS structures with multiple stopbands and passbands [2]. In the context of EMI shielding, **stopband** MSs have predominantly been designed for microwave frequency applications [2, 10, 11]. However, these MS designs primarily targeted the microwave frequency range. At mm-wave frequencies, various passband non-conformal MSs have been introduced in [12– 14]. For instance, in [12], a via-based 2.5D structure was implemented to enhance OIA stability. In [15], a dual-stopband mm-wave MS was developed using a multilayered cross-loop resonator and a wire grid, though it exhibited low OIA stability, and the **stopband** was confined to the 28 GHz frequency band. Additionally, [14] presented a screen-printed, transparent polyethylene terephthalate (PET)-based band-stop MS designed for 28 GHz applications, supporting both normal and conformal configurations. These studies highlight that designing a conformal, miniaturized, single-layer MS capable of dual-band or multi band transmission blocking, polarization insensitivity, wide OIA stability, and reliable performance on curved surfaces remains a significant challenge in the field.

2. Design Analysis

The two-layered geometry of the developed MS-unit cell is presented in Figure 1. ($r1$ = 3.70 mm, fs = 0.20, fl = 0.60 mm, sw = 0.20 mm, al = 1.05 mm, sl = 0.80 mm, gl = 0.52 mm, w = 0.2 mm, cl = 3.0 mm, and cw = 0.15 mm). The top layer of the structure features a fractal square ring, interconnected with an inner square ring via four connecting arms. On the back layer, a cross resonator is incorporated to enhance the design's functionality. Together, the MS array elements are engineered to induce a phase delay when subjected to oblique incidence of EM waves, optimizing their performance for specific applications. As illustrated in Figure 2(a), the $N \times N$ mm^2 MS experiences a (φ) phase difference when the **incident angle** is θ, and the phase difference φ is estimated by Equation (1) [16], where $\omega = 2\pi f$, P_c is the unit cell periodicity, and υ_c is the velocity of light.

$$\varphi = \frac{\omega}{\upsilon_c} (NP_c \sin \theta) \qquad ...(1)$$

The phase difference at the OIA triggers resonance frequency shifting, which can be reduced by a miniaturized MS unit cell [14]. Flexible MS designs are essential for various applications, such as radomes, to enhance adaptability and conform to curved surfaces. This section introduces an ultrathin, miniaturized MS geometry tailored for such requirements. The unit cell of the proposed MS measures 4 × 4 mm^2. It is constructed using a Rogers RT5880 substrate, which has a dielectric constant of 2.2 and a thickness of 0.254 mm. The MS is characterized by dual stopbands at 28 GHz and 38 GHz, offering excellent transmission-blocking performance. Moreover, its symmetrical rotational structure ensures polarization insensitivity, making it suitable for diverse operational environments. The reflection (S_{11}) and transmission (S_{21}) coefficients for both the top-layer and dual-layer MSs are shown in Figure 2(b). The single-layer MS exhibits a resonance at 29.7 GHz with a **bandwidth** percentage (BW%) of 65%, with $S_{21} < -10$ dB over the frequency range of 20–39.4 GHz. In contrast, the dual-layer MS demonstrates two resonance frequencies at 27.3 GHz and 39.2 GHz, with bandwidths of 42.8% and 24.2%, respectively. Additionally, the **passband** frequency for the dual-layer MS spans from 30.9–35.28 GHz.

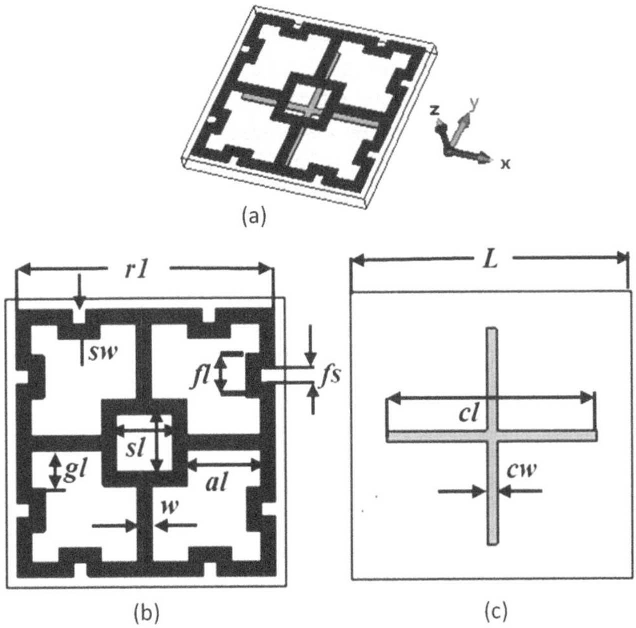

Fig. 1 (a) perspective view, (b) front, and (c) back views of MS-unit cell.

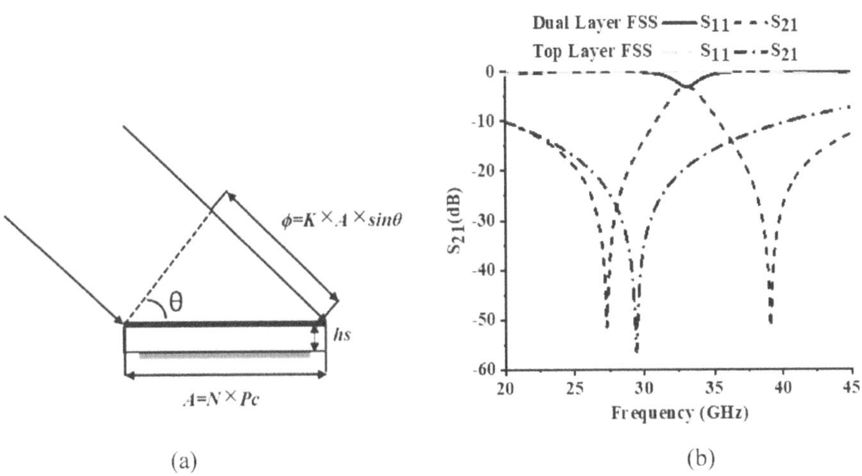

Fig. 2 (a) phase difference between parallel rays at the OIA on the MS array, (b) S-parameters for the top layer and dual layer structure of MS.

3. Results Analysis

3.1 Equivalent Circuit Model (ECM)

The mechanism of a transmission block with an MS is demonstrated using an equivalent circuit model (ECM) in Figure 3(a). The ECM is mode led based on the concept in [10, 13, 17]. The top fractal interconnected resonator generates inductance ($L1$), while its inner segment forms a parallel capacitance ($C1$) in conjunction with $L1$. Additionally, a series capacitance ($C2$) is introduced between adjacent unit cells. The substrate material acts as a transmission line with an impedance defined as $Z_{TL} = Z_o/\sqrt{\varepsilon}$. The cross resonator on the back layer adds inductance ($L2$) and capacitance ($C3$), which arise from the coupling between neighboring unit cells. An Advanced Design System (ADS) is used to simulate the ECM, with parameter adjustments made to achieve the desired **stopband** frequency. For the design and simulation of MS, CST Microwave Studio software is employed. The simulations were conducted using a frequency-domain solver with a default tetrahedral mesh configuration. Unit cell boundary conditions were applied along the x- and y-axes, while an EM wave was propagated along the z-axis using Floquet port excitation. The simulated results for the MS and ECM demonstrate strong agreement, as shown in Figure 3(b).

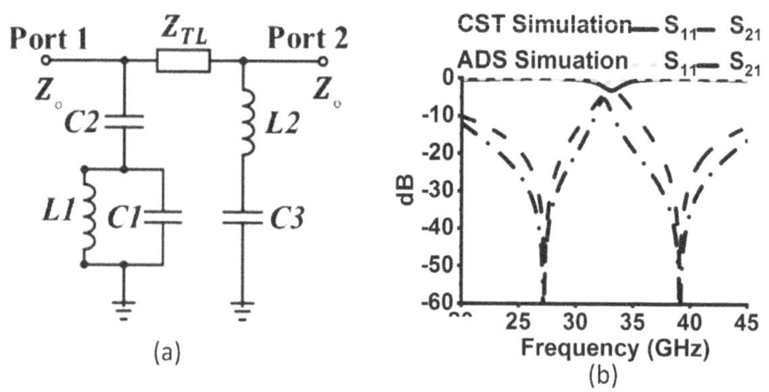

Fig. 3 (a) ECM ($L1 = 0.51$ nH, $C = 0.026$ pF, $C2 = 0.041$ pF, $C3 = 0.051$ pF, $L2 = 0.32$ nH, $Z_o = 377\ \Omega$, and ZTL $= Z_o/\sqrt{\varepsilon}$), and (b) simulated ECM and MS S-parameters.

3.2 Metamaterial Property Analysis

The transmission-blocking property of MS can be understood from the relative permittivity and permeability [6]. Figure 4 shows that the real values of relative permittivity and permeability are alternately negative and produce transmission-blocking properties at the operating frequency. The real part of the relative impedance of MS approaches near zero, indicating the presence of a **stopband** at the corresponding frequency.

3.3 Polarization Insensitivity and Incident Angle Stability

The symmetrical rotational design of the MS ensures polarization insensitivity for both transverse electric (TE) and transverse magnetic (TM) modes, as well as for all

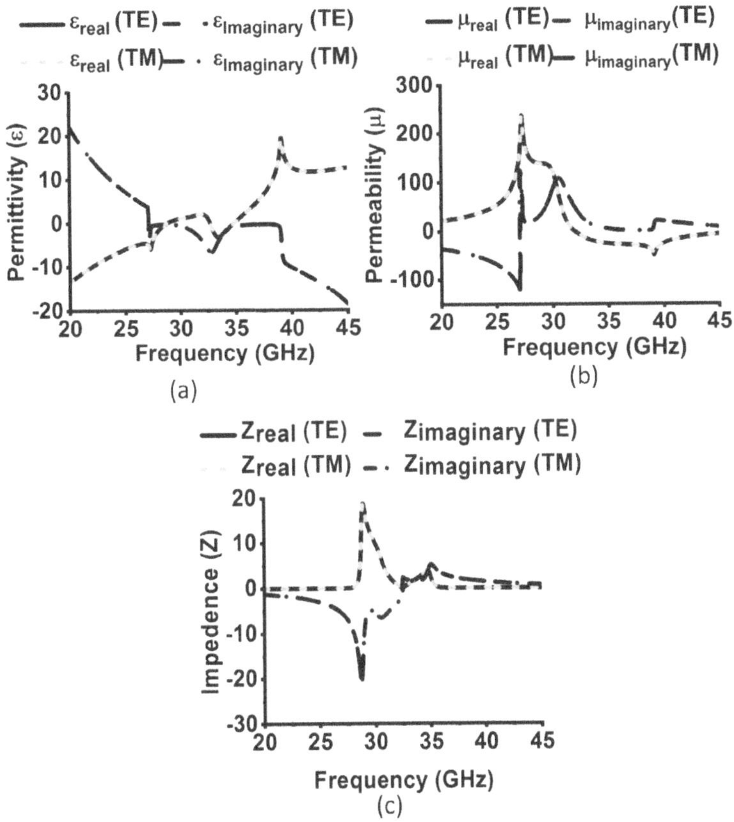

Fig. 4 (a) relative permittivity, (b) relative permeability, and (c) relative impedance of the MS at TE and TM modes.

possible polarization **incident angles.** Figure 5(a) illustrates the distinct transmission-blocking behavior of the MS for TE and TM polarizations. The S_{21} parameters of the MS under TE and TM polarizations at various oblique incidence angles (OIAs, θ) are shown in Figure 5(b, c), respectively. The design achieves high OIA stability up to 60° at the desired frequency, attributed to its compact unit cell size. OIA stability is defined as the ability of the MS to maintain an S_{21} level below – 10 dB at the desired frequency across a range of oblique incident angles of the incoming EM waves. The resonance frequency experiences shifting due to phase variations at different OIAs. The **stopband** filter response remains stable up to a θ angle of 20°. As the θ angle increases, the stability changes differently for TE and TM modes. In the TE mode, the stability increases, while in the TM mode, it decreases as the θ-angle increases. This behavior is attributed to the variation in the MS impedance ($Z_{TE/TM}^{MS}$) for both TE and TM **polarizations** at higher θ angles, which can be described by Equation (2) [16].

$$Z_{TE}^{MS} = \frac{Z_o}{\cos\theta} \text{ and } Z_{TM}^{MS} = Z_o \cos\theta \qquad \ldots(2)$$

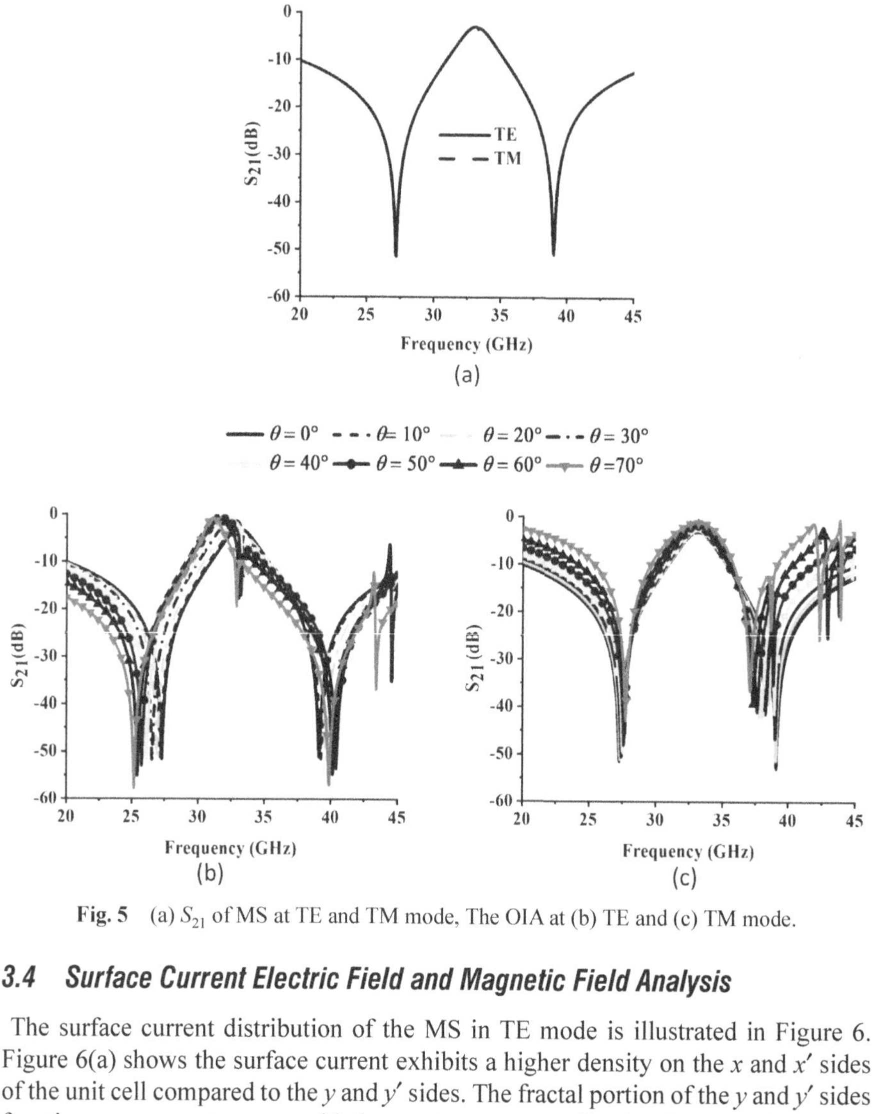

Fig. 5 (a) S_{21} of MS at TE and TM mode, The OIA at (b) TE and (c) TM mode.

3.4 Surface Current Electric Field and Magnetic Field Analysis

The surface current distribution of the MS in TE mode is illustrated in Figure 6. Figure 6(a) shows the surface current exhibits a higher density on the x and x' sides of the unit cell compared to the y and y' sides. The fractal portion of the y and y' sides functions as a current source, with the maximum current flowing downward along the vertical interconnecting arm. Additionally, an antiparallel **current flow** is observed along the horizontal interconnecting arm. Additionally, antiparallel current flow occurs along the horizontal interconnecting arm. Figure 6(b) reveals that the cross resonator on the backside of the MS has a comparatively lower current density than the front **resonator**. Consequently, the upper resonator predominantly contributes to the resonance peaks observed at 27.3 GHz.

On the other hand, a higher current density appears at the back side cross **resonator**, illustrated in Figure 6(c-d). Hence, the 39.12 GHz frequency band is achieved by coupling the front fractal square resonator and backside cross resonator. Figure 7 presents the surface current distribution of MS at TM mode, which shows a

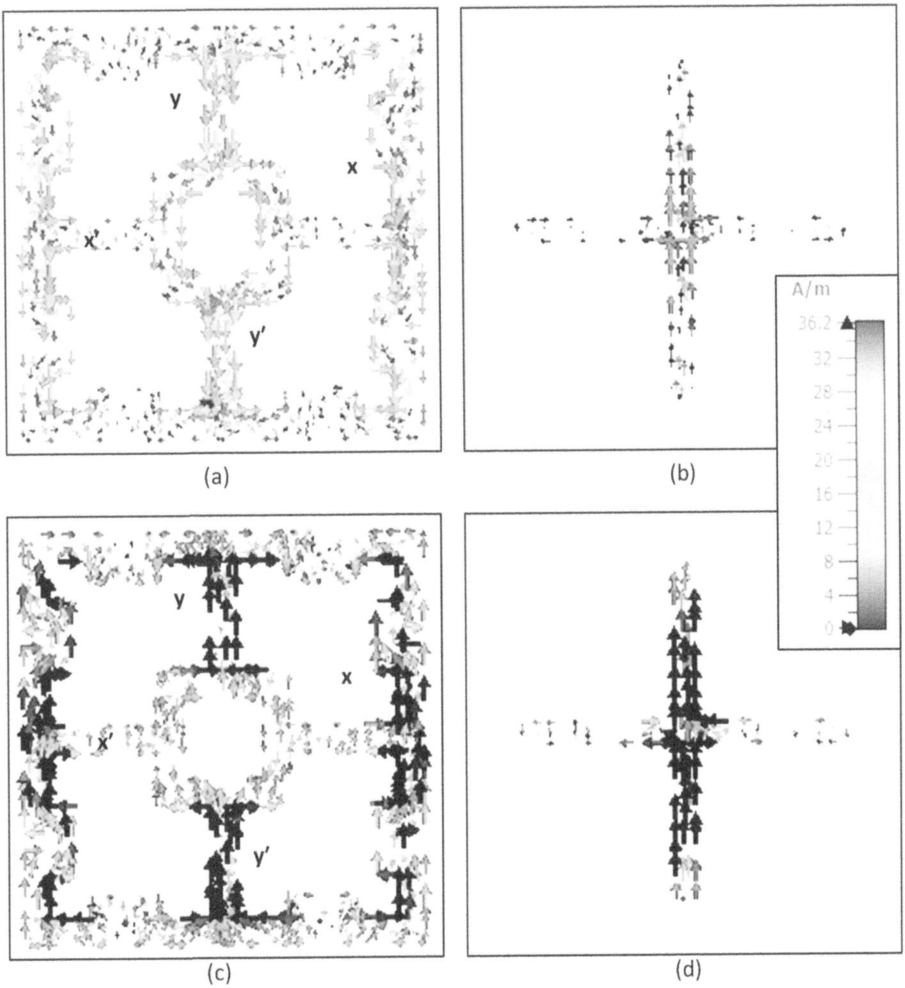

Fig. 6 Surface current distributions of the MS for TE mode at 27.3 GHz – (a) front, (b) back; at 39.2 GHz – (c) front, (d) back.

similar surface current distribution with a 90° rotational form due to the rotation of the field vector at TM mode. The e-field distribution of MS is presented in Figure 8 for both modes. At TM mode, the field intensity is high at the vertical edge of the top resonator at 27.3 GHz frequency, and the horizontal ram of the cross resonator at the backside shows a lower intensity than the top resonator. Therefore, the top resonator mainly achieves a 27.3 GHz frequency **stopband**.

On the other hand, a similar e-field distribution appeared at 37.9 GHz frequency, but the intensity is high on the back resonator, indicating that **stopband** resonance is mainly attain ed by the back resonator. At TE mode, the field intensity is rotated by 90° due to the incident **EM wave** vector polarization. The h-field distribution of the MS is presented in Figure 9, where the h-field at TM mode in Figure 9(a, b, c, d) indicates the

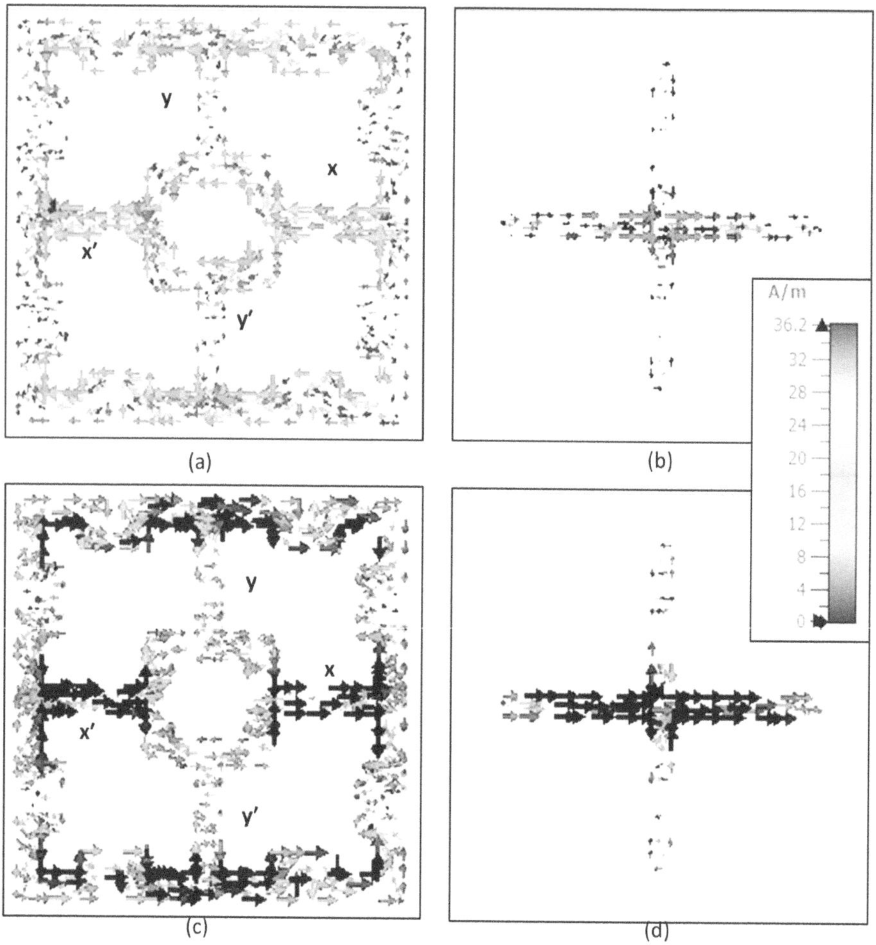

Fig. 7 Surface current distributions of the MS for TM mode: at 27.3 GHz – (a) front, (b) back; at 39.2 GHz – (c) front, (d) back.

top resonator has high intensity at 27.3 GH frequency, and the back resonator has high intensity at 39.2 GHz frequency. Indicated top and back resonators are responsible for 27.3 GHz and 39.2 GHz frequency, respectively. The h-field at TE mode in Figure 9(e, f, g, h) shows a similar field distribution to the h-field at TM mode. But 90° rotational form due to the rotation of the field vector at TE mode.

4. Measurement Results

The developed MS, consisting of a 20 × 20 unit cell array, is fabricated with an overall size of 80 × 80 mm². The transmission coefficient (S_{21}) of the MS is measured for both normal and curved surfaces. The front and back side views of the fabricated MS are presented in Figure 10(a, b), with an extended view. The measurement setup is presented in Figure 11(a) to determine the S_{21} of the MS. The experiment is conducted

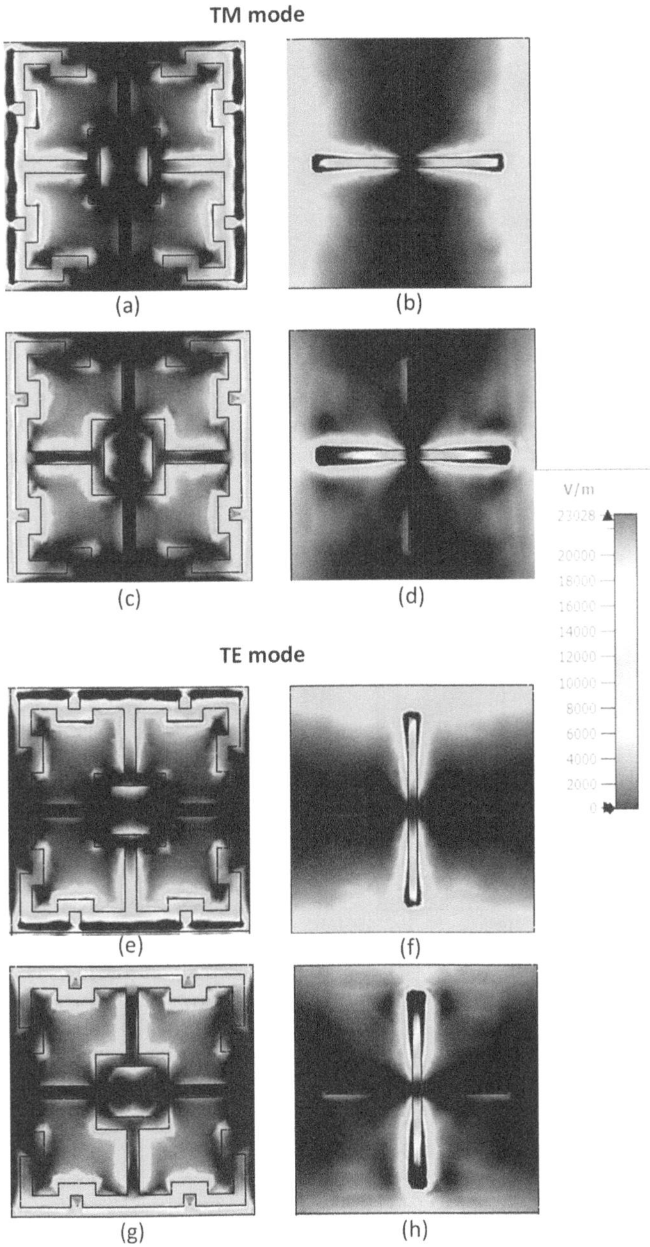

Fig. 8 E-field of the MS for **TM mode**: at 27.3 GHz – (a) front, (b) back; at 39.2 GHz – (c) front, (d) back. and **TE mode**: at 27.3 GHz (e) front, (f) back; at 39.2 GHz – (g) front, (h) back.

in an anechoic chamber, where two horn **antennas** are used to transmit and receive the **EM wave**. The MS is placed between the **antennas** during the measurement The plane wave incident on the MS is ensured by the distance of $>2D^2/\lambda$ between the MS and horn **antenna**, where λ is the center frequency, and D is the MS length.

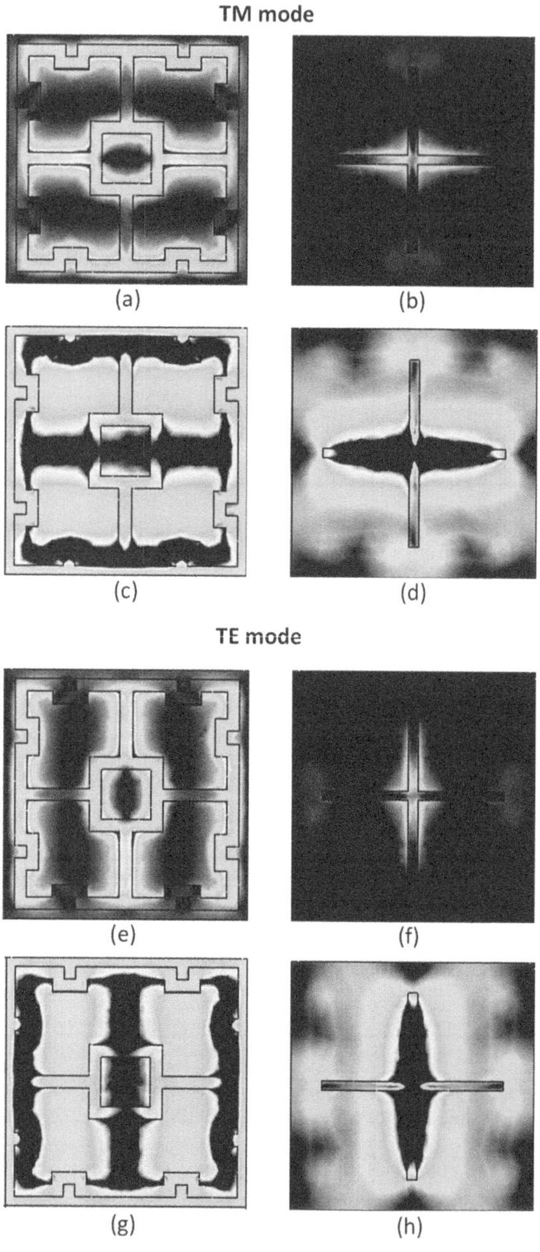

Fig. 9 H-field of the MS for TM mode: at 27.3 GHz (a) front, (b) back; at 39.2 GHz (c) front, (d) back. And TE mode: at 27.3 GHz (e) front, (f) back; at 39.2 GHz (g) front, (h) back.

The simulated and measured S_{21} of the normal MS are presented in Figure 11(b). The measurement results agree well with the simulated results. However, a slight shift of the upper-frequency band toward higher frequencies is observed, which is attributed to edge diffraction caused by the finite number of unit cells, as well as fabrication

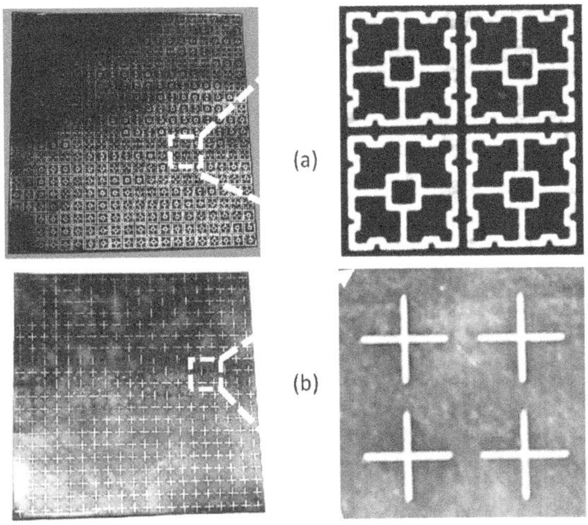

Fig. 10 Fabricated 20 × 20 MS array: (a) front view, and (b) back view.

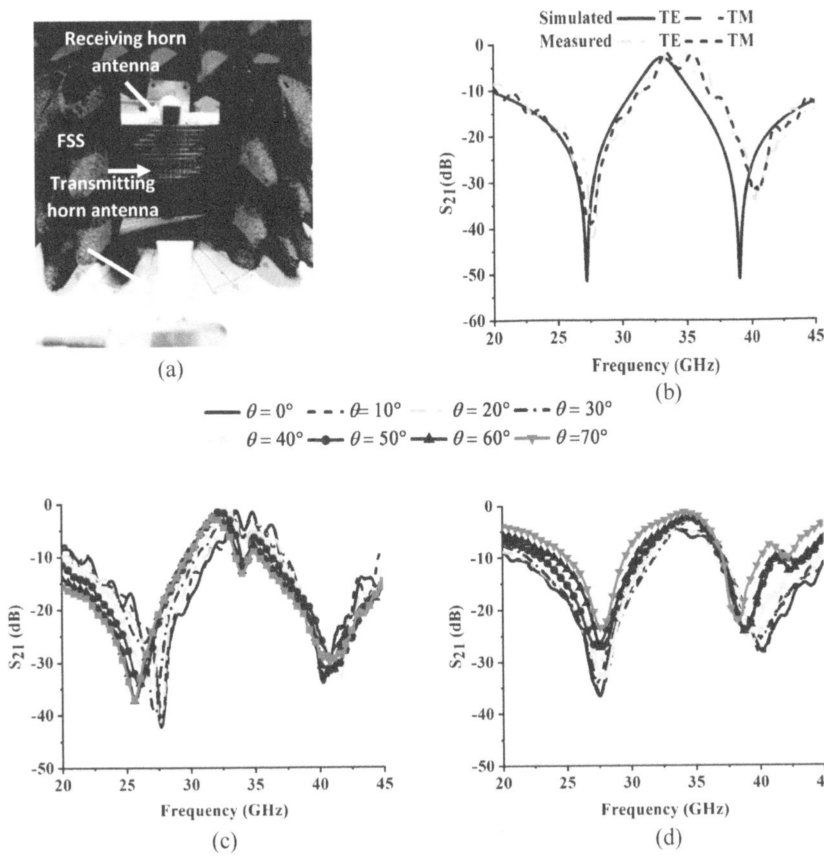

Fig. 11 (a) measurement setup, (b) simulated and measured S_{21} at TE and TM polarized waves, (c) measured S_{21} at different OIAs at TE polarized wave, and (f) measured S_{21} at different OIAs at TM polarized wave.

and measurement tolerances. The measured (– 10 dB) stopbands are achieved from 20–30.9 GHz and 35.28–45 GHz, where the measured resonances are shown at 27.3 GHz and 40.4 GHz, respectively. Figure 11(c, d) presents the measured S_{21} parameters of the designed MS for both TE and TM polarizations across varying incidence angles (θ). The results indicate that the stopband broadens for TE polarization while narrowing for TM polarization as θ increases, consistent with the behaviour described by Equation (2). These measured outcomes align closely with the simulated results shown in Figure 5 (b, c). The developed MS demonstrates potential for EMI shielding applications within the frequency ranges of 20–30.9 GHz and 35.28–45 GHz [12].

5. Conformal Applications

The MS was bent to a radius of R = 25.5 mm for conformal MS measurements, with the setup depicted in Figure 12(a). The midpoint of the conformal MS was aligned with the centerline of the receiving and transmitting horn **antenna** apertures. A microwave absorber was placed at the edges of the MS to eliminate reflections between the receiving and transmitting **antennas**. The simulated and measured S_{21} parameters of the conformal MS are shown in Figure 12(b), Both results exhibit stable performance. The simulated – 10 dB S_{21} bandwidths were 20–30.9 GHz and 35.3 GHz, while the measured results were 20–31.2 GHz and 36.6–45 GHz, demonstrating good agreement.

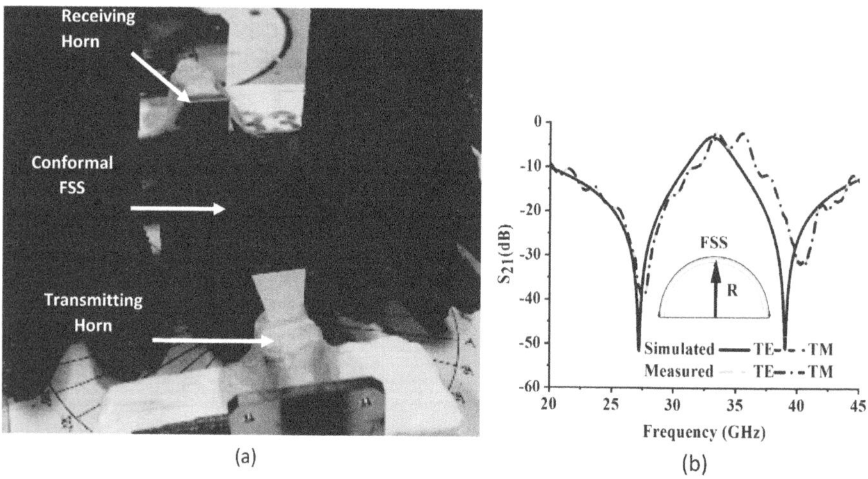

Fig. 12 (a) Conformal MS with measurement setup; (b) S_{21} of simulated and measured conformal MS at the TE and TM polarized normal incident EM wave.

6. Performance Comparison of MS with Existing Polarization Insensitive MS

Table 1 compares the developed MS with existing works, where [10, 11, 16, 18] are designed for the **stopband** application and [12, 13] are developed for the passband application. In [16], a stopband MS was designed for a 5G mmWave frequency but

Table 1 Comparison of MS with existing polarization-insensitive MS.

Ref/	*Periodicity (λ_o)*	*Substrate h(mm)/εr*	*Filter Type*	*Frequency (GHz)*	*BW (%)*	*Covered 5G Band*	*OIA Stability*	*Conformal*	*Resonator Configurations*
[18]	0.096 λ_o	0.25/4.4	Stop band	4, 6.96, 11.3	13.9, 25, 6.9	–	60°	Yes	Twisted spiral triangle cross dipole
[10]	0.26 λ_o	0.127/2.2	Stop band	10	40	–	60°	Yes	Polygon shape
[12]	0.26 λ_o	0.38/2.2	Pass band	27.75	9.6	n261	60°	No	Vertical via based 2.5 D mandarin line
[13]	0.23 λ_o	5.0/2.45	Pass band	28	7.017	n261	50°	No	Dual loop resonator
[16]	0.4 λ_o	0.51/2.2	Stop band	27.2, 33.1	8.3, 13.8	n261	35°	No	Cross loop resonator with a wire grid.
[14]	0.3 λ_o	0.125/3	Stop band	28	29.3	n257	60°	Yes	Meshed resonators
Developed MS	0.26 λ_o	0.254/2.2	Stop band	28, 38	42.8 24.2	n257 n260	60°	Yes	Back-to-back resonators

did not cover the 38 GHz frequency band. The design did not support the conformal structure configuration and achieved a small oblique incident angle stability of 35° because of its higher periodicity.

Moreover, the listed works were polarization insensitive The MSs in [10, 11] can be used for conformal application due to their thin and flexible substrate material. Finally, the developed MS is designed on a thin (0.254 mm) flexible substrate with a low periodicity (0.26 λ) for the unit cell. This design feature enables conformal application and ensures high stability at oblique **incident angles**. Additionally, the symmetrical rotational design of the MS provides polarization insensitivity. Most importantly, the MS is designed for 5G mmWave **stopband** applications at 28 GHz and 38 GHz, achieving a wide bandwidth percentage (BW%) of 42.8% at 27.3 GHz and 24.2% at 39.12 GHz.

7. Applications Investigations of MS

7.1 EMI Shielding

The developed MS can potentially be used in EMI shielding applications at frequencies ranging from 20–30.9 GHz and 35.28–45 GHz [12]. The quantity of EMI shielding is defined in terms of shielding effectiveness (SE), which is calculated by the ratio of transmitted (*Et*) and incident (*Ei*) **EM waves** using Equation (3).

$$\mathrm{SE(dB)} = -20\log|E_t/E_i| = -S_{21}\,\mathrm{(dB)} \qquad ...(3)$$

The simulated and measured SEs at normal and different θ angles for TE and TM-polarized **EM waves** are presented in Figures 13 and 14, respectively. The simulated SE of the MS are 50 dB at 27.3 GHz and 39.2 GHz, respectively. On the other hand, the measured SE is 40 dB and 35 dB at 27.3 GHz and 39.2 GHz, respectively. A slight deviation in the measured SE at the resonance frequency is observed, which is attributed to the scattering of transmitted waves from the edge of the MS. Additionally, the receiving horn is linked to a portion of the incoming **EM wave**, influencing the SE measurements. The measured SE can be improved by positioning a dielectric lens in front of the horn **antenna** to focus and aggregate the incident **EM wave** onto the MS surface or by fabricating a larger MS array to minimize edge effects.

7.2 Gain Enhancement of Defected Ground Antenna by MS

The reflected wave from the developed MS is characterized as an AMC [19–21]. The AMC represents the properties of the perfect magnetic conductor (PMC). The AMC replicates the behaviour of a PMC, reflecting waves with zero-degree phase at its resonant frequency [19]. This distinctive feature makes the AMC an ideal choice for enhancing **antenna** performance and optimizing various microwave devices [22]. The AMC characteristics of the developed MS can be understood from the reflection phase value shown in Figure 15(a). The phase value of the reflected wave between +90° to −90° is considered as AMC bandwidth. The AMC bandwidth of

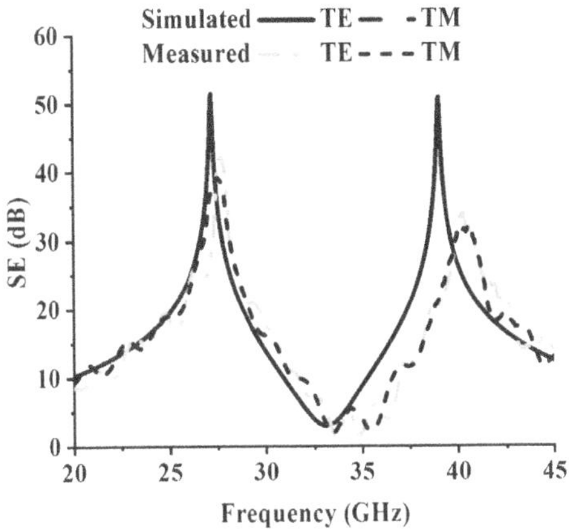

Fig. 13 SE of the developed MS for TE and TM modes at the normal incident of EM. wave

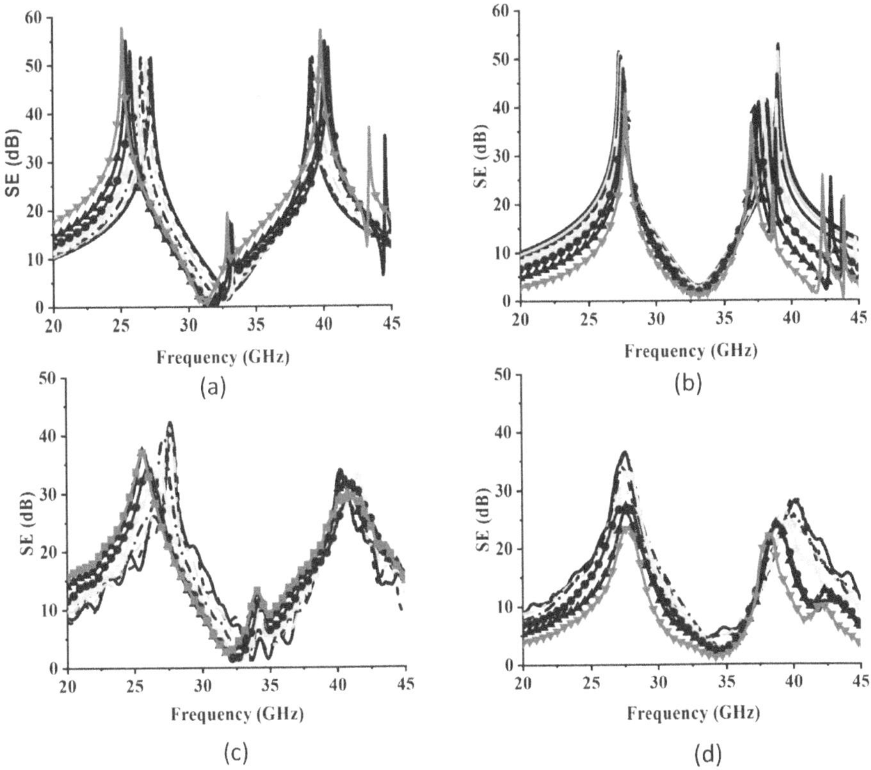

Fig. 14 Simulated SE at OIA: (a) TE, and (b) **TM polarized** EM waves; measured SE at OIA: (c) TE, and (d) TM polarized EM waves.

MS ranges from 20–31.96 GHz and 33.96–40.3 GHz, respectively. Figure 15(a) shows the reflected phase of 0° at 29.4 GHz frequency as the resonance frequency [23]. The constructive interference of radiated waves from the **antenna** and reflected waves by MS is presented in Figure 15(b). The **EM wave** reflecting characteristic of MS is investigated by designing a dual-band four-element MIMO antenna that operates at n257 and n260 bands. Figure 16(a) shows the single **antenna** geometrical parameters, and Figure 16(b) shows the four-element MIMO **antenna** is designed on a 22 × 22 mm^2 substrate in rotational form. The substrate material (Rogers RT5880) with loss tangent 0.0009, dielectric constant 2.2, and thickness 0.79 mm is used to design the four-element MIMO antenna.

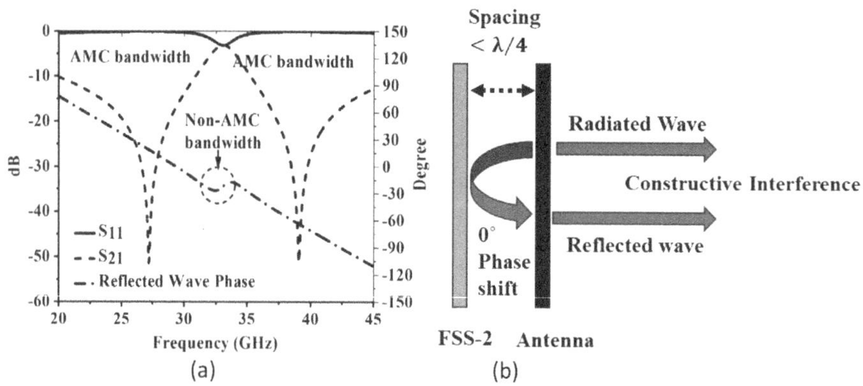

Fig. 15 (a) MS AMC characterisation of reflected wave or **stopband** frequency, (b) constructive interference by a reflected and radiated wave from MS/AMC/FSS.

The defected ground four-element MIMO antenna gain is increased significantly by positioning MS at the backside of the antenna from a 2.5 mm distance (Figure 16(c)). The designed antenna was fabricated, shown in Figure 17, and measured in an anechoic chamber to validate the application of MS. The simulated s-parameters of the MS base MIMO antenna show a – 10 dB operational bandwidth in both the n257 and n260 bands, illustrated in Figure 18(a). The isolation of the MIMO antenna among its four elements is less than – 18 dB, revealed in Figure 18(b). The measured scattering parameters of the fabricated antenna agree well with the simulated one shown in Figure 18(c, d). Finally, the wave reflecting behaviour and improvement in the gain due to the constructive interference can be understood from the antenna radiation pattern and broadband gain plot at operational bandwidth.

The reflecting behavior of EM waves by the MS structure is illustrated through the two-dimensional radiation patterns of the **antenna**, both with and without MS (Figure 19). The inclusion of MS results in a higher front-to-back ratio, significantly reducing back radiation at 28.5 GHz and 38 GHz compared to the regular antenna, as depicted in Figure 19(b, d). Figure 20 provides the three-dimensional radiation patterns of the MIMO antenna with and without MS at 28.25 GHz and 38 GHz. After integrating the MS, the patterns in Figure 20(b, c) demonstrate a substantial reflection of back-propagated EM waves. These reflected waves constructively interfere with forward-propagated waves, enhancing the overall performance. Furthermore, the

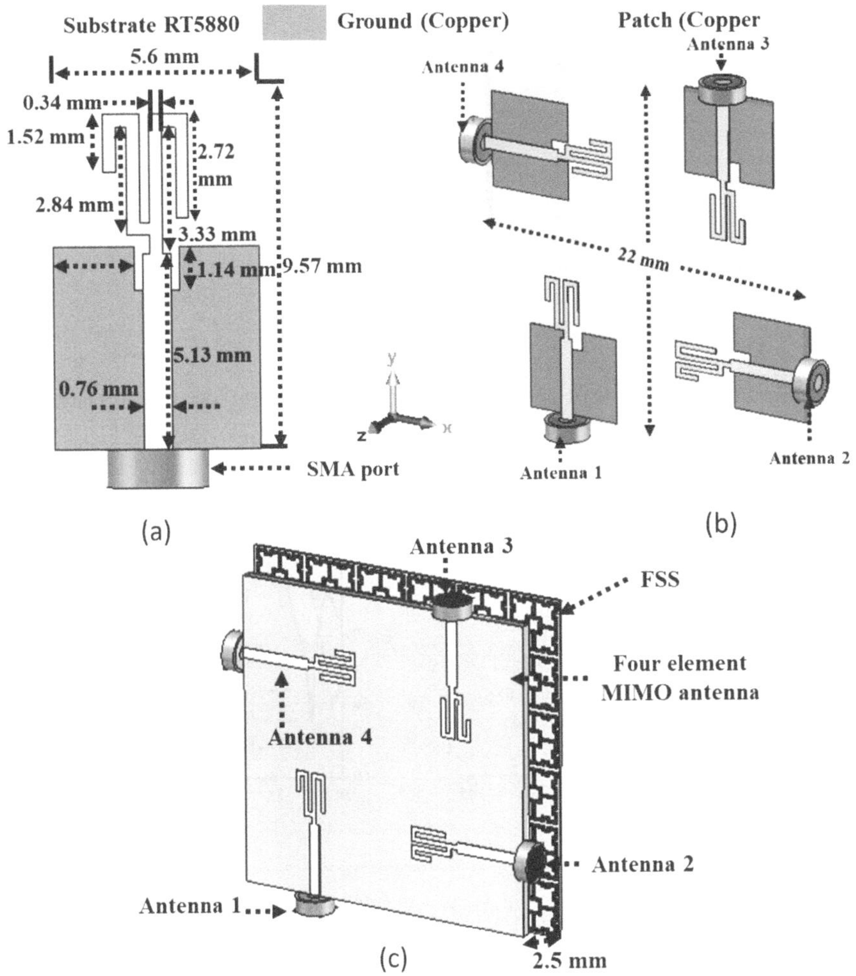

Fig. 16 (a) Single **antenna** geomaterial parameters; (b) four-element MIMO configuration of the designed antenna; (c) four-element MIMO antenna with developed MS back layer.

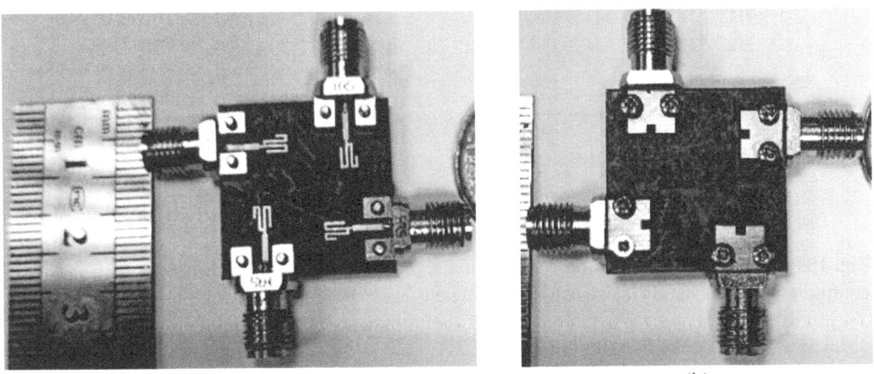

(a) (b)

(c)

Fig. 17 The regular **fabricated MIMO** antenna: (a) front view, (b) back view, (c) fabricated MS backed MIMO antenna.

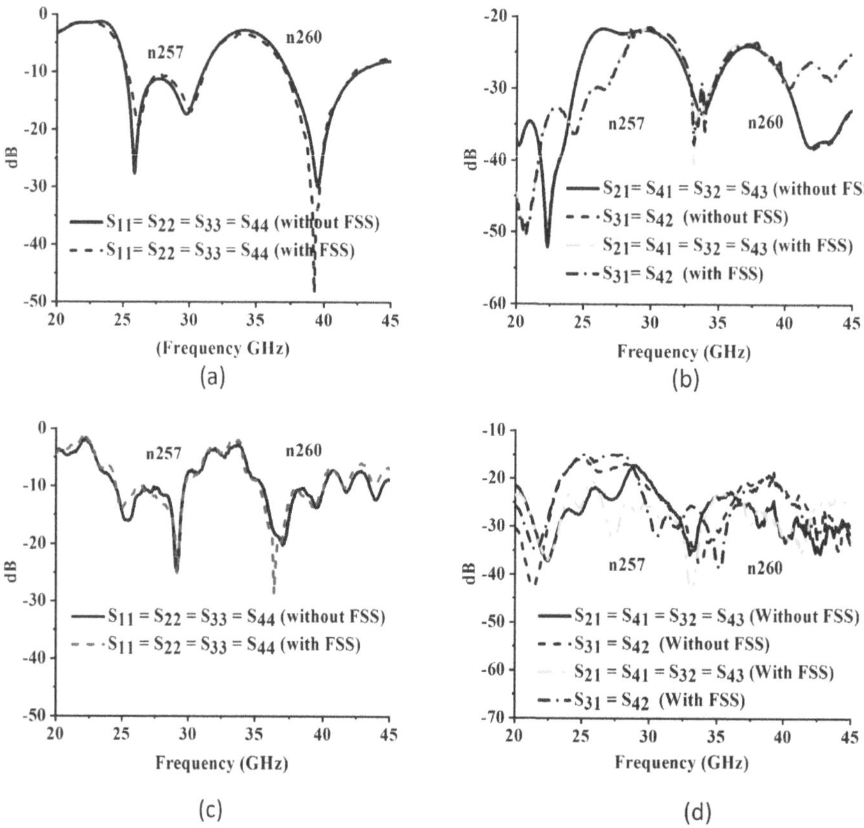

Fig. 18 Simulated: (a) **reflection** coefficient and (b) isolation of the **antenna** elements in MIMO configuration; measured: (c) reflection coefficient, and (b) isolation of the antenna elements in MIMO configuration.

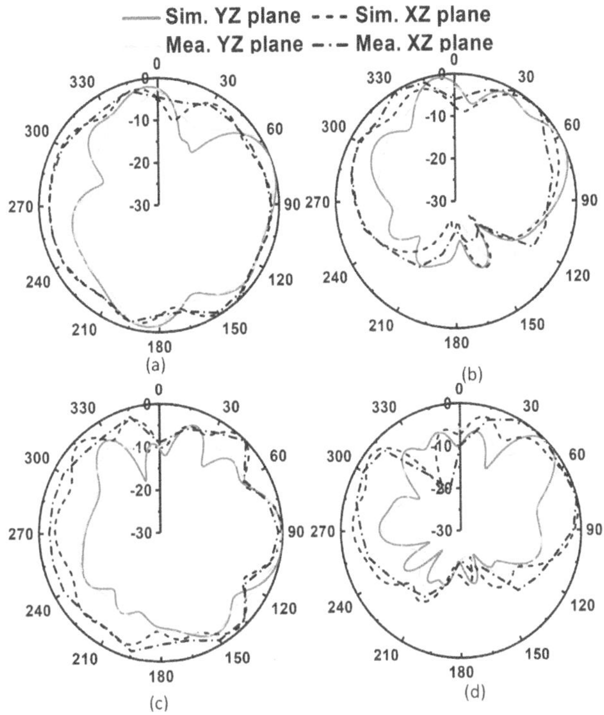

Fig. 19 Two-dimensional radiation pattern of the MIMO antenna at 28.25 GHz: (a) Without MS, (b) With MS; and at 38 GHz frequency: (c) Without MS, (d) With MS.

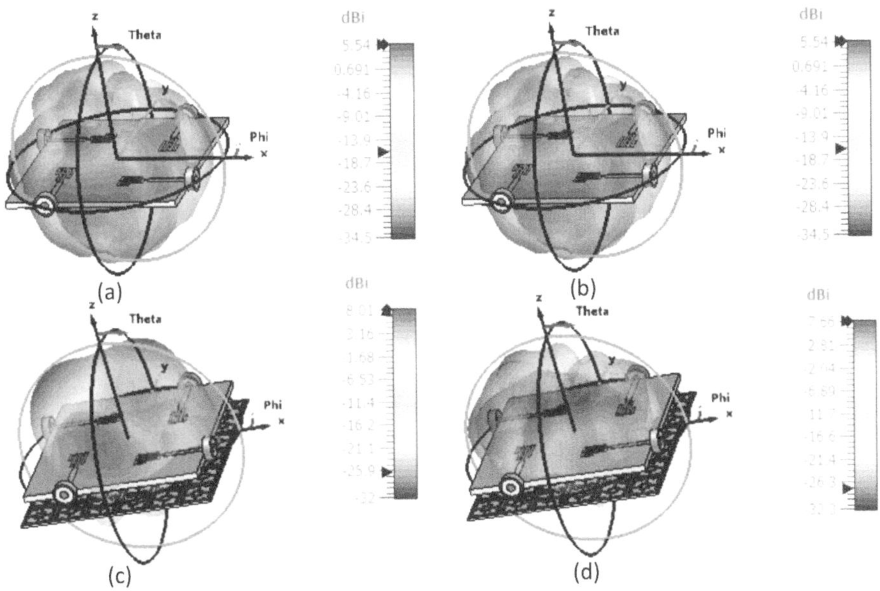

Fig. 20 Three-dimensional radiation pattern of the MIMO antenna (excitation of single port) without MS: (a) at 28.25 GHz and (b) at 38 GHz frequency; with MS: (c) at 28.25 GHz, and (d) at 38 GHz frequency.

broadband gain of the MIMO antenna with and without MS is presented in Figure 21. The integration of the MS increases the maximum gain to approximately 3.5 dBi in both frequency bands. As previously discussed, the reflected **EM waves** from MS, exhibiting a phase variation between +90° and −90°, create constructive interference with the forward-radiated waves along the antenna's off-axis (z-axis), thereby boosting the antenna's gain.

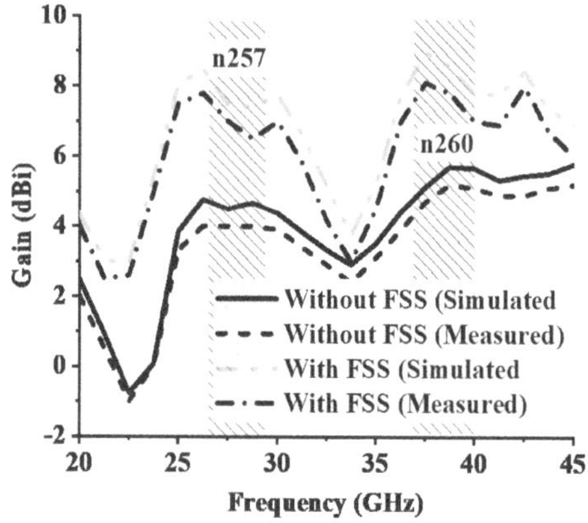

Fig. 21 Simulated and measured broadband gain of the MIMO antenna with and without MS/FSS

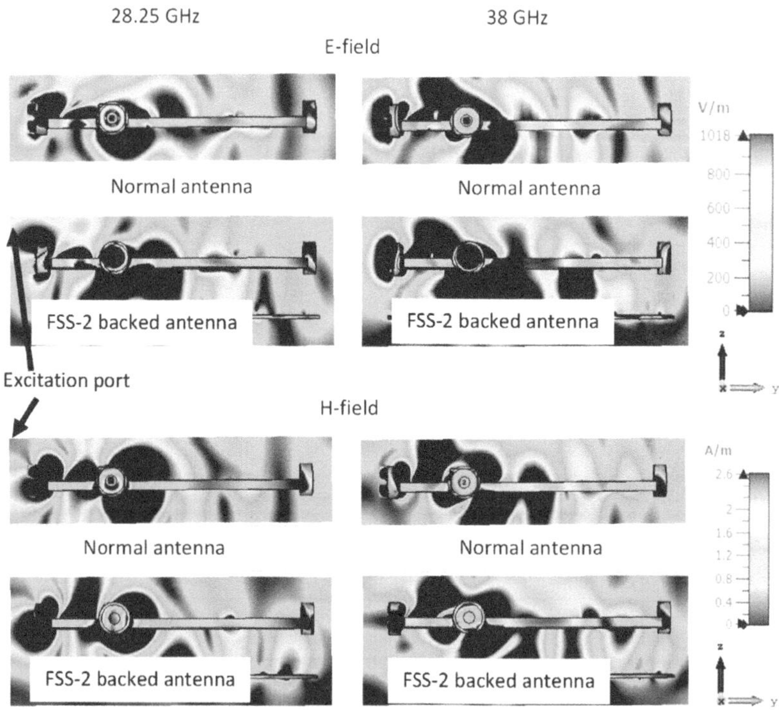

Fig. 22 E-field and H-field of the normal and MS/FSS backed MIMO antenna

Figure 22 shows the developed MS-based MIMO **antenna** and normal MIMO **antenna** electric and magnetic fields at 28.25 and 38 GHz frequencies. The MS structure completely blocks the **EM wave** towards the negative *z*-axis and reflects towards the positive *z*-axis. Figure 2.23 shows the retrieved ECC and DG values of the MS-based MIMO **antenna**, calculated by Equation (4.8 and 4.9). The ECC <0.001 and DG > 9.99 at the n257 and n260 bands, which is near ideal.

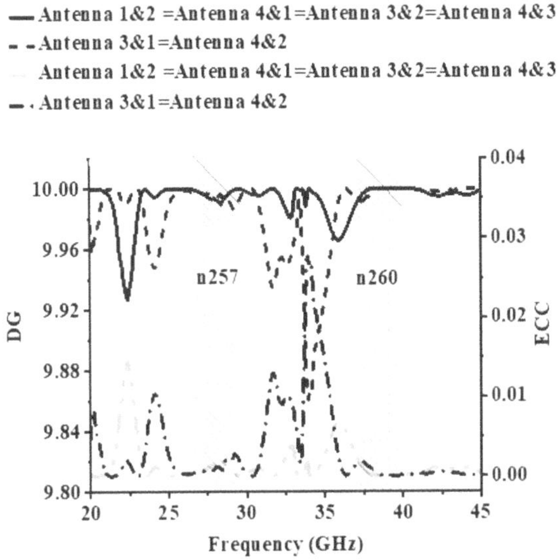

Fig. 23 ECC and DG value between the MS based MIMO elements.

A comprehensive comparison of the developed MS-based MIMO antenna with existing mmWave antennas is provided in Table 2. Various planar printed log-periodic dipole array (PLPDA) antennas have been introduced in [24–26] for broadband 5G mmWave applications, demonstrating significant gain enhancement through the integration of different metamaterial structures. Additionally, other antenna designs such as a shared aperture antenna [28], Bow-tie antenna [29] and Antipodal Vivaldi antenna [30] incorporating metasurfaces were developed for 5G applications, where the use of **metamaterial** structures markedly improved their gain performance. Similarly, multilayer microstrip antennas [27] and dual-layer two-element slotted patch antennas [31] have been presented, featuring metamaterials positioned above the antenna with an air gap to enhance gain and overall efficiency. The MIMO parameters analyses were performed only [27, 31]. The developed MS–2 based MIMO **antenna** achieved dual operational band bandwidth at 5G mmWave frequency compared to the other listed antenna. Also, 3.5 dBi gain enhancement is done by MS, which is higher than the existing antennas. The investigation of the MIMO parameters analysis, dual-band at 5G mmWave frequency, and higher gain prioritize the designed antenna over the other works listed in Table 2.

Table 2 Comparison of developed MS-based MIMO antenna with existing works.

Ref	*Size (mm^2)*	*Gain Enhancement Technique*	*Operation BW (GHz)*	*Gain Enhances dBi*	*MIMO*	*ECC/DG*
[24]	13 × 22	Semi-circular splits ring	25.7–40	1.9	No	–
[25]	15 × 22	Semi ring patch	26–40	1.6 dBi	No	–
[26]	10 × 17.5	Bow tie directors	40–50	0.73	No	–
[27]	10 × 10	Cross shape Resonator MS	28–34	N/A	Yes	N/A
[28]	31 × 31	Splits ring resonators	27–30	1.6	No	–
[29]	30 × 30.5	Meander and triple I shape	24.5–27.5	2.9	No	–
[30]	40 × 24	Inverted V shape	24.8–34.52	1.78	No	–
[31]	17 × 11.7	Splits ring resonators MS	24.5–26.5	2.2	Yes	$<0.1e^{-6}$
Our work	22 × 22	Back-to-back resonator MS	28 (n257) 38 (n260)	3.5	Yes	<0.001/> 9.99

8. Summary

This chapter discusses the design of a polarization-insensitive dual-band planar metamaterial surface (MS) tailored for antenna applications and electromagnetic interference (EMI) shielding, particularly in 5G millimeter-wave systems. The MS is designed with a two-layer geometry, combining a fractal square ring on the top layer and a cross resonator on the back to achieve dual stopbands at 28 GHz and 38 GHz, offering effective frequency filtering. The design ensures polarization insensitivity and incident angle stability up to 60°, making it versatile for dynamic environments and conformal applications like radomes. Simulations and experimental results validate its low-profile, compact structure, and flexibility for use on curved surfaces. Additionally, the MS enhances antenna performance by improving gain and radiation patterns, particularly in MIMO antenna configurations, through the artificial magnetic conductor (AMC) effect. The MS demonstrates significant potential for 5G EMI shielding and antenna performance improvement, offering a promising solution for next-generation wireless communication systems.

References

[1] T. Hong, S. Guo, W. Jiang and S. Gong. "Highly Selective Frequency Selective Surface With Ultrawideband Rejection," *IEEE Transactions on Antennas and Propagation,* vol. 70, no. 5, pp. 3459–3468, 2022.

[2] S. Dey and S. Dey. "Conformal Miniaturized Angular Stable Triband Frequency Selective Surface for EMI Shielding," *IEEE Transactions on Electromagnetic Compatibility,* 2022.

[3] M. Pazokian, N. Komjani and M. Karimipour. "Broadband RCS reduction of microstrip antenna using coding frequency selective surface," *IEEE Antennas and Wireless Propagation Letters,* vol. 17, no. 8, pp. 1382–1385, 2018.

[4] W.-J. Liao, W.-Y. Zhang, Y.-C. Hou, S.-T. Chen, C.Y. Kuo and M. Chou. "An MS-integrated low-RCS radome design," *IEEE Antennas and Wireless Propagation Letters,* vol. 18, no. 10, pp. 2076–2080, 2019.

[5] F.A. Dicandia and S. Genovesi. "Linear-to-Circular Polarization Transmission Converter Exploiting Meandered Metallic Slots," *IEEE Antennas and Wireless Propagation Letters,* vol. 21, no. 11, pp. 2191–2195, 2022.

[6] M.A. Shukoor and S. Dey. "Novel dual-mode polarization insensitive wide angular stable circular ring based deca-band absorber for RCS and EMI shielding applications," *IEEE Transactions on Electromagnetic Compatibility,* vol. 64, no. 5, pp. 1337–1345, 2022.

[7] FCC. "America's 5G Future." fcc.gov. https://www.fcc.gov/5G (accessed 16 December 2022).

[8] D. Brenner. "Global 5G spectrum update," in *Spectrum Strategy & Technology Poli cy*: SVP, 2020.

[9] V. Chaudhary and R. Panwar. "Hybrid ECM blended whale optimization derived frequency-selective conformal EMI shielding structure using ferrite substrate," *IEEE Transactions on Magnetics,* vol. 57, no. 8, pp. 1–10, 2021.

[10] M. Bilal, R. Saleem, Q.H. Abbasi, B. Kasi and M.F. Shafique. "Miniaturized and flexible MS-based EM shields for conformal applications," *IEEE Transactions on Electromagnetic Compatibility,* vol. 62, no. 5, pp. 1703–1710, 2020.

[11] M.H. Nisanci and F. de Paulis. "Easy-to-design-and-manufacture frequency selective surfaces for conformal applications," *IEEE Antennas and Wireless Propagation Letters,* vol. 20, no. 5, pp. 753–757, 2021.

[12] D. Li, T.-W. Li, E.-P. Li and Y.-J. Zhang. "A 2.5-D angularly stable frequency selective surface using via-based structure for 5G EMI shielding," *IEEE Transactions on Electromagnetic Compatibility,* vol. 60, no. 3, pp. 768–775, 2017.

[13] H. Chen, H. Chen, X. Xiu, Q. Xue and W. Che. "Transparent MS on Glass Window for Signal Selection of 5G Millimeter-Wave Communication," *IEEE Antennas and Wireless Propagation Letters,* vol. 20, no. 12, pp. 2319–2323, 2021.

[14] D.J. King, K. Hettak, M.R. Chaharmir and S. Gupta., "Flexible Ink-Minimized Screen-Printed Frequency Selective Surfaces with Increased Optical Transparency for 5G Electromagnetic Interference Mitigation," *IEEE Transactions on Components, Packaging and Manufacturing Technology,* vol. 13, no. 1, pp. 110–119, 2023.

[15] S. Dey, S. Dey and S.K. Koul. "Second-Order, Single-Band, and Dual-Band Bandstop Frequency Selective Surfaces at Millimeter Wave Regime," *IEEE Transactions on Antennas and Propagation,* 2022.

[16] S. Dey, S. Dey and S.K. Koul. "Second-order, single-band, and dual-band bandstop frequency selective surfaces at millimeter wave regime," *IEEE Transactions on Antennas and Propagation,* vol. 70, no. 8, pp. 7282–7287, 2022.

[17] T.-W. Li et al. "A novel miniaturized multiband strong coupled-MS structure insensitive to almost all angles and all polarizations," *IEEE Transactions on Antennas and Propagation,* vol. 69, no. 12, pp. 8470–8478, 2021.

[18] S. Dey and S. Dey. "Conformal miniaturized angular stable triband frequency selective surface for EMI shielding," *IEEE Transactions on Electromagnetic Compatibility,* vol. 64, no. 4, pp. 1031–1041, 2022.

[19] R. Dewan et al. "Artificial magnetic conductor for various antenna applications: An overview," *International Journal of RF and Microwave Computer-Aided Engineering,* vol. 27, no. 6, p. e21105, 2017.

[20] H. Malekpoor. "Analysis of 2 × 2 MIMO array loaded by artificial magnetic conductor with improved gain, isolation, and bandwidth using equivalent transmission line model for indoor wireless base station," *International Journal of RF and Microwave Computer-Aided Engineering,* p. e23307, 2022.

[21] A. Kumar, A. Kumar and A. Kumar. "Gain enhancement of a wideband rectangular ring monopole millimeter-wave antenna using artificial magnetic conductor structure," *International Journal of RF and Microwave Computer-Aided Engineering,* p. e23319, 2022.

[22] M.A. Shahzad et al. "An Artificial Magnetic Conductor-Backed Compact Wearable Antenna for Smart Watch IoT Applications," *Electronics,* vol. 10, no. 23, p. 2908, 2021.

[23] Z. Jiang, Z. Wang, L. Nie, X. Zhao and S. Huang. "A Low-Profile Ultrawideband Slotted Dipole Antenna Based on Artificial Magnetic Conductor," *IEEE Antennas and Wireless Propagation Letters,* vol. 21, no. 4, pp. 671–675, 2022.

[24] Y. Chen et al. "Landstorfer Printed Log-Periodic Dipole Array Antenna With Enhanced Stable High Gain for 5G Communication," *IEEE Transactions on Antennas and Propagation,* vol. 69, no. 12, pp. 8407–8414, 2021.

[25] G. Zhai et al. "Gain-enhanced planar log-periodic dipole array antenna using non-resonant metamaterial," *IEEE Transactions on Antennas and Propagation,* vol. 67, no. 9, pp. 6193–6198, 2019.

[26] Q.-X. Chu, X.-R. Li and M. Ye. "High-gain printed log-periodic dipole array antenna with parasitic cell for 5G communication," *IEEE Transactions on Antennas and Propagation,* vol. 65, no. 12, pp. 6338–6344, 2017.

[27] M. Akbari, H.A. Ghalyon, M. Farahani, A.-R. Sebak and T.A. Denidni. "Spatially decoupling of CP antennas based on MS for 30-GHz MIMO systems," *IEEE Access,* vol. 5, pp. 6527–6537, 2017.

[28] K.G. Sadananda, M.P. Abegaonkar and S.K. Koul. "Gain equalized shared-aperture antenna using dual-polarized ZIM for mmWave 5G base stations," *IEEE Antennas and Wireless Propagation Letters,* vol. 18, no. 6, pp. 1100–1104, 2019.

[29] H. Jiang, L.-M. Si, W. Hu and X. Lv. "A symmetrical dual-beam bowtie antenna with gain enhancement using metamaterial for 5G MIMO applications," *IEEE Photonics Journal,* vol. 11, no. 1, pp. 1–9, 2019.

[30] A.S. Dixit and S. Kumar. "The enhanced gain and cost-effective antipodal Vivaldi antenna for 5G communication applications," *Microwave and Optical Technology Letters,* vol. 62, no. 6, pp. 2365–2374, 2020.

[31] S. Tariq, S.I. Naqvi, N. Hussain and Y. Amin. "A metasurface-based MIMO antenna for 5G millimeter-wave applications," *IEEE Access,* vol. 9, pp. 51805–51817, 2021.

[32] M.L. Hakim, T. Alam and M.T. Islam, "Polarization-insensitive and oblique incident angle stable miniaturized conformal FSS for 28/38 GHz mm-wave band 5G EMI shielding applications," IEEE Trans. Antennas Propag., vol. 22, no. 11, pp. 2644–2648, Nov. 2023.

CHAPTER 3

Miniaturization Mechanism of Planar Antenna using Metamaterial Structure

Mohammad Tariqul Islam[1*], *Md Mushfiqur Rahman*[2] and *Md Shabiul Islam*[3]

1. Introduction

The backbone of microwave integrated circuit (MIC) is the microstrip patch antenna (MPA). Since MIC can be referred to as the mother of all modern wireless portable devices, the design of MPA will have a large impact on the future market of these devices. But the conventional microstrip patch antenna works on the principle of half wavelength resonance which has a limitation of size reduction. A lot of techniques have been applied to improve this limitation and the level of compactness achieved by these methods is in general in the scale of one fifth of the operating wavelength. But it is true that some techniques achieve very promising **miniaturization** but with the cost of very complex structure, for example, multilayer structure. This type of structure can cause additional cost in designing compact portable devices. On the other hand, Metamaterial structures are specially designed to give resonance at low frequency. Especially, the structures with single or multiple splits cause strong gap capacitance along the splits. Since the value of resonant frequency is inversely proportional to the square root of the capacitance, the large value of gap capacitance gives low frequency resonance. In this way, miniaturization is achieved with very simple design steps.

2. Background

Metamaterial structures have the potential to give electromagnetic properties rather than natural materials. Besides that, they have some structural benefits. The presence

[1] Department of Electrical, Electronic and Systems Engineering, Universiti Kebangsaan Malaysia, Bangi, Selangor, Malaysia

[2,3] Faculty of Engineering, Multimedia University, Cyberjaya, Selangor, Malaysia.
Email: mushfique398@gmail.com[2]; shabiul.islam@mmu.edu.my[3]

* Corresponding author: tariqul@ukm.edu.my

of one or more capacitive gaps in a loop structure will result in a very high value of capacitance which will cause single or multiple resonances at very low frequency regions. For this reason, metamaterial structures are often used partially with the half-wavelength resonators to achieve a multiband property where in most cases one resonance will be found at a low frequency region. In this case, the resonance occurs even before the fundamental mode of resonances of half-wavelength structures which are referred to as **negative order resonances**. But this partial loading will not reduce the overall dimension of the structure. Thus, these types of antennas can cover multiple service bands so that one antenna can be sufficient for a particular device. But they are not appropriate for the compact modules used for Internet of things (IoT) devices or for other compact portable devices. For this reason, if a single metamaterial unit cell is considered as the only radiating patch of the antenna, then the overall dimension will be compact and at the same time by using multiple splits the multiband property can also be achieved. For this purpose, the unit cell needs to be excited by an appropriate feed structure to form an antenna which will cause an even or odd mode of resonances similar to the resonances of the single unit cell.

3. Metamaterial Inspired Techniques for Miniaturization

Many researches are done on partial loading of metamaterial structure to get a compact design. Wang et al., 2011 [1] proposed an antenna where the low frequency resonance is characterized by the coupling among a small loop ring on the top side and two **split ring resonators** (SRRs) on the back side (Figure 1).

The small loop ring forms the radiating element of the antenna which is inductive in nature and this inductive reactance is matched by the **capacitive loads** realized by the split-ring oscillator. These inductive and capacitive reactance form a pair of LC resonators. The separation of SRR gaps and the circumference of the small ring can be tuned to control the resonant frequencies. Further investigation finds that the larger square shaped SRR on the back side is responsible for the lower frequency band and the rectangular shaped SRR causes the higher frequency band. The substrate used here is FR4. The dimension of the antenna with the substrate is 30 mm × 30 mm × 1.8 mm.

A similar approach is proposed by [2], where two SRRs are placed on the same plane as shown in Figure 2. In this case the inner rectangle shaped closed ring is used as the feeding ring and the outer square shaped ring is used as the coupling element.

When two square-shaped SSRs with equal space between them are positioned in such a way that the slits are placed in opposite sides and the inner ring is used as the radiator whereas the outer ring acts as a coupling element, this topology can cause **multi-frequency response**. When the space between them is varied, the positions of the multi-frequency resonances can be tuned. Among the multi-mode operation, the lowest mode is found at a wavelength which is far less than $\lambda/4$. The antenna substrate used here is FR4 which is 1 mm thick. The dimension of the antenna is 50 mm × 50 mm × 1 mm [3], which follows this approach in a different way. Here two SRRs are loaded on the top and bottom faces of the antenna in such a way that each encloses a printed dipole arm as shown in Figure 3. This combination causes three resonant modes.

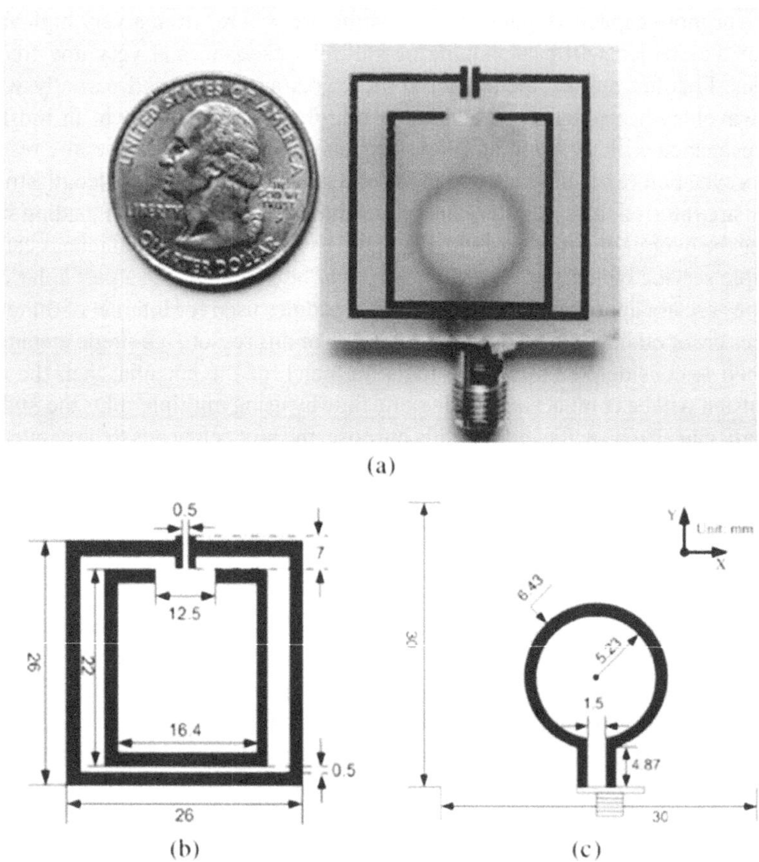

Fig. 1 Fabricated Prototype and Geometry of the Antenna Loaded by Two Split-Ring Resonators: (a) Fabricated Antenna, (b) Bottom View, (c) Top View.

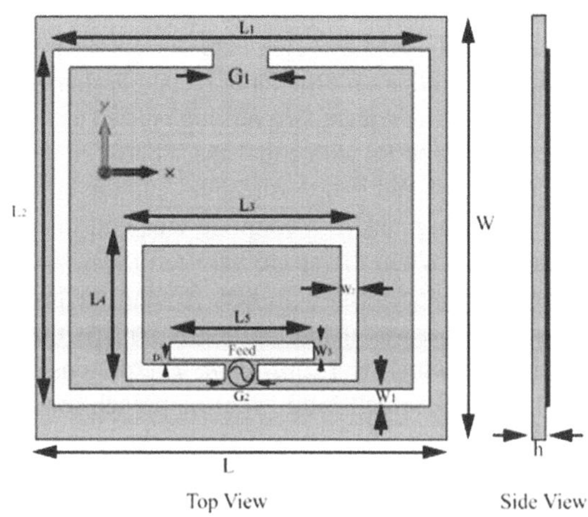

Fig. 2 Geometry of the Antenna with **Degraded Multimodal SRRs.**

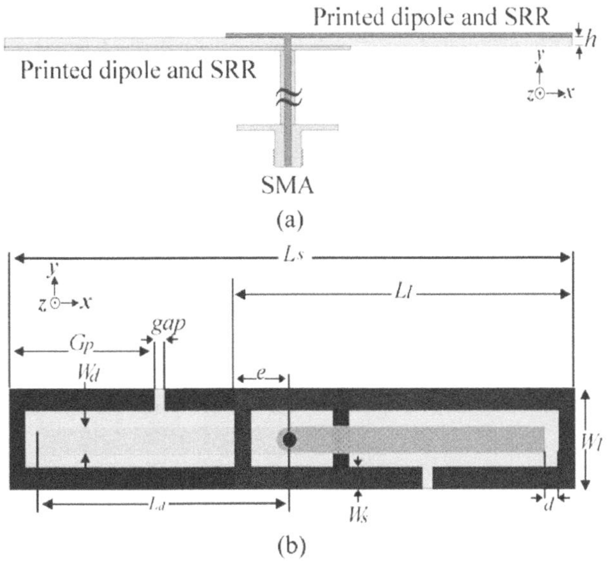

Fig. 3 Detailed Geometry of the **Dipole Antenna** Embodied with Two SRRs.

Fig. 4 Printed **Meander Line Antenna** Incorporated with Two SRRs.

After investigating the resonant modes, it is found that the dipole without the SRRs causes the highest order mode of the band. The placement of one SRR on the top side generates another lower mode of resonance. But when another SRR is placed on the back side it produces the lowest resonance mode. The antenna is printed on a Rogers 5880 substrate with a thickness of 0.508 mm. The dimension of the antenna is 9.6 mm × 74.4 mm × 0.508 mm.

Some researchers [4] propose a meander line antenna incorporating two square-shaped SRRs as depicted in Figure 4. The splits on the SRRs are placed in a perpendicular position. The antenna has been designed in three steps. In the first step, no SRRs are incorporated and the meander line patch is the only resonator which causes dual band resonance. After that, one SRR is incorporated which causes the resonant frequencies to shift to the lower region. Finally, another SRR is added which

(a) (b)

Fig. 5 Printed **Ring Monopole Antenna** with Two SRRs. (a) Top Side (b) Bottom Side.

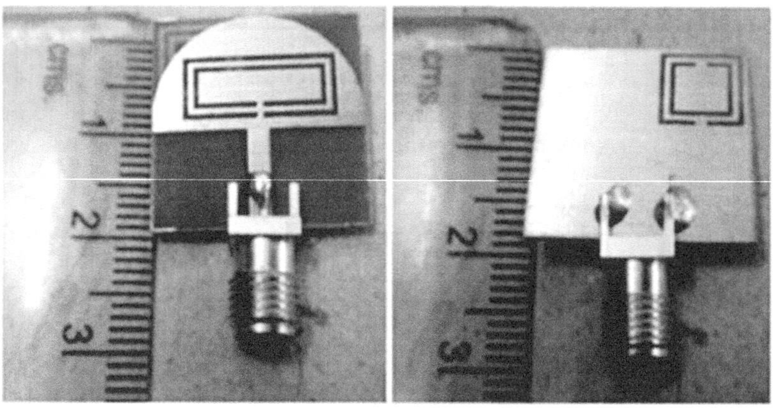

(a) (b)

Fig. 6 Printed **Semi-Circular Monopole Antenna** Loaded with **Complementary SRR:** (a) Top View, (b) Bottom View.

shifts the resonances further to the lower region. The optimized antenna is fabricated on the FR4 substrate with a thickness of 1.59 mm. The dimension of the structure is 22 mm × 24 mm.

Ali and his team [5] applied a similar approach where, in this case, the main resonator is the ring monopole rather than the meandered patch as shown in Figure 5. Inside the ring monopole, two SRR structures are loaded where the larger SRR is directly connected to the ring by a stub at the ring-feed junction and the smaller one is placed inside the larger SRR. This modification gives resonance at lower frequency than the ring monopole alone. In addition, a parametric analysis is done on the length of the external SRR which reveals that when the length is varied from 8.8 mm – 7 mm, the best impedance matching is found at the value of 7.5 mm. The antenna is fabricated on the FR4 substrate which is 1.6 mm thick. The dimension of the antenna is 30 mm × 24.8 mm ×1.6 mm.

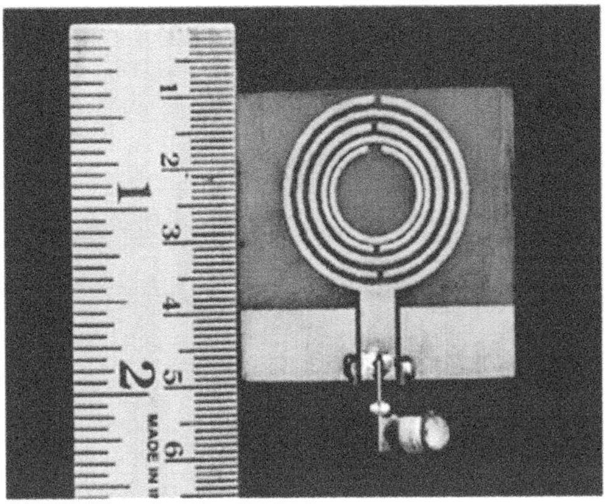

Fig. 7 Printed **Annular Split Ring Antenna** with Incorporation of Four SRRs.

Another identical type of approach is adopted by [6], where the semi-circular monopole is the main radiator as depicted in Figure 6. This radiator with a complementary SRR loaded ground plane causes single resonance at high frequency and a spike at the low frequency region. When two **C-shaped SRRs** are cut on the semicircle, multiple resonances occur including one resonance near the previous spike. The antenna is fabricated on a FR4 substrate with a thickness of 0.5 mm. The dimension of the antenna is 20 mm × 20 mm × 0.5 mm.

Thamil Selvi and his group [7] applied the previous methods in a different fashion. They utilized one unit cell of a metamaterial structure as the main radiating element. Without the split in the ring the resonance at low frequency suffers from poor impedance matching. When the split is added it causes low frequency resonance with acceptable impedance matching. After that they added additional four split rings which are not connected to each other as well as not with the main radiating element to get a multiband antenna as shown in Figure 7. Since strong coupling occurs between them, they give a multi band performance which is justified by designing the same antenna but with no splits. The same design is also investigated with different substrate materials and found that decreasing the value of permittivity shifts the resonance to the higher frequency region. Finally, the antenna is fabricated using FR4 substrate with a thickness of 0.8 mm. The dimension of the antenna is 40 mm × 40 mm × 0.8 mm.

Another group of researchers [8] proposed a Complementary SRR loaded monopole antenna where the main radiating element is a patch of rectangular shape as depicted in Figure 8. The solid rectangle with a closed end U-shaped slot gives a high frequency resonance at 5 GHz. A single split is cut at the left arm of the U-shaped slot which gives a low frequency resonance at 2.1 GHz. Finally, the CSRR structure is deployed inside the U-shaped slot to get another additional resonance. They also used FR4 substrate which is 1.6 mm thick. The dimension of the antenna is 19.18 mm × 22.64 mm × 1.6 mm.

Another approach to getting an electrically small antenna is the application of a zeroth order resonator which consists (ZOR) of a periodic structure of composite

(a) (b)

Fig. 8 Printed **Rectangular Monopole Antenna** Loaded with Complementary SRR: (a) Top View, (b) Bottom View.

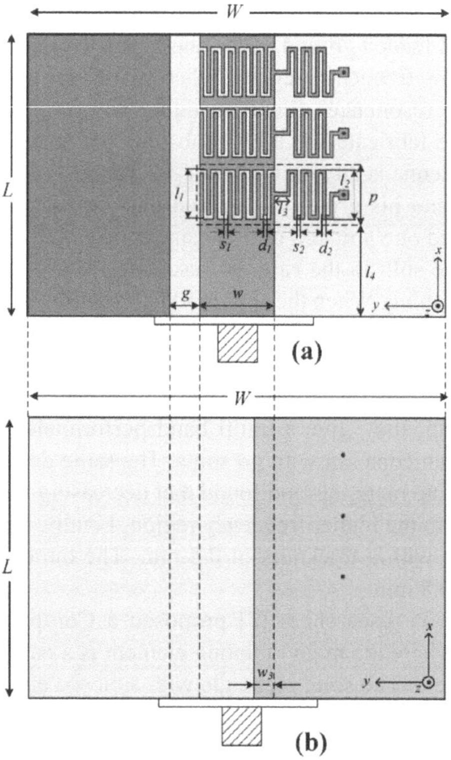

Fig. 9 Detailed Geometry of the **Meander Line Loaded ZOR** (Zero Order Resonance) Antenna with CRLH-TL: (a) Top View, (b) Bottom View.

right/left-handed (CRLH) transmission lines (TLs). The antenna using this structure can generate resonance which has zero propagation constant, i.e., for a finite and nonzero frequency it can support infinite wavelength. As a result, the resonance of these types of antennas does not depend on the physical length. Chen & Chen and his research team [9] developed one ZOR antenna where a meander line structure shorted to ground plane induces a shunt inductance and the interdigital capacitor induces a series capacitance and the combination of these two components are used to form the CRLH-TL as depicted in Figure 9. Another important design characteristic of this type of structure is the use of a ground plane on both sides, i.e., one is the coplanar ground, and another ground plane is used at the back side. This configuration causes a zero-order resonance at 1.94 GHz. The antenna is fabricated on a Rogers 5880 substrate with a thickness of 1.57 mm. The dimension of the antenna is 28 mm × 50 mm × 1.57 mm.

In another research, some researchers [10] developed a CRLH-TL based ZOR antenna where two units of CRLH-TL unit cell are connected to the **coplanar ground** plane and cascaded side by side above the Centre patch as shown in Figure 10. The unit cells are also joined to the ground plane at the back side through some vias. The ZOR of this configuration is found at 2.59 GHz. The substrate used here is Rogers 5880. The dimension of the antenna is 33 mm × 20 mm × 1.57 mm.

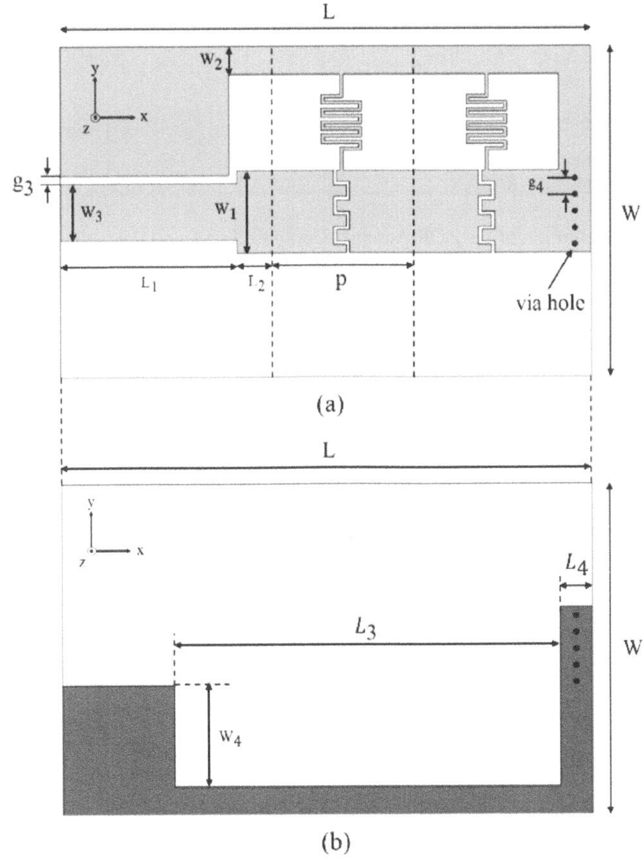

Fig. 10 Detailed Geometry of the Meander Line Loaded ZOR (Zero Order Resonance) Antenna with CRLH-TL: (a) Top View, (b) Bottom View.

Chi and his team [11] propose another type of ZOR antenna in which they only use the co-planar ground as presented in Figure 11. Here, the meander lines of the unit cells are not connected to the ground plane through vias like the previous structures. Rather, they are used as a connection between the ground and the patch, i.e., the antenna is short-circuited. However, the operation of this configuration is almost the same as the previous structures. The only difference is that in this case, the **interdigital capacitor** is the most influential part for the resonance. They also use the Rogers 5880 substrate. The size of the structure is 32.2 mm × 20 mm × 1.57 mm.

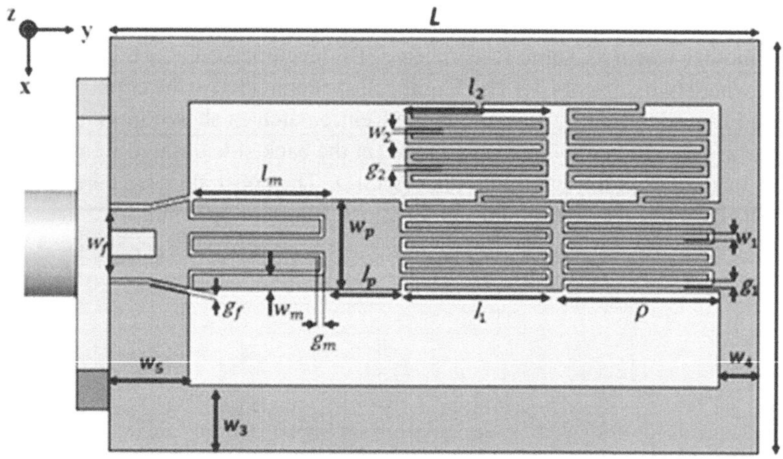

Fig. 11 Configuration of the ZOR Antenna Composed of Two-Unit Cells of the Composite Right/Left-Handed (CRLH) Artificial Structure.

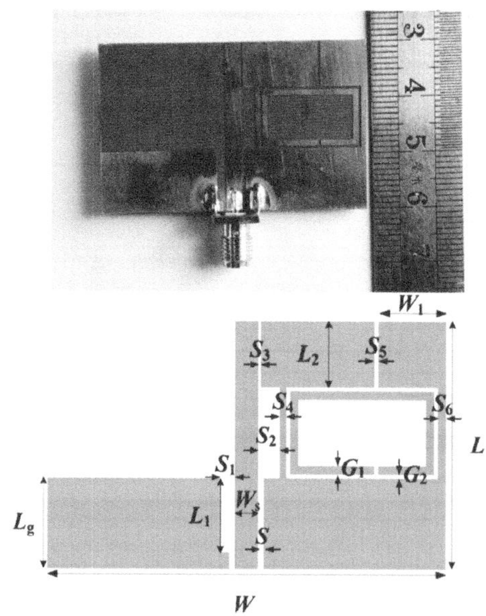

Fig. 12 Configuration of Single CRLH-TL Unit Cell Loaded ZOR Antenna.

Another group of researchers [12] designed a different type of ZOR antenna without an interdigital capacitor and meander line. In this case, a strip is added between a rectangle-shaped patch and coplanar ground plane to induce the **shunt inductance** (Figure 12). On the other hand, the shunt capacitance is realized by the **coupling** between the feeding-line ground and the patch. And the series inductance is formed by the coupling between the **CPW (Coplanar Waveguide**) ground plane and the patch. In this way the CRLH-TL is constructed. They have only used one unit cell. The material used here is F4B-2 which is 1.5 mm thick. The overall size of the antenna is 50 mm × 30 mm × 1.5 mm.

4. Methodology

The overall methodology of this mechanism starts with the simulation of a metamaterial unit cell. When the required resonance characteristics are achieved then the unit cell can be used to form an antenna by modelling an appropriate feeding network with proper boundary conditions.

4.1 Metamaterial Unit Cell Simulation

First of all, the radiation characteristics of the unit cell is determined by designing and simulating the unit cell structure. In this case, CST simulation software is used. In simulation, the transmission coefficient of the unit cell is specially observed because the unit cell will be used as the main radiator of the antenna. The permittivity and permeability are also checked to confirm the presence of strong electrical polarization which is important to get a low frequency resonance. If the structure is used as the parasitic element or as the superstrate to improve the radiation of another radiator then its surface property, namely, permittivity and permeability or more precisely, **single and double negative characteristics** need to be observed as well. In this chapter, these applications are not focused. For this reason, the simulation of the unit cell structure will be sufficient rather than simulating the arrays of that unit cell structure. However, the solver used here is the Transient Time Domain Solver rather than the Frequency Domain Solver. More specifically, the Finite Integration Technique (FIT) of Transient solver is used to simulate the structure. The FIT is actually a numerical method which provides a universal spatial discretization scheme. It is suitable for various electromagnetic problems ranging from static field calculations to high frequency applications in time or frequency domain. The following design specifications need to be maintained during the simulation of the unit cells.

- The dimension of the structure has been chosen initially from the selection of the resonant frequency.
- The frequency range should be large enough compared to the expected required bandwidth of the simulated unit cell.
- The properties of the background material are listed in Table 1. Since the ports are always connected along the positive and negative *Z*-axis, the distances from the center of the unit cells along the *X*, *Y* and *Z*-axis are set to zero.
- Onc of thc most important factors to considcr in simulating thc unit cclls is thc selection of appropriate boundary conditions. Since the unit cells are anisotropic

in nature and the microstrip feed line is used to excite the unit cell antenna, the perfect electric wall and the perfect magnetic wall are not defined along the *Z*-axis. Since the ports are connected along the *Z*-axis for all orientations, The boundary condition "Open (add space)" is always applied along the positive and negative *Z*-axis. The extra space along the positive and negative sides of the *Z*-axis is chosen to be half of the length of the longer side (rectangular structure) or half of the length of the equal side from the surface of the substrate.

- Two ports are connected along the positive and negative *Z*-axis. The size of the ports is automatically determined by CST. The distance of the reference plane from the ports is an important factor to determine the permittivity and permeability of the metamaterial structure but has no effect on the determination of the resonant frequency.
- After considering all the specifications, the metamaterial structure is simulated after selecting the source type to all ports and choosing all mode options. The accuracy is chosen to be −40 dB.

Table 1 The Properties of the Background Material of the Unit Cell Simulation.

Type	*Epsilon*	*Mu*	*Rho [kg/m^3]*	*Thermal Conductivity [W/K/m]*	*Heat Capacity [kJ/K/kg]*	*Diffusivity [m^2/s]*
Normal	1.0	1.0	1.204	0.026	1.005	2.14872e-05

4.2 Antenna Design and Simulation

When the unit cell design and simulation are completed, the antenna is formed by exciting the unit cell structure by a 50 Ohm microstrip feed line. The width of the microstrip line for the 50-ohm impedance matching is determined by using the macros function of CST. First of all, the setup is chosen as a thin microstrip and then the parameters which are needed to be given are as follows: permittivity of the substrate, the height of the substrate and the width. The width is varied until the output impedance becomes almost 50 Ohm. The frequency range is selected the same as the unit cell simulation. The background properties for the antenna simulation are the same as the unit cell simulation to achieve similar radiation characteristics. The boundary condition needs to be chosen carefully to simulate the antenna. Since far field radiation patterns need to be observed, the Open (add space) boundary condition is applied in all directions except the negative *Y*-axis where the feed line is located. An SMA is connected at the center of the feed line and then a waveguide port is connected between the positive and ground terminals of the SMA connector. The dimension of the port is automatically calculated by the software. The solver parameters are the same as the unit cell simulation.

5. Antenna with Symmetrical Metamaterial Unit Cell [26]

In this segment, a patch antenna is proposed where the main radiator is an **ELC (Electric inductive and capacitive) resonator** which can be compared to a unit cell of an epsilon negative metamaterial array structure. The structure is highly symmetrical in nature. However, the split width and the size of a simple partial ground plane is modified to achieve resonance at lower frequency with wider bandwidth. Moreover, two metallic blocks are etched on the right and left sides of the feeding line to increase the gain but at the same time the bandwidth is reduced slightly. The overall electrical measurement of the miniaturized antenna is 0.26 $\lambda L \times$ 0.26 $\lambda L \times$ 0.017 λL, where λL is the operating wavelength in free medium at the lowest − 10 dB cut-off frequency. The proposed design is fabricated and measured. It is found from the measurement that the operating bandwidth of the antenna for |S11| less than − 10 dB is 19.74% and the peak realized gain is 1.15 dBi. Moreover, the horizontal plane radiation pattern is found to be **omnidirectional** and in the elevation plane a dumbbell-shaped pattern is observed. The levels of cross-polarization in the elevation and horizontal planes are both less than − 12 dB.

5.1 Antenna Configuration

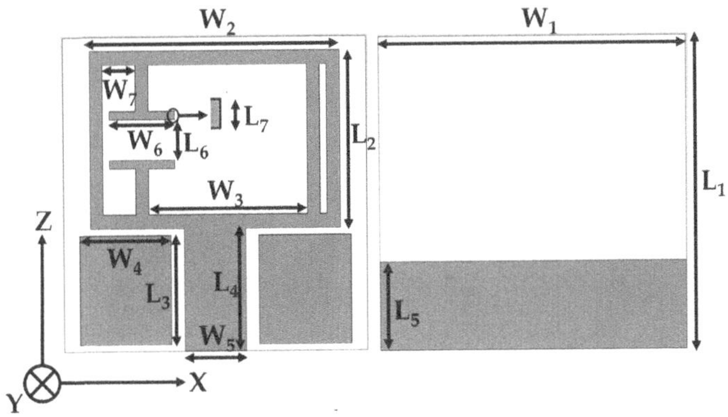

Fig. 13 The Configuration of the Final Optimized Antenna: (a) Top View, (b) Bottom View.

Table 2 Detailed Measurements of the ELC Resonator Loaded Antenna with Partial Ground Plane.

Parameters	*Values (mm)*	*Parameters*	*Values (mm)*
L_1	23	W_2	19
L_2	13	W_3	12
L_3	08	W_4	7.0
L_4	09	W_5	4.6
L_5	6.5	W_6	5.0
L_6	3.0	W_7	2.5
L_7	0.6	W_8	1.52
W_1	23		

RT Duroid 5880 substrate with a thickness of 1.52 mm is used as the dielectric material of the antenna. The value of the **dielectric constant** of this material is 2.2 and the value of loss tangent is 0.0009. The proposed optimized design of the antenna is depicted in Figure 13, and its detailed measurements are listed in Table 2. The design of the antenna is started by the analysis of the ELC (electric LC) resonator with **appropriate boundary conditions**. Then, the initial antenna is formed which is fed by a fifty-ohm microstrip feed structure with a finite ground plane size. Further, some evolutionary steps are applied for better antenna performance. A detailed description of the entire analysis is given in the subsequent sections.

5.2 Design and Performance Analysis

5.2.1 Modelling of the ELC Resonator

The ELC resonator is 19 mm wide and 13 mm long while each side is 1 mm wide. The capacitive gap is 5 mm wide and 0.6 mm thick. The substrate material used here is RT Duroid 5880 which is 1.52 mm thick and has a dielectric constant of 2.2. Since the resonator is **anisotropic**, i.e., it shows different resonance characteristics for different **applied field orientations**, it needs to be investigated for different boundary conditions [13–16].

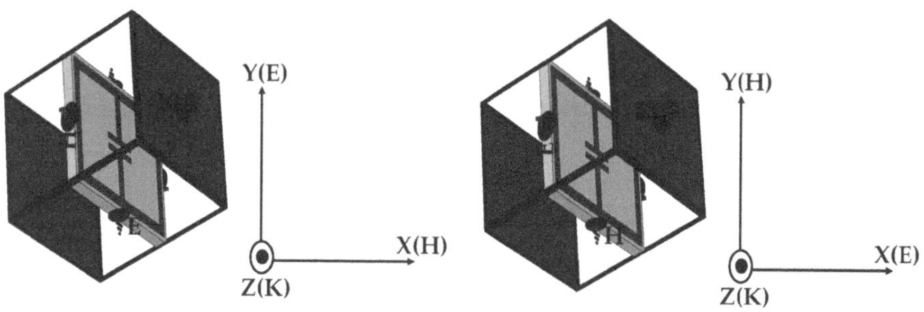

Fig. 14 Simulation Setup of the ELC Unit Cell: (a) Setup 1, (b) Setup 2.

However, the boundary conditions where the magnetic field is applied perpendicularly to the resonator are not discussed here. This is because the symmetry of the ELC structure belongs to the C_{2h} group, which does not possess **magnetoelectric coupling** [17]. For this reason, the presented structure is modelled by two boundary conditions. In one boundary condition, the electric field is applied along the *y*-axis and the magnetic field along the *x*-axis (Setup 1), whereas in another boundary condition the field orientations are opposite to the former one (Setup 2) (Figure 14).

The simulation of this resonator with these two setups is done by a time-domain solver of CST Microwave Studio software. Since in Setup 1 the **electric field polarization** is found along the split (Figure 15(a)), it induces a current in the structure that circulates in two loops as shown in Figure 15(b), resulting in two different current

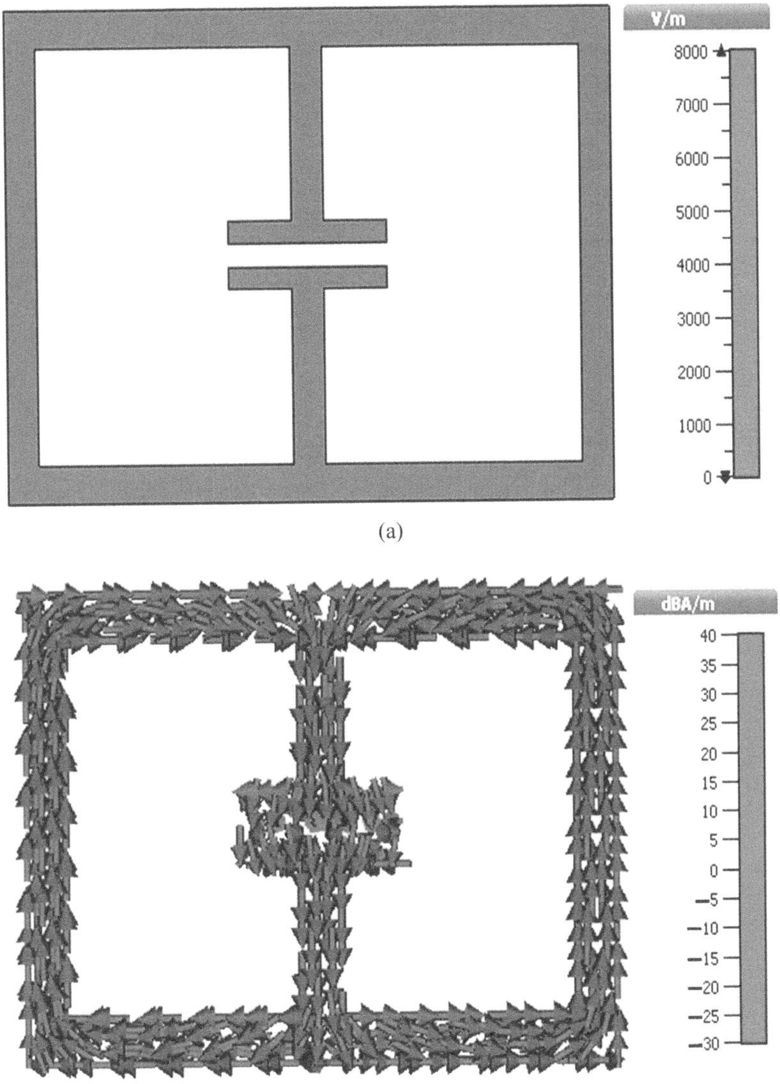

(a)

(b)

Fig. 15 Distribution of: (a) Electric Field, and (b) Surface Current at 2.67 GHz for Setup 1.

paths. Due to these current paths, two resonances are found at 2.67 GHz and 8.03 GHz frequencies (Figure 16(a).

On the contrary, when the electric field polarization is found along the non-split bearing sides (Figure 17(a)), the induced current is confined to the opposite sides of the resonator. For this reason, it cannot complete a loop and results in a reduced electrical length as shown in Figure 17(b). For this reason, a single resonance occurs at 4.23 GHz (Figure 16(b)), which is higher than the previous orientation (2.67 GHz).

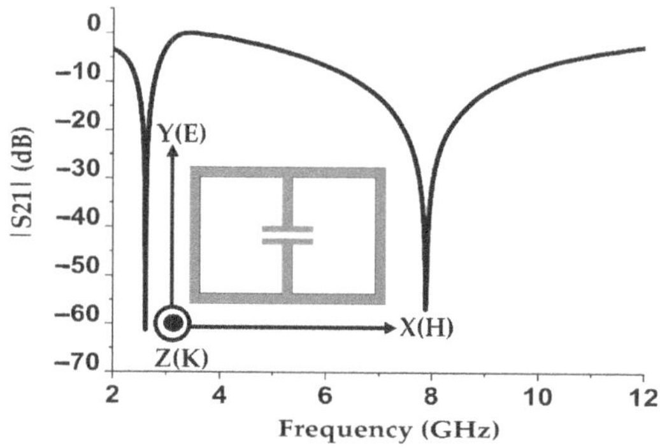

(a)

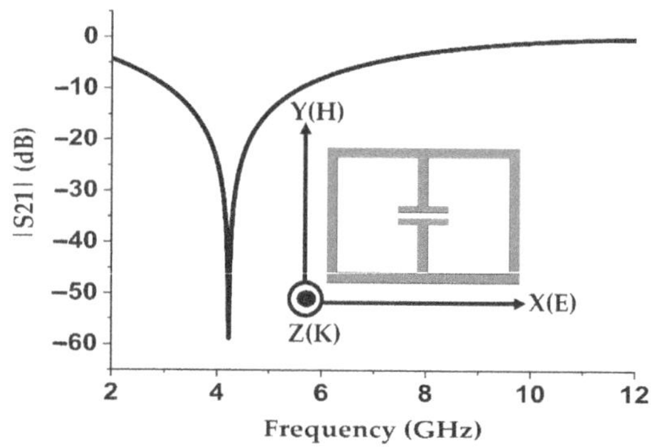

(b)

Fig. 16 S21 Spectrum of the ELC Resonator for: (a) Setup 1, and (b) Setup 2.

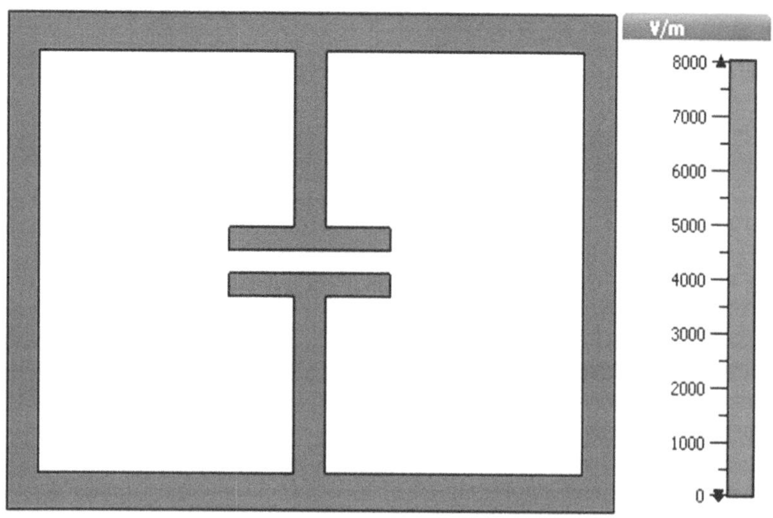

(a)

Contd.

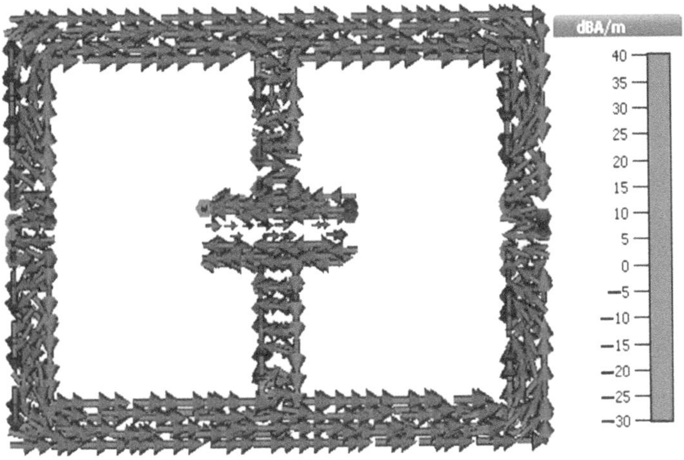

(b)

Fig. 17 Distribution of: (a) Electric Field, and (b) **Surface Current** at 4.23 GHz for Setup 2.

5.2.2 Design Evolution of the Antenna

From the above analysis, it is clear that when the resonator is used as the main radiating structure of an antenna, it will exhibit the same resonance characteristics provided the substrate material and its thickness, as well as their **orientation**, are the same for both cases. The same orientation can be achieved when the structure is excited by a microstrip feed line in a way so that the electric field polarization will be found along the split. Since Setup 1 causes low-frequency resonance, the antenna will be miniaturized if the ELC is excited in a way as depicted in Figure 18 because then the electric field polarization will be found along the capacitive gap as in Setup 1. However, antenna simulation shows that when the ELC is excited as shown in Figure 18, then two resonances occur at almost the same frequencies (2.88 GHz and

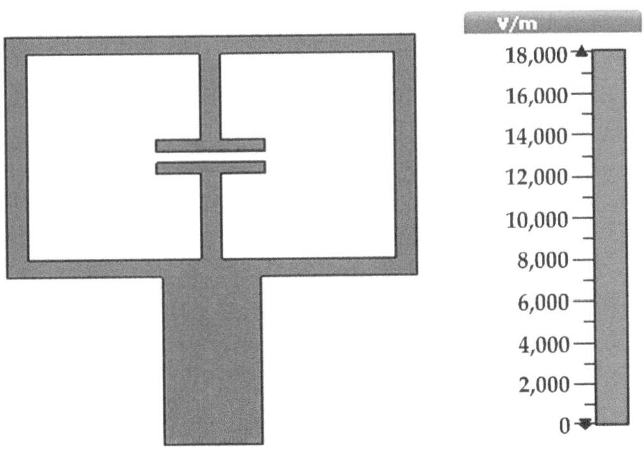

Fig. 18 **Electric Field Distribution** of Antenna 1.

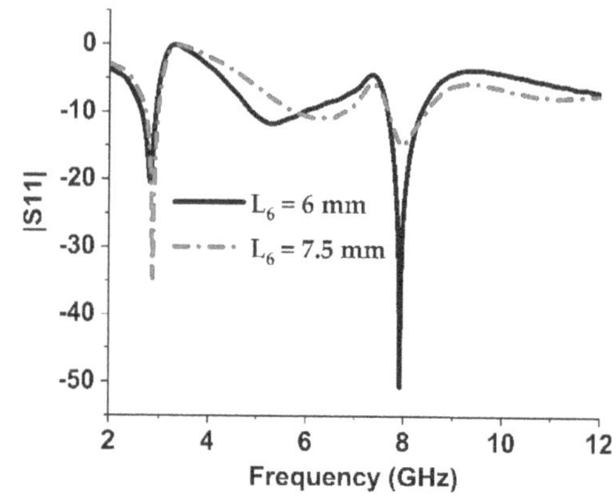

Fig. 19 Simulated Resonance Characteristics of Antenna 1 for Different Ground Plane Sizes.

7.96 GHz) as in Setup 1, but they are not matched for a unique ground plane dimension as presented in Figure 19. For compactness, the ground plane dimension, L_5 = 7.5 mm is chosen. However, this antenna results in a narrow bandwidth (10.41%) and poor gain (0.07 dBi).

The performance of Antenna 1 is improved at the desired frequency region by some evolutionary steps as shown in Figure 20. Since the antenna is designed for the 5G C band applications, the resonant frequency needs to be shifted to the higher

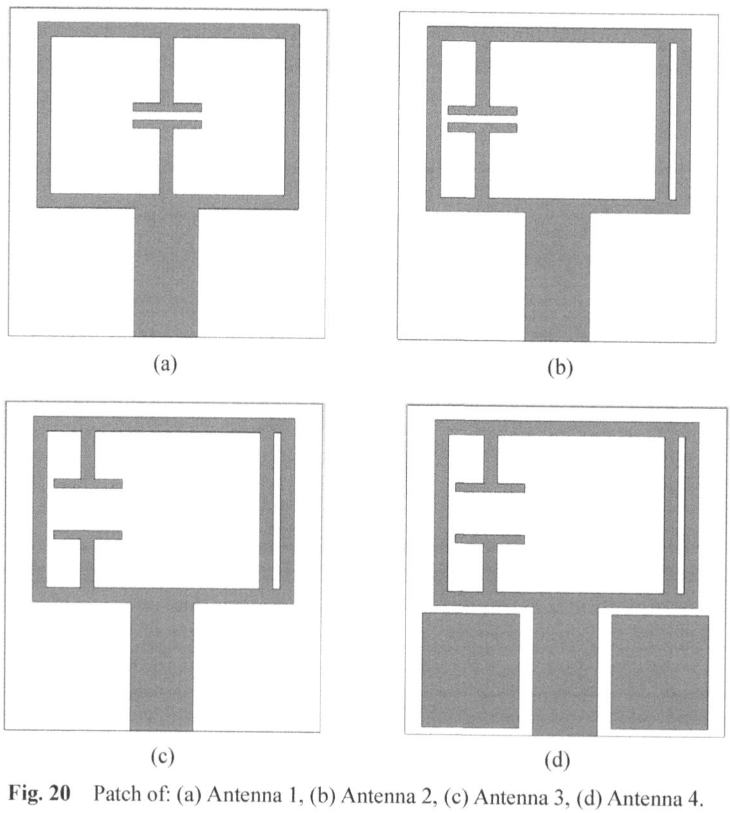

Fig. 20 Patch of: (a) Antenna 1, (b) Antenna 2, (c) Antenna 3, (d) Antenna 4.

frequency region. It can be accomplished either by reducing the length and width or by reducing the electrical length or by decreasing the **gap capacitance** of Antenna 1 (Figure 20(a)). The first method is avoided since it can cause an even less efficient radiator. However, the second method is implemented in two steps.

Firstly, the capacitive gap is moved near to the left side, which changes the current path inside the structure, and, secondly, an inductive strip is connected in parallel inside the structure which is near to the right side (Figure 20(b)). As a result, the electrical length, responsible for the low-frequency resonance, is reduced, which shifts the resonance to the higher frequency region as shown in Figure 21(a). However, the second method has a negligible effect on bandwidth, but the gain is increased considerably as shown in Figure 21(b).

The third method is adopted by increasing the capacitive gap width of the resonator (Antenna 3 in Figure 20(c)), which causes the gap capacitance to decrease. For this reason, the resonance is shifted to the higher frequency region (Figure 21(a)). A **parametric study** on the gap width is shown in Figure 22 to demonstrate the effect of gap capacitance on reflection coefficients. The study shows that increasing the capacitive gap shifts the resonance to the higher frequency region and increases the bandwidth as well.

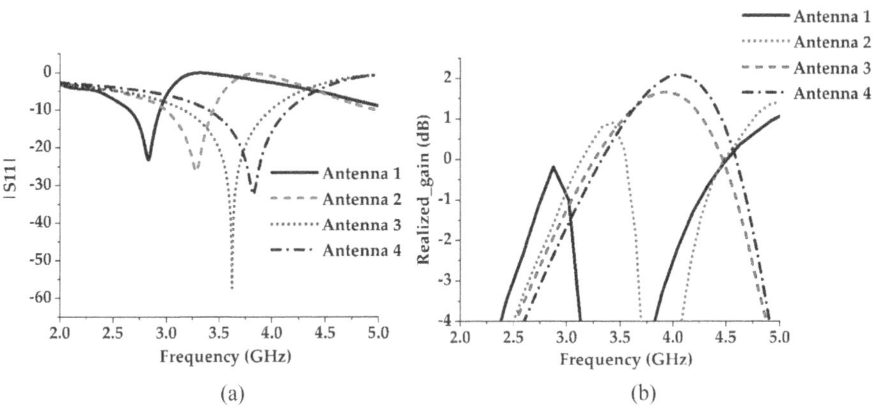

Fig. 21 (a) **Comparison of S-Parameters** of the Antennas, (b) Comparison of Gains of the Antennas.

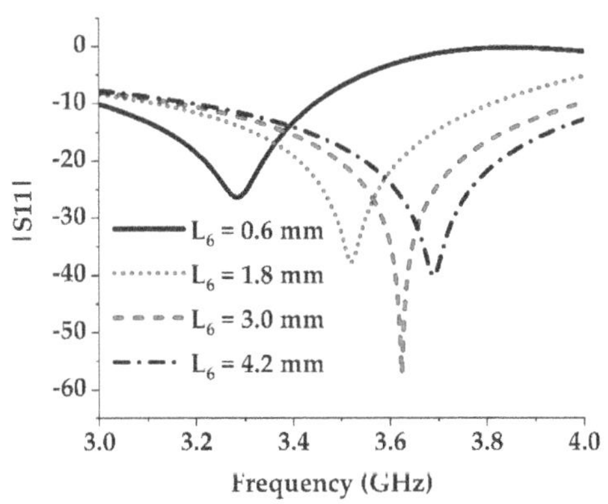

Fig. 22 Variation of Reflection Coefficients with Different Gap Widths.

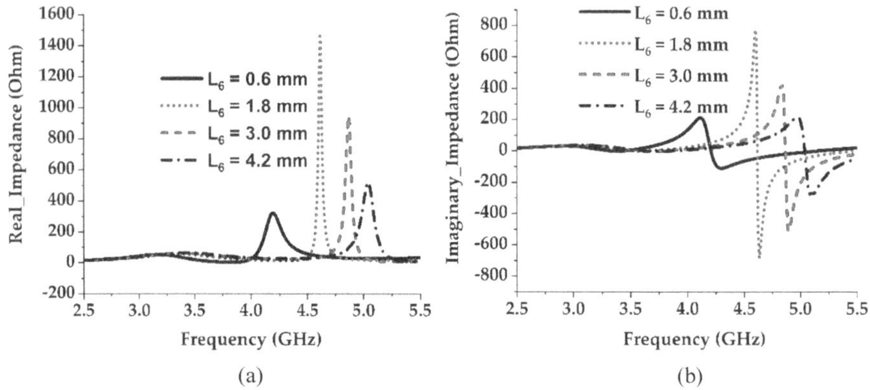

Fig. 23 Variation of **Real and Imaginary Impedance** Values with Different Gap Widths: (a) Real Impedances, (b) Imaginary Impedances.

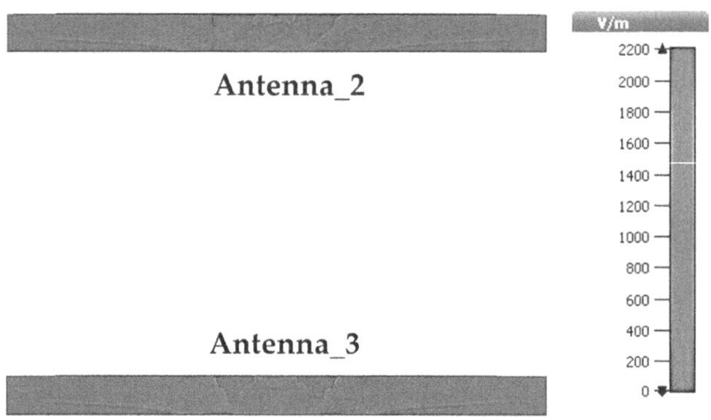

Fig. 24 Simulated Electric Field Distribution in the Cross-Section of Antenna 2 and Antenna 3.

However, the effect of gap width on antenna bandwidth can be easily realized by the parametric studies of various gap widths on real and imaginary values of the input impedance (see Figure 23(a and b)). As the gap width widens, a decreasing real impedance and an increasing imaginary impedance are observed at high frequency which causes a wider bandwidth. This happens because the electric field between the metallic ground plane and feed structure decreases as the gap width increases, which in turn decreases the coupling capacitance between the feed and the ground, which is shown in Figure 24. The optimized value of the gap width (L_6) is 3 mm. Further increment of the gap width results in poor impedance matching and the bandwidth enhancement is also negligible. Finally, the gain of the antenna is improved by adding two rectangular metallic blocks of the same size on the right and left sides of the feed structure (Antenna 4 in Figure 20(d). These **parasitic elements** increase the current density at the left side of the capacitive gap and at the lower side of the structure, which causes the gain to increase as presented in Figure 21(b). A slight reduction in bandwidth is also observed.

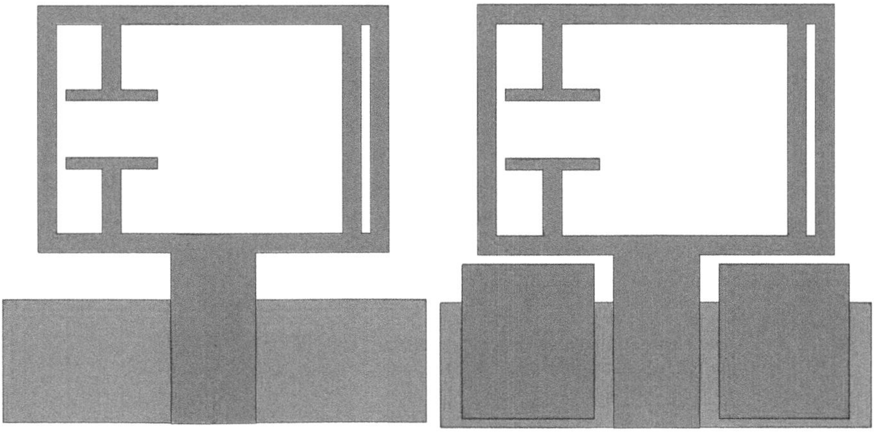

(a) (b)

Fig. 25 Electric Field Distribution of the Patch of: (a) Antenna 3, and (b) Antenna 4 at 3.6 GHz and 3.8 GHz, respectively.

The effect of adding the **rectangular blocks** can be easily detected from the electric field distribution of the antenna patch. As seen in Figure 25, the blocks increased the electric field on the overall patch, especially at the lower part of the radiator and the left part of the radiator near the gap. That's why the antenna gain is considerably enhanced. The equivalent lumped element circuit of the structure can be approximated as shown in Figure 26. Here, Cse1 and Cse2 = the capacitance between the edges of the capacitive gap and the left arm. Cs = The capacitance due to the gap of L_6. Ca = the capacitance between the two arms on the right side. Cp1 and Cp2 = The capacitances between the parasitic blocks and the patch. Cg1 And Cg2 = The coupling capacitances between the patch and parasitic blocks with the ground.

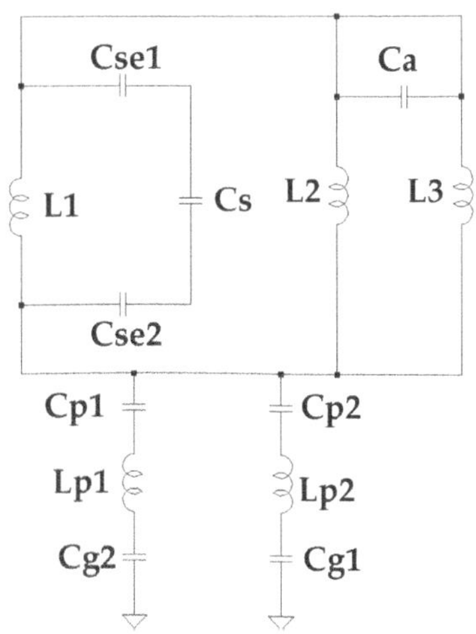

Fig. 26 **Equivalent Lumped Element Circuit** of the Proposed Antenna (Antenna 4 in Figure 8d).

5.3 Results Analysis

The antenna of Figure 20(d) is fabricated after optimization and measured to validate the above analysis. The results of simulation agree well with the measurement, where a slight variation is observed. It is actually due to the cable loss and fabrication tolerance. In addition, the cable used in the **Satimo StarLab measurement system** is very long. The reflection coefficient is measured using a PNA series (**power network analyser**) VNA (Agilent N5227A) (10 MHz to 67 GHz), whereas the radiation patterns and gain are observed using the Satimo StarLab antenna pattern measurement system. The photography of the prototype is given in Figure 27.

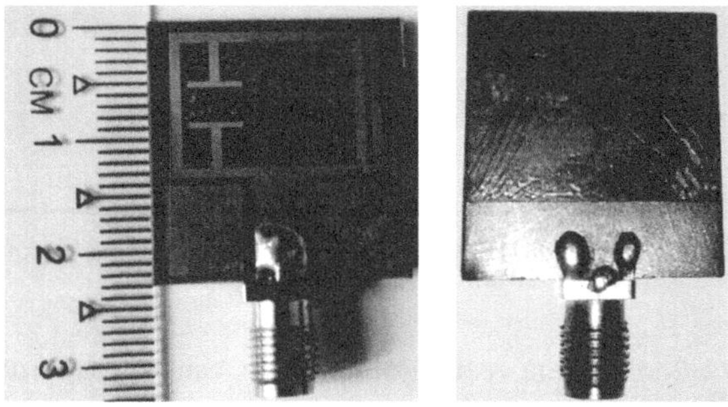

(a) (b)

Fig. 27 (a) Top View and (b) Bottom View of the Fabricated Antenna.

The measurement setup of VNA and Satimo Star LAB are shown in Figure 28 (a and b). The reflection coefficient and realized gain in simulation and measurement of the prototype are shown in Figure 29(a and b), respectively. The measured reflection coefficient for |S11| < −10 dB is from 3.4 – 4.2 GHz, and a peak gain of 1.16 dBi is realized.

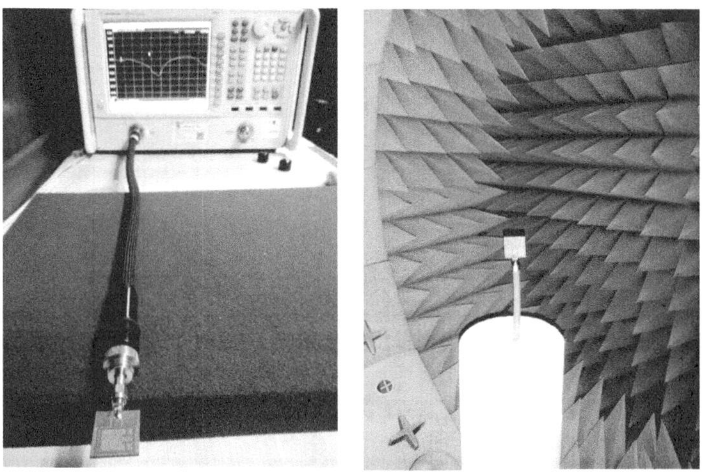

(a) (b)

Fig. 28 Measurement Setup of: (a) VNA, and (b) Satimo Star Lab.

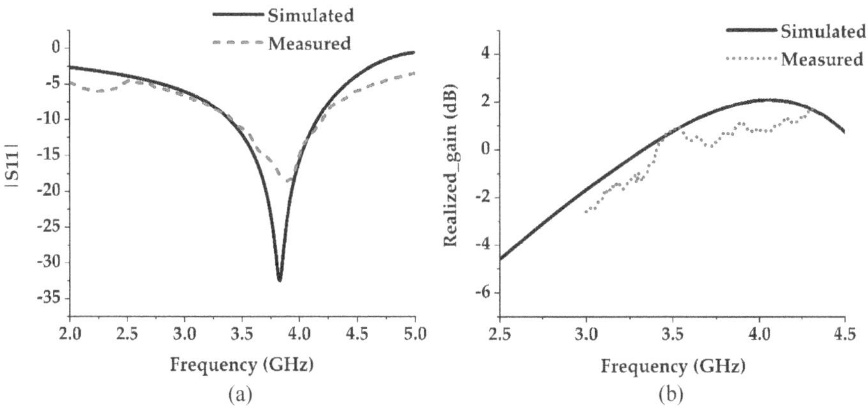

Fig. 29 (a) **Reflection Coefficient** and (b) Realized Gain of the Final Antenna.

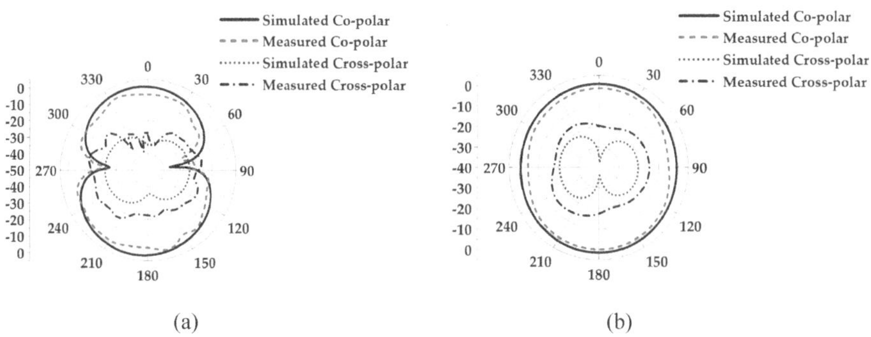

Fig. 30 **Radiation Patterns** of the Final Antenna: (a) YZ-Plane (Elevation Plane), (b) XY-Plane (Horizontal Plane).

The elevation plane and the horizontal plane radiation patterns extracted from simulation and measurement are presented in Figure 30(a and b), respectively, which are in good agreement. The measured radiation pattern in the E-plane is dumbbell-shaped while the radiation pattern in the H-plane is omnidirectional, and both patterns are stable. However, the measured levels of cross-polarization in the elevation and horizontal planes are not more than − 12 dB. The simulated 3d gain plots at three different frequencies are shown in Figure 31.

6. Antenna with Asymmetrical Metamaterial Unit Cell and DGS [27]

In this section, an **asymmetric metamaterial unit cell** structure is used as the main radiator of a patch antenna and then slightly modified by adding a meander line structure inside the unit cell. Without modification, the antenna loaded with a single unit cell, depending on the microstrip feed structure orientation, gives similar **even and odd modes of resonances** as the unit cell structure when examined as a metamaterial surface. Since compactness is the main target, the feed line alignment

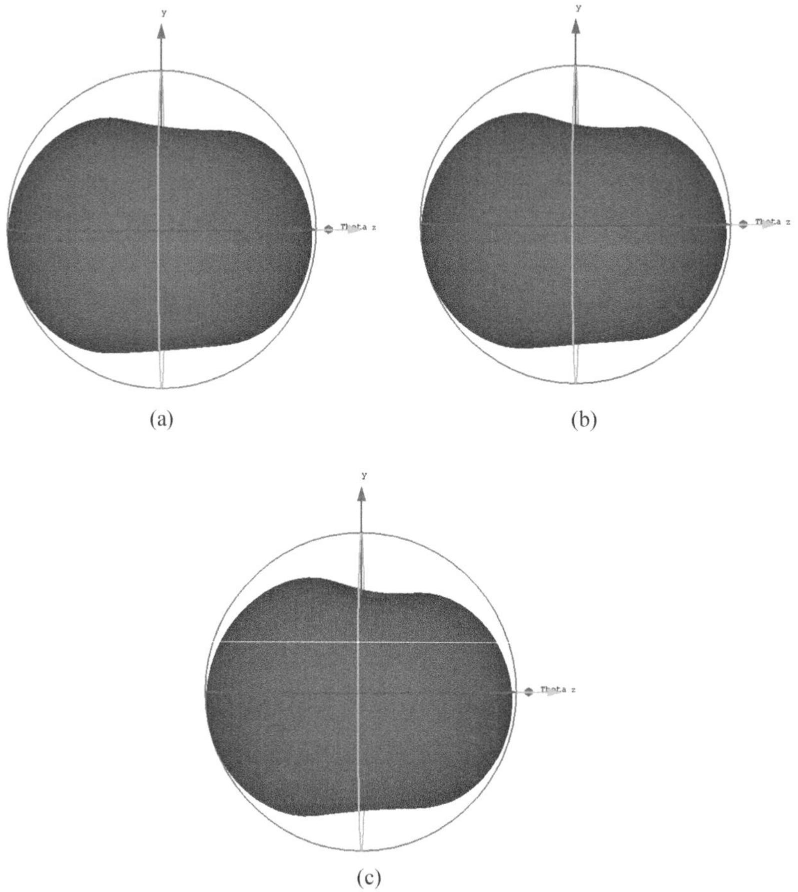

Fig. 31 Simulated 3d Gain at: (a) 3.4 GHz, (b) 3.9 GHz, and (c) 4.2 GHz.

which causes resonance at lower frequency regions, is accepted here. After that a meander line structure is added inside the unit cell antenna. The meander structure is parallelly placed by the side where the split is located and joins the sides which are perpendicular to the feed line structure. This modification of the structure causes another resonant mode at higher frequency region when the metallic ground plane is defected by another meander line structure. Specific dimensions of the DGS structure, the meander line structure, and the metamaterial unit cell structure are parametrically optimized to locate these modes of resonance dips at two distinct frequencies having appropriate intervals to achieve wide bandwidth. Last of all, the feed structure is moved from the centre position of the patch to improve the impedance matching in such a way that the bandwidth remains unaffected. The overall dimension of the miniaturized antenna is 0.25 λ_L × 0.23 λ_L × 0.02 λ_L, where λ_L is the operating wavelength at the centre frequency of the impedance bandwidth in free space. The optimized antenna is fabricated and measured. Measurement results show that the novel design exhibits similar radiation patterns like a monopole antenna. Moreover, the prototype gives a percentage bandwidth of 26.6%, ranging from 3.26 – 4.26 GHz where |S11| is less than − 10 dB and a pick realized gain of 1.26 dBi is observed. In addition, the levels of the cross-polarization recorded from the simulation and measurement are lower than − 15 dB in the h-plane.

6.1 Antenna Geometry

The antenna is fabricated on Rogers RT Duroid 5880 substrate of thickness 1.52 mm which has a dielectric constant of 2.2 and loss tangent of 0.0009. The configuration of the proposed antenna is depicted in Figure 32 and the dimensions are listed in Table 3.

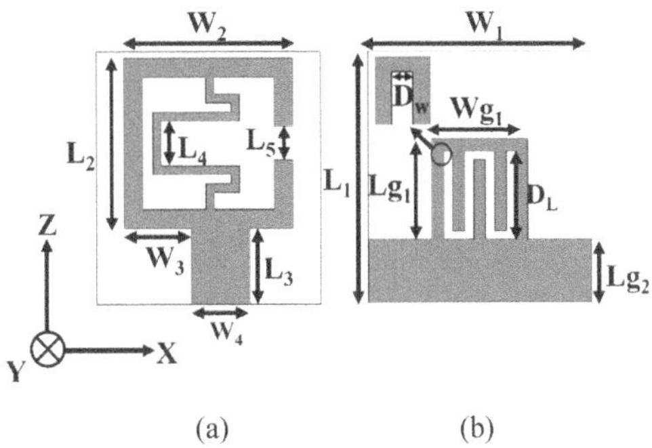

Fig. 32 The Configuration of the Final Optimized Antenna: (a) Top View. (b) Bottom View.

Initially, the resonance characteristics of the SRR unit cell with a single split are investigated. Then the microstrip feed line is utilized to excite the resonator with proper orientation and boundary condition to get similar resonance characteristics as investigated. After that, some techniques are applied to get enhanced bandwidth with a compact structure. The numerical simulations are done with CST microwave studio software in the time domain.

Table 3 Detailed Measurements of the Meander Line Loaded SRR Antenna with DGS.

Parameters	*Values (mm)*	*Parameters*	*Values (mm)*
L_1	20	W_1	18
L_2	13.5	W_2	13.5
L_3	6	W_3	5.44
L_4	3.6	W_4	4.62
L_5	2.7	Wg_1	7.8
Lg_1	8	D_L	7
Lg_2	5	D_W	0.7

6.2 Design and Performance Analysis

6.2.1 Resonance Characteristics of the Unit Cell

Initially, a square-shaped ring with a length of 13.5 mm on each side and a width of 1.5 mm is selected. Then a split of length 0.9 mm is cut in the middle of the right

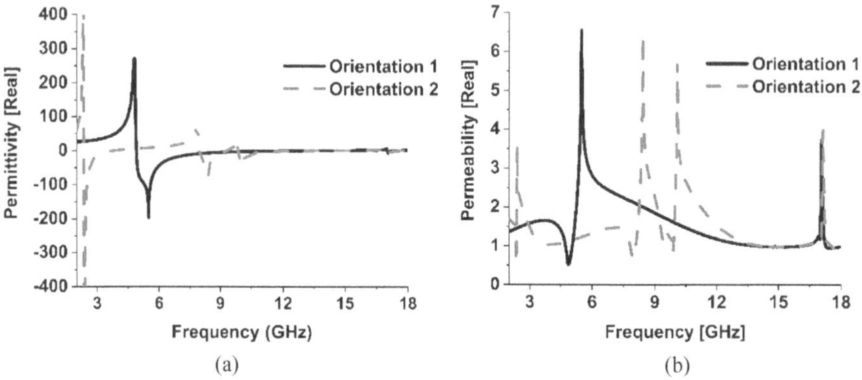

Fig. 33 (a) **Permittivity,** and (b) **Permeability** of Configuration 2 for Different Orientations.

side of this ring. The resonator formed in this way is anisotropic in nature, i.e., it will exhibit different resonance characteristics for different orientations of incident electric field vector, $\vec{E}$ and magnetic field vector, $\vec{H}$ and wave vector, $\vec{K}$. Originally, the SSR exhibits **mu-negative characteristic** when the magnetic field is applied orthogonally to the unit cell plane, otherwise it will show epsilon-negative property [18–20] as shown in Figure 33. However, there can be three possible configurations. Each configuration further has two different orientations after fixing the *z*-axis.

The configurations of the field vectors are given below:

1. In configuration 1, the $\vec{H}$ vector is normal to the unit cell plane, i.e., it is aligned with the *z*-axis (fixed), where in orientation 1, $\vec{k}$ and $\vec{E}$ are applied to the *x*-axis and the *y*-axis respectively and in orientation 2, $\vec{k}$ and $\vec{E}$ are applied to the *y*-axis and the *x*-axis respectively.
2. In configuration 2, the $\vec{k}$ vector is normal to the unit cell plane, i.e., it is applied to the *z*-axis (fixed), $\vec{E}$ and $\vec{H}$ are applied to the *x*-axis and the y axis respectively (Orientation 1) and vice versa (Orientation 2).
3. In configuration 3, the $\vec{E}$ vector is normal to the unit cell plane, i.e., it is applied to the *z*-axis, $\vec{k}$ and $\vec{H}$ are applied to the *x*-axis and the *y*-axis respectively and vice versa.

No noticeable resonance characteristics (is not shown here) is observed for both orientations of configuration 3 in the desired frequency range. On the other hand, configuration 1 is not investigated here since the electric polarization found in this case is not stronger than configuration 2. Moreover, in this research microstrip feed structure is utilized to give excitation to the resonator where the perpendicular component of the magnetic field of the incident electromagnetic radiation is negligible [21]. For these reasons, this excitation will not reflect the results of configuration 1.

On the contrary, the resonance behaviour of the unit cell for configuration 2 can be achieved easily by exciting the unit cell antenna with a microstrip feed line. The orientations of configuration 2 and their transmission spectra are shown in Figure 34. In both orientations, the wave vector $\vec{K}$ is applied to *z*-axis. However, in orientation 1, $\vec{E}$ is applied to the *x*-axis and $\vec{H}$ to the *y*-axis. In this case, the electric field polarization

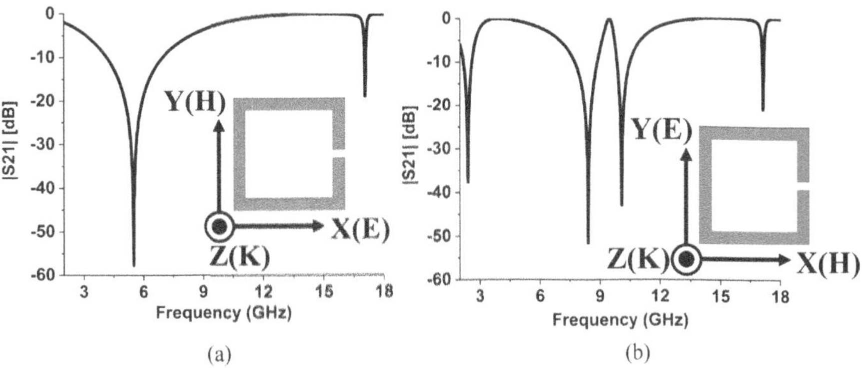

Fig. 34 Transmission Spectrum of the Single-Split Resonator: (a) Orientation 1, (b) Orientation 2.

is along the sides of the unit cell and two successive dips are found at high frequency regions in the transmission spectra which can be referred to as the odd modes of resonances. These are similar to short-wire-like electric resonances [22]. The first excitation mode occurs at f_0 = 5.5 GHz. Whereas, the higher order mode occurs approximately at $3f_0$ position in the frequency range of 2 – 18 GHz.

In orientation 2, $\vec{E}$ is applied to the y-axis and $\vec{H}$ to the x-axis. In this case, strong electric polarization occurs along with the split, and an annular current is induced due to the external electric field which causes **magnetic resonance**. This occurrence is referred to as the electric field coupling to the magnetic resonance [23].

Four dips are observed in the transmission spectra which can be referred to as the even modes of resonances. The first dip is the lowest excitation mode which occurs at f_0 = 2.4 GHz. The second and third dips both correspond to the second excitation mode (since they have similar current density distribution) occur at approximately $4f_0$ positions. The third excitation mode occurs at approximately $8f_0$ position.

The distribution of current on the **unit cell structure** for these orientations at the frequencies of resonance is also investigated to clarify the resonance characteristics (Figure 35). For orientation 1, an odd number of current nodes are found in the current density distribution at the frequency of resonance where the number of nodes increases at higher frequencies of resonances. Three and seven current nodes are observed in the successive **excitation modes** (a current node is referred to as the junction of two current paths or the starting or ending of a current path). Whereas an even number of current nodes are found in the current density distributions at frequencies resonances of orientation 2. Like orientation 1 the number of nodes also increases at higher frequencies of resonances except for the second and third dips. That's why they correspond to the same excitation mode. However, two, four, and eight current nodes are observed at successive frequencies of resonances.

6.2.2 Unit Cell Antenna Formation and Analysis

After investigating the resonance characteristics of the split ring resonator unit cell, it can be said that if the unit cell is excited in a way such that strong electric polarization

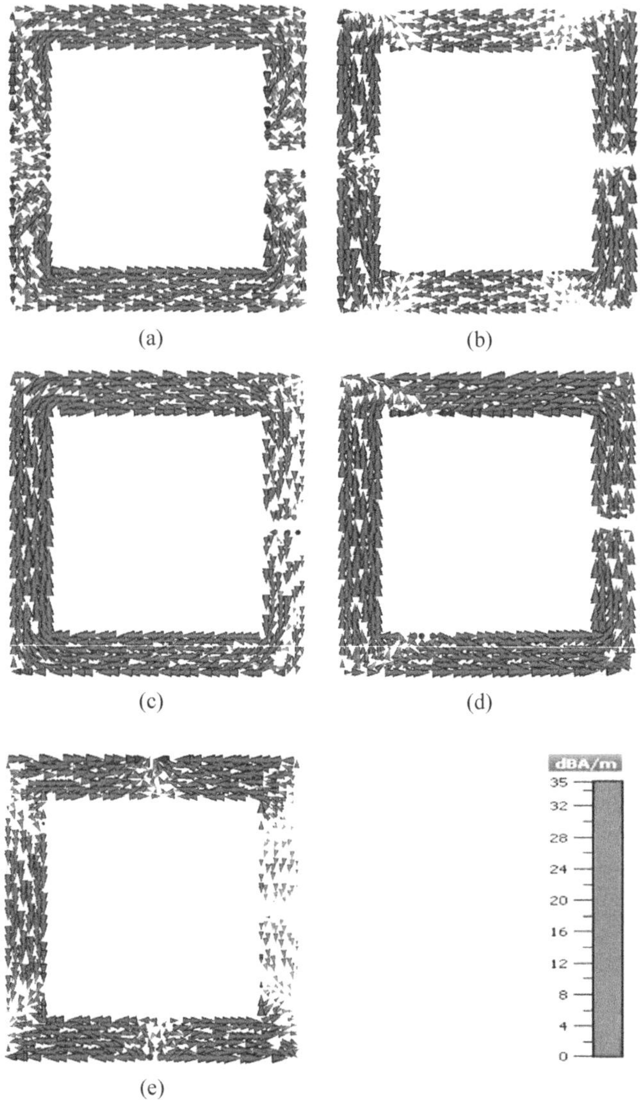

Fig. 35 Current Density at: (a) 5.5 GHz for Orientation 1, (b) 17 GHz for Orientation 1, (c) 2.4 GHz for Orientation 2, (d) 8.4 GHz for Orientation 2, (e) 17 GHz for Orientation 2.

is observed along the split gap as orientation 2, then similar resonance characteristics will be observed.

For this purpose, a **microstrip feed line** is connected at the centre of the side which is orthogonal to the split bearing side to excite the unit cell. In the simulation, strong electric polarization is observed along the split (Figure 36(a)) and the three dips can be obtained easily at the same frequency positions as the unit cell within 2–12 GHz frequency range (Figure 37(a)) and for different ground plane sizes. The frequency positions are similar to orientation 2 of configuration 2.

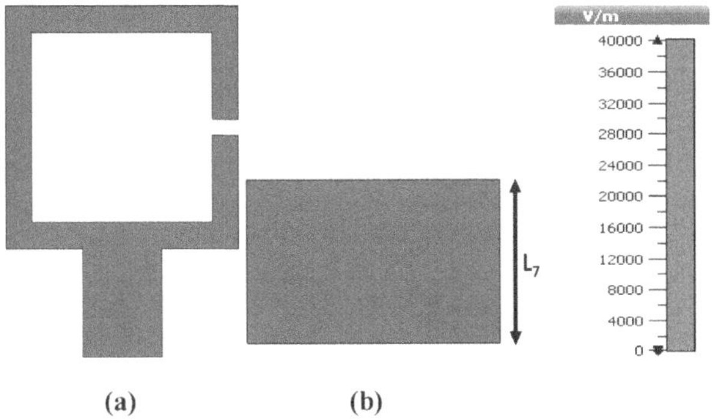

Fig. 36 (a) Electric Field Distribution on the: (a) Patch, and (b) Ground of Antenna 1 at 2.5 GHz.

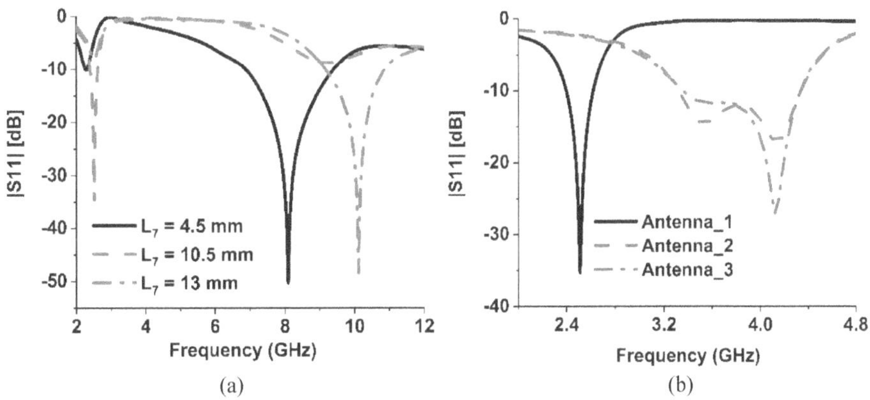

Fig. 37 (a) Reflection Coefficient of Antenna 1 for Different Ground Plane Sizes, (b) Comparison of Reflection Coefficient of 3 Stages of the Proposed Antenna.

It should be noted that in this configuration because of the strong electric polarization along the split, only even modes of resonances will be excited where the resonances at high frequency will cause an increased even number of current nodes in the patch. On the other hand, if the feed is connected at the middle of the non-split bearing side [not shown here] then electric polarization will not be observed along the split. And in this case, odd modes of resonances will be excited. However, in both cases, the creation of additional current nodes in the patch, which will cause higher even or odd mode resonances, can be controlled using different ground plane structures.

Now the antenna (Antenna 1) formed using the feed orientation as shown in Figure 38(a) and ground structure as shown in Figure 38(b) causes low frequency resonance as shown in Figure 37(b) but the bandwidth is very low at the lowest mode of resonance. Since compactness is required, techniques are applied to enhance the bandwidth of the resonance at 2.4 GHz. First of all, a meander line is placed inside the unit cell which is parallelly located with the side where the split is located. This structure causes the resonance to shift to the higher frequency region with a conventional ground plane. But when the novel meander structure is cut on the ground plane the unit cell loaded with the meander line excites another higher mode of resonance. Parameters are optimized for the proper placement of these two resonances to enhance the bandwidth.

The width of the meander line and the split width are optimized to control the position of the first mode of resonance. And the length and the width of the meander structure cut on the metallic ground plane are parametrically optimized to control

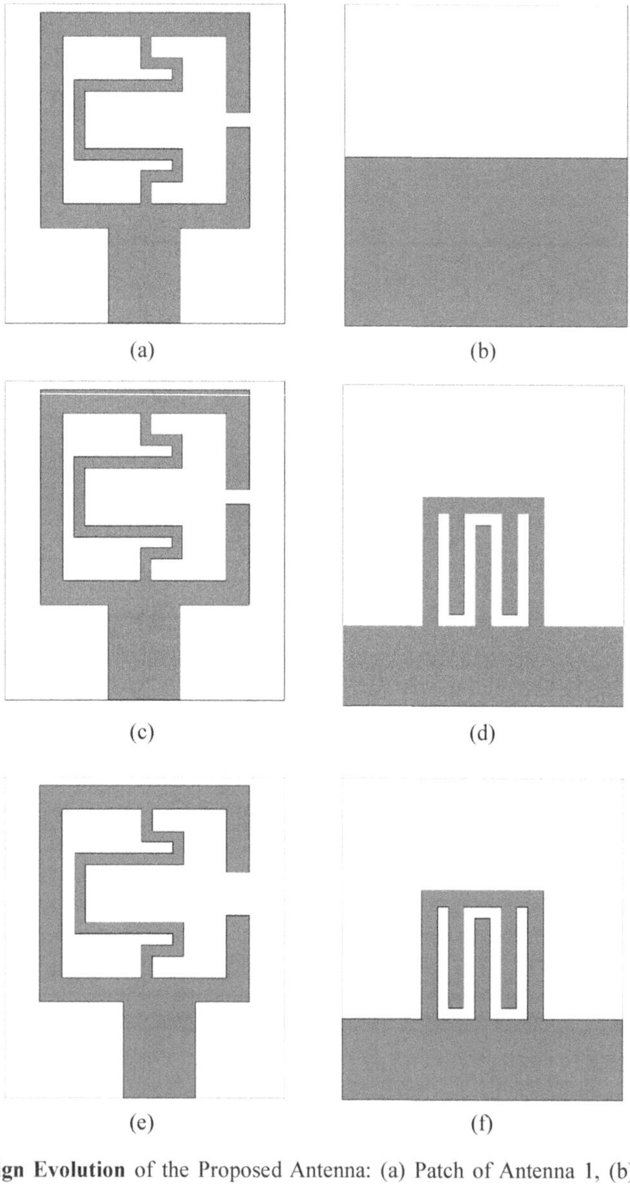

Fig. 38 Design Evolution of the Proposed Antenna: (a) Patch of Antenna 1, (b) Ground of Antenna 2, (c) Patch of Antenna 2, (d) Ground of Antenna 2, (e) Patch of Antenna 3, (f) Ground of Antenna 3.

the position of the second mode of resonance. Last of all, the feed line is moved from the centre position to improve the matching without affecting the bandwidth. The design evolution of the final antenna is given in Figure 38. A comparison of the reflection coefficient of various stages of the proposed antenna is given in Figure 37(b). A detailed analysis of the techniques is given in the successive sections where parametric analysis is done to optimize the values of the parameters. It should be noted that during the parametric analysis on a particular parameter the values of other parameters are kept constant at their optimized values.

6.2.3 Meander Line Analysis

The addition of a **meander line** inside the resonator forms two loops inside the unit cell. Since the loops are connected parallelly to each other they can be compared to two parallel loop inductances. As a result, the total inductance of the structure is reduced, and the **fundamental resonance** is shifted to the higher frequency region. It is well known that the inductance of a flat metal sheet depends on its width, length, and thickness [24]. So, the total inductance of the structure depends on these parameters of the meander line and the ring structure. For simplicity, only the thickness of the meander line is optimized to achieve desired resonance characteristics.

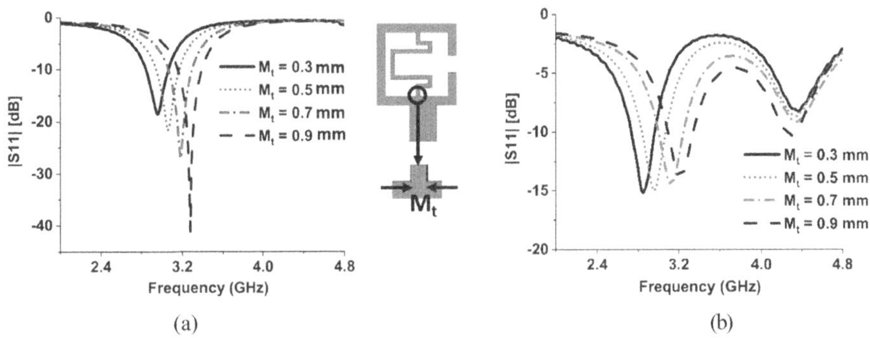

Fig. 39 Parametric Analysis of the Meander Line Width: (a) Without DGS, and (b) With DGS Structure on Reflection Coefficient.

A parametric analysis is done on the width of the meander line, M_t with DGS and without DGS structure (Figure 39). It is evident from these analyses that increasing the width of the meander line will shift the first or fundamental mode of resonance to the higher frequency region for both cases (DGS and without DGS) and does not affect the second mode of resonance for the case of DGS loaded ground. The optimized value of the width is 0.7 mm. The further increment will degrade the impedance matching of the resonance. For simplicity parametric analysis is not performed separately on the other dimensions of the meander line structure such as L_4. But the effect of the length L_4 can be described based on the parametric analysis of the meander line width. For example, increasing M_t will decrease the value of L_4 and vice versa. So, the effect of L_4 on the reflection coefficient of the structure will be reciprocal to the effect of meander line width Mt.

6.2.4 DGS Structure Analysis

A Defective ground structure is a compact structure which creates a defect inside the metallic ground plane as a **periodic or non-periodic arrangement**. They exhibit resonance behaviour and its existence under a microstrip feed structure can interrupt the current path inside the metallic ground plane [25]. That's why, when this meandered geometry is loaded on the metallic ground plane, the current is accumulated along the edges of the meander defects. This current induces additional current nodes at high frequency in the ring as well as in the meander line of the patch which causes an additional even mode of resonance in the high frequency.

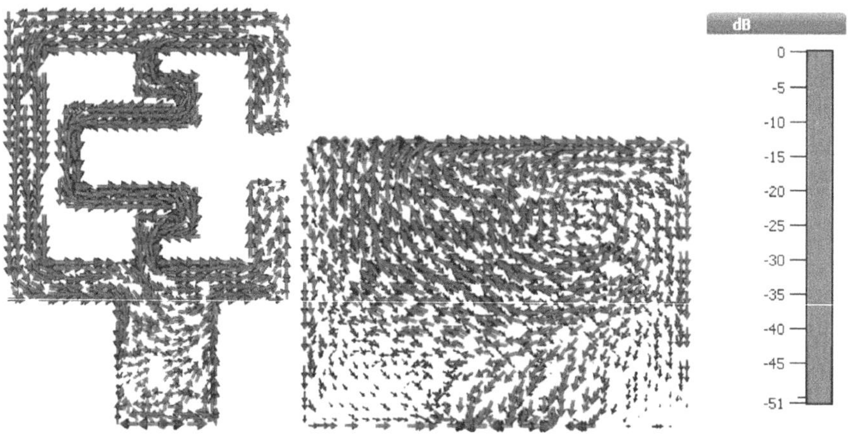

(a) (b)

Fig. 40 Current Distribution without DGS: (a) On the Patch, and (b) On the Ground at 3.5 GHz.

The current distributions on the patch and the ground plane without and with DGS are shown in Figure 40 and Figure 41, respectively. Before adopting DGS, the current on the patch follows a **single direction** which has only two nodes at the two ends of the split. On the other hand, the current in the ground plane which is below the patch also follows a single direction but at the corners which is not underneath the patch follows two different directions (Figure 40). However, the opposite flows of the current at the corners of the ground do not affect the current on the patch.

On the contrary, after adopting DGS on the ground plane, at low frequency the currents on the patch and the ground follow the same mechanism as without DGS (Figure 41(a and b)). But at a high frequency, the current follows different directions in the patch which forms two additional **current nodes** which in turn creates four different loops (Figure 41(c)). As a result, at high frequency the total inductance of the unit cell further reduces which gives rise to another higher even mode of resonance.

The inductance of the loops depends on the location of the current nodes on the patch. Whereas the location of the current nodes on the patch relies on the location of the current nodes on the ground plane. That means, by controlling the position

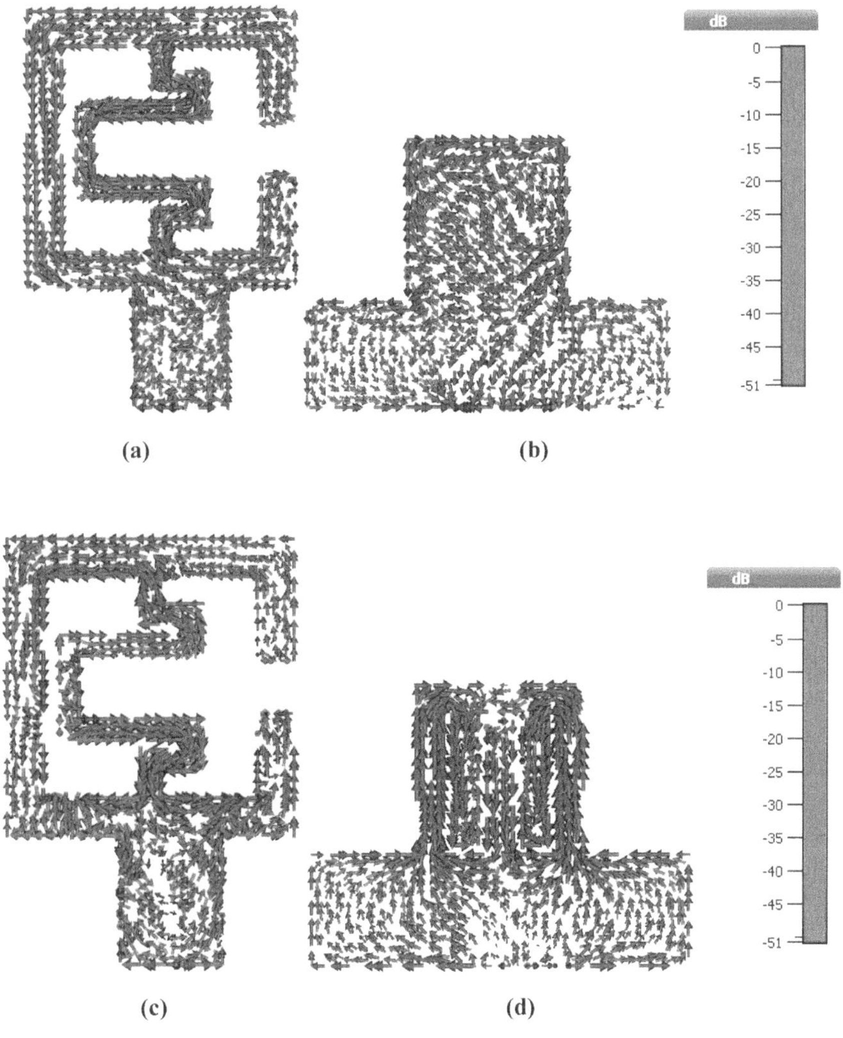

Fig. 41 Current Distribution without DGS: (a) On the Patch, and (b) On the Ground at 3.5 GHz, and with DGS, (c) On the Patch, and (d) On the Ground at 4.13 GHz.

of the current node on the metallic ground plane, the position of the resonance at higher frequency can be controlled. At high frequency, the currents flow from the opposite sides of the DGS to the centre causing a current node to generate at the middle top position of the structure (Figure 41(d)). And it is observed from the current distribution that the positions of the current nodes on the left side of the ring as well as on the meander line are generated just above the current node position of the DGS. The length and width of the DGS are optimized to control the position of the current node on the DGS which in turn controls the position of current nodes on the patch.

A parametric analysis is done on the length and width of DGS to see their effect on the second resonance (Figure 42). It is found that when the DGS length is increased the second resonance shifts to the lower frequency region with improved **impedance**

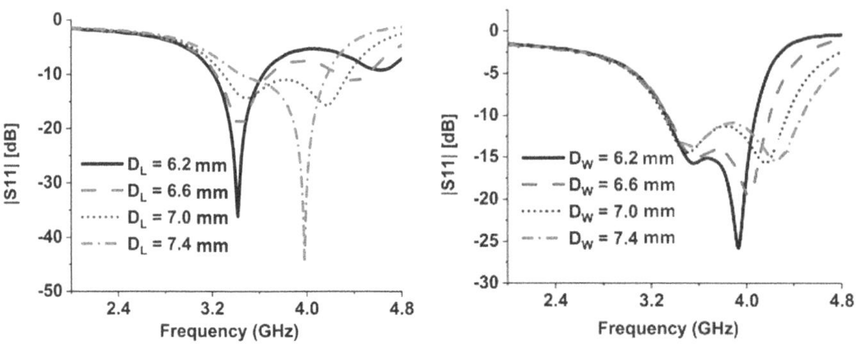

(a) (b)

Fig. 42 Parametric Analysis of the DGS: (a) Length, and (b) Width on Reflection Coefficient.

matching but degrades the impedance matching of the first resonance (Figure 42 (a)). Conversely, when the DGS width is decreased (Figure 42 (b)) the second resonance shifts to the lower frequency region with improved impedance matching but also improves the impedance matching of the first resonance. The parameters are optimized at values (DL = 7 mm, DW = 0.7 mm) for which the bandwidth is maximum.

6.2.5 Split Width Analysis

As mentioned earlier, increasing the width of the meander line shifts the first resonance to the higher frequency but the impedance matching degrades for higher values of width. The **split width** is another parameter that can shift the first resonance to the higher frequency region. A parametric analysis is done on the split width to see its effect on both resonances as shown in Figure 43. It is found that as the value of width is increased the first resonance shifts to the higher frequency region and at the same time improves the impedance matching of the second resonance.

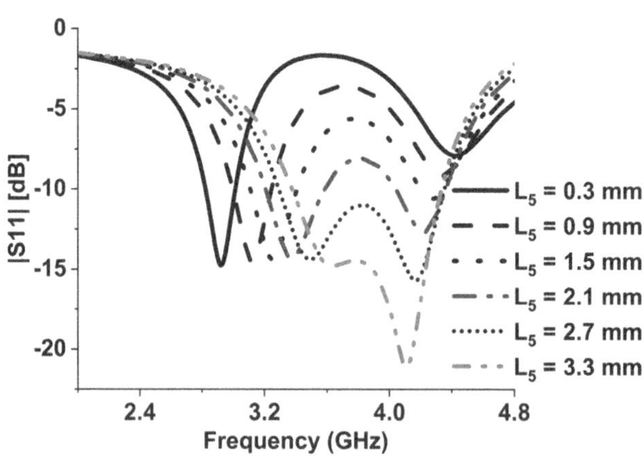

Fig. 43 The Parametric Analysis of the Split Width on Reflection Coefficient.

On the other hand, it also enhances the impedance bandwidth of both bands. Another parametric analysis is done on split width to see its effect on real and imaginary values of input impedance (Figure 44). It is found that increasing the value of the width flattens the real value of the input impedance of the bands toward 50 ohms at higher frequency (Figure 44(a) and at the same time, flattens the imaginary value of them towards 0 ohm at the same frequency region (Figure 44(b)). For this reason, two bands are placed closely with maximum **impedance bandwidth**. However, the width is optimized at a value (0.7 mm) where the bandwidth is maximum.

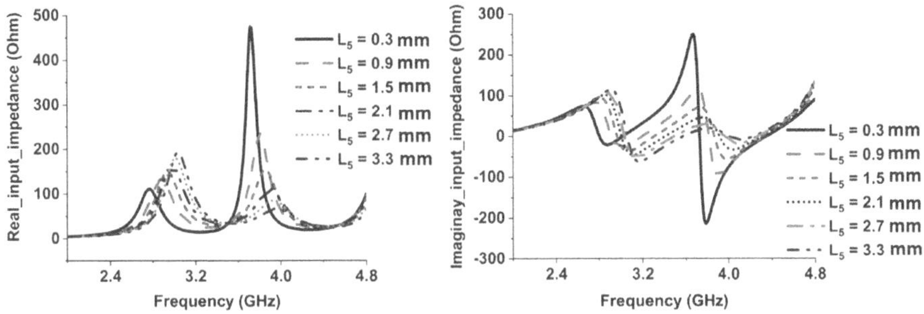

(a) (b)

Fig. 44 The Parametric Analysis of the Split Width on: (a) Real Input Impedance, (b) Imaginary Input Impedance.

6.2.6 Results Analysis

To verify the above analysis, the proposed antenna is fabricated (Figure 45) and measured. The reflection coefficients extracted from the simulation and the **measurement** of the final antenna agrees well as shown in Figure 46(a).

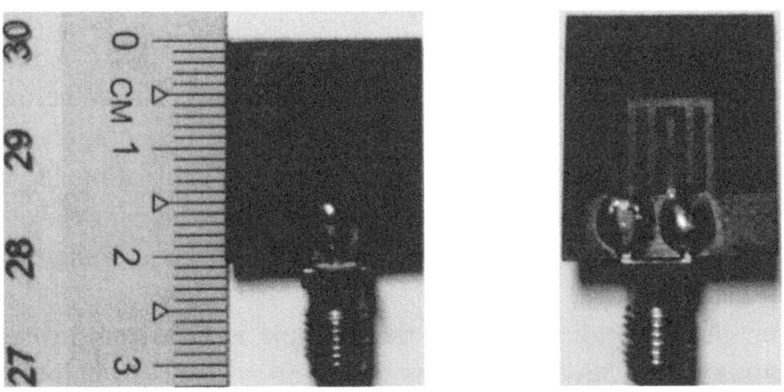

(a) **(b)**

Fig. 45 (a) Top View and (b) Bottom View of the **Fabricated Antenna.**

The antenna achieves an impedance bandwidth of approximately 26.3% for $|S11|$ less than -10 dB varying from 3.26 GHz – 4.26 GHz, where only a small region within this bandwidth is found where the $|S11|$ value fluctuates between -9 dB to -10 dB. On the other hand, the simulated result shows a percentage bandwidth of 25.75% varying from 3.35 GHz – 4.34 GHz. This small mismatch may be caused due

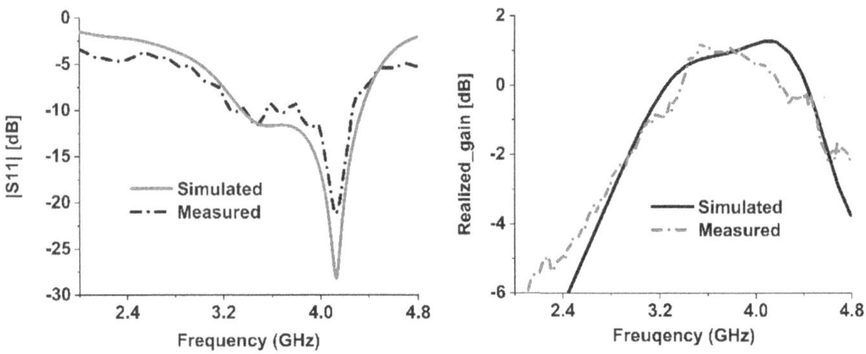

(a) (b)

Fig. 46 (a) Reflection Coefficient, and (b) **Realized Gain** of The Final Antenna.

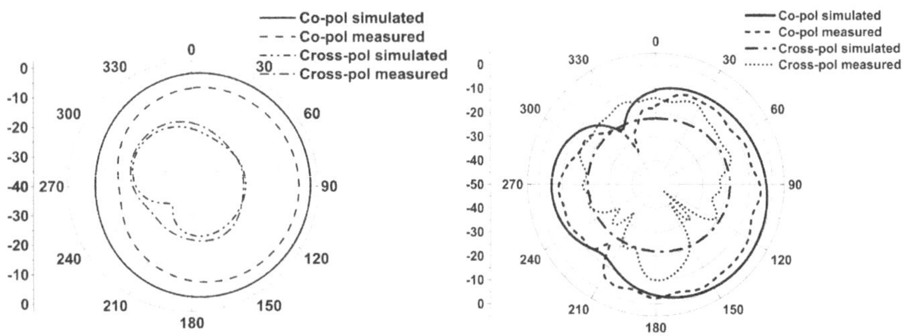

(a) (b)

Fig. 47 Radiation Patterns of the Final Antenna: (a) XY-Plane, (b) XZ-Plane.

to the error on soldering or fabrication tolerance (Figure 46(a)). The realized gain in simulation varies from 0.37 dBi – 1.35 dBi within the bandwidth. Whereas the realized gain in measurement ranges from −0.7 dBi – 1.26 dBi which agrees well with the simulated data (Figure 46(b)). The radiation patterns in simulation and measurement at 4 GHz of the final antenna are given in Figure 47 which shows well agreement. From the radiation patterns in both planes, it is evident that maximum radiation occurs near the split bearing side.

However, the radiation pattern in the H-plane is **quasi-omnidirectional** with **low cross-polarization** level. The cross-polarization levels in both simulation and measurement are recorded less than −15 dB in this plane.

7. Summary

In this chapter, a detailed discussion on the mechanism of miniaturization using metamaterial structures is made. Two different types of unit cell structures are designed namely symmetric and asymmetric structures. In the first case, the antenna is formed using an electrically coupled LC resonator where the metallic ground plane on the back side is used partially for better impedance matching. The internal structure of the resonator causes low frequency resonance which is expected and after that it is modified to achieve desired resonance behavior. Later, two additional parasitic elements are placed on both sides of the microstrip feed line, and the split width is optimized to get enhanced bandwidth and better realized gain. The measured fractional bandwidth is 19.74% for |S11| less than − 10 dB ranging from 3.39 GHz −4.15 GHz where a gain of 1.16 is realized. The final prototype shows good cross polarization levels in both planes. In the second case, a patch antenna is formed using an asymmetrical and single metamaterial unit cell is described where the ground plane is defected by a meander line structure. A meander line added split ring resonator is used as the main radiator of the antenna. Without the meander line, the antenna causes a low frequency resonance. The addition of the meander line causes the resonance to shift to the 5G 3GPP bands. A novel DGS structure is loaded on the metallic partial ground plane to excite another mode of resonance with the presence of the meander line inside the unit cell structure. Then the entire structure is modified to locate the two resonance dips at distinct frequencies having appropriate intervals and this helps to widen the bandwidth. A bandwidth of 26% for |S11| less than − 10 dB varying from 3.26 GHz − 4.26 GHz is experimentally recorded. The levels of cross-polarization are not more than − 15 dB in the horizontal plane which is excellent in terms of any DGS loaded structures. Before the analysis of these structures a detailed discussion is made on the simulation methodology of the simulation software to achieve the accurate resonance characteristics. Considering all the results of these compact antennas it can be said that the miniaturization mechanism of antennas using metamaterial structures is cost- effective and simple compared to the other available miniaturization techniques.

References

[1] L. Wang, Yuan, M.Q. and Liu, Q.H. A Dual-Band Printed Electrically Small Antenna Covered by Two Capacitive Split-Ring Resonators. *IEEE Antennas and Wireless Propagation Letters*, vol. 10, 824–826, 2011.

[2] K. Xu, Liu, F., Peng, L., Zhao, W.S., Ran, L. and Wang, G. Multimode and Wideband Printed Loop Antenna Based on Degraded Split-Ring Resonators. *IEEE Access*, vol. 5, 15561–15570, 2017.

[3] K.E., Kedze, Wang, H. and Park, I. Compact Broadband Omnidirectional Radiation Pattern Printed Dipole Antenna Incorporated with Split-Ring Resonators. *IEEE Access*, vol. 6, 49537–49545, 2018.

[4] K. Kumar Naik. Asymmetric CPW-Fed SRR Patch Antenna for WLAN/WIMAX Applications. *AEU - International Journal of Electronics and Communications*, vol. 93, 103–108, 2018.

[5] T. Ali, Mohammad Saadh, A.W., Biradar, R.C., Anguera, J. and Andújar, A. A Miniaturized Metamaterial Slot Antenna for Wireless Applications. *AEU - International Journal of Electronics and Communications*, vol. 82, 368–382, 2017.

[6] R. Samson Daniel, Pandeeswari, R. and Raghavan, S. Multiband Monopole Antenna Loaded with Complementary Split Ring Resonator and C-Shaped Slots. *AEU - International Journal of Electronics and Communications*, vol. 75, 8–14, 2017.

[7] N. Thamil Selvi, Thiruvalar Selvan, P., Babu, S.P.K. and Pandeeswari, R. Multiband Metamaterial-Inspired Antenna Using Split Ring Resonator. *Computers & Electrical Engineering*, vol. 84, 106613, 2020.

[8] R. Samson Daniel, , Pandeeswari, R. and Raghavan, S. A Compact Metamaterial Loaded Monopole Antenna with Offset-Fed Microstrip Line for Wireless Applications. *AEU - International Journal of Electronics and Communications*, vol. 83, 88–94, 2018.

[9] P.W. Chen and Chen, F.C. Asymmetric Coplanar Waveguide (ACPW) Zeroth-Order Resonant (ZOR) Antenna with High Efficiency and Bandwidth Enhancement. *IEEE Antennas and Wireless Propagation Letters*, vol. 11, 527–530, 2012.
[10] H. Lee, H., Woo, D.J. and Nam, S. Compact and Bandwidth-Enhanced Asymmetric Coplanar Waveguide (ACPW) Antenna Using CRLH-TL and Modified Ground Plane. *IEEE Antennas and Wireless Propagation Letters*, vol. 15, 810–813, 2016.
[11] Chi, P.L. and Shih, Y. Sen. Compact and Bandwidth-Enhanced Zeroth-Order Resonant Antenna. *IEEE Antennas and Wireless Propagation Letters*, vol. 14, 285–288, 2015.
[12] G. Wang and Feng, Q. A Novel Coplanar Waveguide Feed Zeroth-Order Resonant Antenna with Resonant Ring. *IEEE Antennas and Wireless Propagation Letters*, vol. 13, 774–777, 2014.
[13] J.D. Baena, Jelinek, L., Marquás, R. and Zehentner, J. Electrically Small Isotropic Three-Dimensional Magnetic Resonators for Metamaterial Design. *Applied Physics Letters*, vol. 88, no.13 , 134108, 2006.
[14] C.L. Holloway, Love, D.C., Kuester, E.F., Salandrino, A. and Engheta, N. Sub-Wavelength Resonators: On the Use of Metafilms to Overcome the $\lambda/2$ Size Limit. *IET Microwaves, Antennas, and Propagation*, vol. 2, no. 2 , 120–129, 2008.
[15] J.B. Pendry, , Holden, A.J., Robbins, D.J. and Stewart, W.J. Magnetism from Conductors and Enhanced Nonlinear Phenomena. *IEEE Transactions on Microwave Theory and Techniques*, vol. 47, no. 11 , 2075–2084, 1999.
[16] S. Zahertar, Yalcinkaya, A.D. and Torun, H. Rectangular Split-Ring Resonators with Single-Split and Two-Splits under Different Excitations at Microwave Frequencies. *AIP Advances*, vol. 5, no. 11 , 117220, 2015.
[17] D. Schurig, Mock, J.J. and Smith, D.R. Electric-Field-Coupled Resonators for Negative Permittivity Metamaterials. *Applied Physics Letters*, vol. 88, no. 4 , 041109, 2006.
[18] N. Engheta, and Ziolkowski, R. (2006). *Metamaterials: Physics and Engineering Explorations*. Wiley (Online Library).
[19] I.K. Kim and Varadan, V.V. Electric and Magnetic Resonances in Symmetric Pairs of Split Ring Resonators. *Journal of Applied Physics*, vol. 106, no. 7, 074504, 2009.
[20] T. Koschny, Soukoulis, C.M. and Zhou, J. Magnetic and Electric Excitations in Split Ring Resonators. *Optics Express*, vol. 15, no. 26, pp. 17881–17890, . 2007.
[21] Constantine A. Balanis. *Antenna Theory: Analysis and Design*, 4th Edition, Wiley. 1104. 2016. https://www.wiley.com/en-us/Antenna+Theory%3A+Analysis+and+Design%2C+4th+Edition-p-9781118642061.
[22] T. Koschny, Kafesaki, M., Economou, E.N. and Soukoulis, C.M. Effective Medium Theory of Left-Handed Materials. *Physical Review Letters*, vol. 93, no. 10 , 107402. 2004.
[23] N. Katsarakis, Koschny, T., Kafesaki, M., Economou, E.N. and Soukoulis, C.M. Electric Coupling to the Magnetic Resonance of Split Ring Resonators. *Applied Physics Letters*, vol. 84, no. 15, 2943, 2004.
[24] F.E. Terman. *Radio Engineers' Handbook*, 1943. McGraw-Hill Book Co. NY & London.
[25] F. Yang and Rahmat-Samii, Y. *Electromagnetic Band Gap Structures in Antenna Engineering: Frontmatter*, pp. 1–266. 2008. doi:10:1017/C8O9780511754531.
[26] M.M. Rahman, M.S. Islam, M.T. Islam and T. Alam, "Electric-Field-Coupled Resonator Antenna for 5G Applications," Materials 2022, vol. 15, Page 5247, vol. 15, no. 15, p. 5247, Jul. 2022, doi: 10.3390/MA15155247.
[27] M.M. Rahman, M.S. Islam, M.T. Islam, S.S. Al-Bawri and W.H. Yong, "Metamaterial-Based Compact Antenna with Defected Ground Structure for 5G and Beyond," Computers, Materials & Continua, vol. 71, no. 2, pp. 2383–2399, Dec. 2021, doi: 10.32604/CMC.2022.022150.

CHAPTER 4

High Refractive Index Metamaterial-based Radiation Aggregation for Gain Enhancement

Mohammad Lutful Hakim[1], *Mohammad Tariqul Islam*[2*] and *Touhidul Alam*[3]

1. Introduction

The upcoming **wireless communications** system is the mm-wave fifth-generation (5G) technology. This paradigm shift in connectivity is driven by the exponentially growing demand for high data rates, ultra-low latency, and an increasingly congested sub-6 GHz frequency spectrum. To address these limitations, 5G systems will primarily operate in the mm-wave frequency region, ranging from 24–100 GHz, where abundant bandwidth is available to support advanced communication technologies [1]. The 5G mm-wave communications are expected to revolutionize daily life through its excellent connectivity with modern technology like smart factories, automotive cars, smart homes, telemedicine, virtual reality, etc. [2–6]. These advancements will enable faster and more reliable data transmission, transforming various sectors, including healthcare, transportation, and urban infrastructure. Recognizing its transformative potential, standardization organizations, government bodies, and telecom industries are actively working on deploying and standardizing mm-wave 5G communication systems. For instance, specific mm-wave frequencies such as 26 GHz, 28 GHz, 37 GHz, and 38 GHz have been allocated by different countries for 5G new radio (NR), ensuring global compatibility and deployment [7–9]. As mm-wave frequencies come with higher bandwidth, they also face inherent challenges such as higher

[1,2] Department of Electrical, Electronic and Systems Engineering, Universiti Kebangsaan Malaysia, Bangi, Selangor, Malaysia.
[3] Pusat Sains Angkasa (ANGKASA), Institut Perubahan Iklim, Universiti Kebangsaan Malaysia, 43600 UKM Bangi, Selangor, Malaysia.
Email: lutful@ukm.edu.my[1]; tariqul@ukm.edu.my[2]
* Corresponding author: tariqul@ukm.edu.my, lutful@ukm.edu.my

free-space path loss, increased atmospheric absorption, and reduced communication range compared to the fourth-generation (4G) systems operating in lower frequency bands. Consequently, developing antennas capable of mitigating these losses is critical. High-gain antennas are required on both ends of communication devices to counteract the high propagation loss and atmospheric attenuation in the mm-wave frequency range [10]. Patch antennas are attractive for their low profile, low cost, and planar structure, but they have disadvantages such as low gain and narrow bandwidth [11–14]. The planar **antenna** with high gain in communication systems is attractive for satellite or base station applications. The upcoming 5G wireless communication systems will be implemented in the mm-wave frequency, but these frequencies go through high free space path loss. Hence, the communication range degrades than the 4G frequency band. High-gain antennas are required on both ends of the communication device to overcome atmospheric and high propagation loss in mm-wave frequency [15, 16]. From this perspective, the metamaterial is a strong candidate for designing an antenna with this required feature [17–21]. By leveraging metamaterial (MTM) properties, researchers have demonstrated significant improvements in antenna functionality. Typically, metamaterials are positioned either on the radiating element of an antenna or behind a defected ground plane. While these arrangements improve performance, they often introduce design challenges, such as increased bulkiness, mechanical complexity, and fabrication costs [22], where the gap between the antenna and meta-surface layer increases the bulkiness and complexity of the antenna. An antenna integrated with a MTM superstrate, particularly one where a lens is positioned on the radiating patch, offers an effective means of enhancing gain by concentrating the radiation beam. This configuration capitalizes on the unique electromagnetic properties of MTMs, enabling the focusing of emitted energy for improved directivity and performance. Such antennas are particularly advantageous for achieving wideband operational frequency, a critical feature for modern communication systems like mm-wave 5G. However, they often encounter notable drawbacks, including low radiation efficiency due to the bulky size of the superstrate and inherent losses within the thick dielectric material. These limitations challenge their feasibility for compact and high-performance applications. In research [23–25], the use of a MTM superstrate on the antenna patch significantly boosted gain performance. However, these improvements were accompanied by increased mechanical complexity and a higher-profile design, primarily caused by the presence of an air cavity between the superstrate and the antenna. This air gap, while beneficial for some electromagnetic properties, compromises structural simplicity and can hinder miniaturization efforts. For instance, the vertical placement of MTM structures has been successfully employed to improve the gain of bow-tie antennas [26]. In [27, 28], metamaterial was vertically positioned in the direction of the electromagnetic **(EM) wave** propagation to enhance the gain. Despite these successes, configurations without an air gap, such as those utilizing multiple layers of meta-surfaces, have shown promise in improving the performance of planar antennas. The multi layered approach minimizes the need for bulky cavities and air gaps, leading to more streamlined designs. However, this method introduces other challenges, including increased design complexity, narrow operational bandwidth, and elevated fabrication costs. Such trade-offs necessitate careful consideration during the design process, especially when scalability and affordability are key concerns.

Alternatively, antenna array configurations are another approach to achieving high gain. These designs employ multiple radiating elements to boost overall performance. However, array antennas often fail to achieve significant channel capacity improvements compared to single antennas. This is primarily because the arrays are typically fed by a single port, similar to single-element antennas. Moreover, the intricate power divider networks required for arrays lead to additional power losses, reducing overall efficiency [29]. As a result, while array configurations are useful for certain applications, they fall short of addressing the challenges posed by high-frequency mm-wave systems. On the other hand, meta-surfaces have been extensively employed with patch and slot antennas due to their remarkable ability to manipulate EM waves [7]. These meta-surfaces enable enhanced performance by tailoring the wave propagation characteristics, such as refraction, reflection, and polarization, to optimize the antenna's radiating properties. While single-layer antennas incorporating MTMs have been demonstrated in [20–23], these designs are predominantly in the microwave frequency range, where gain and bandwidth improvements are relatively modest. Additionally, these studies largely neglect the investigation of MIMO (Multiple-Input Multiple-Output) performance, which is critical for applications in the mm-wave frequency range. The analysis of MIMO performance is particularly crucial for mm-wave systems, as MIMO configurations provide multiple independent channels, facilitating increased data traffic, improved spectral efficiency, enhanced link reliability, and greater overall capacity [30–32]. Hence, the antenna with a MIMO configuration is the key requirement for mm-wave 5G applications. The ability of MIMO systems to mitigate path loss and multipath interference makes them indispensable for the high-speed, high-reliability demands of mm-wave 5G applications. Considering these discussions, it becomes clear that a high-gain antenna with a wide operational bandwidth is not merely desirable but a necessity for mm-wave 5G applications. Furthermore, a comprehensive investigation of MIMO performance in such antenna designs is vital to unlocking their full potential. By addressing these challenges through innovative MTM designs, antenna engineers can pave the way for the next generation of high-performance, compact, and efficient wireless communication systems [33].

This chapter introduces a MTM-incorporated planar dipole antenna designed for mm-wave 5G applications. The designed high-refractive-index MTM dipole antenna (HRIMDA) achieves a gain improvement of 3.5 dBi using a horizontal H-shaped MTM structure and offers an operational bandwidth of 23–38 GHz. Additionally, the MIMO configuration of HRIMDA demonstrates excellent isolation (< -30 dB) between two adjacent antennas, near-ideal ECC (<0.0001), and DG (>9.99). These features make the HRIMDA an attractive solution for advanced mm-wave 5G applications requiring compact, high-performance antennas with reliable MIMO capabilities.

2. Metamaterial Unit Cell Design

Metamaterials (MTMs) are a class of artificial materials engineered to exhibit unique EM properties that are rarely found in natural materials. These properties, such as negative permittivity, negative permeability, or even a negative refractive index, enable unparalleled control over EM wave propagation. By manipulating their structure rather than their composition, MTMs can achieve behavior that transcends the constraints of conventional materials, making them particularly appealing for a wide range of applications, including antenna design, wave manipulation, cloaking, and sensing [34–39]. The electric and magnetic quantities, also known as essential parameters, explain a medium's EM wave behavior. In this work, a horizontal H-shaped MTM resonator is developed for gain enhancement across the mm-wave frequency range. Unlike conventional materials, this MTM is designed to operate effectively within the constraints of compact antenna systems while maximizing performance. The developed horizontal H-shaped MTM is designed on the Rogers RT5880 substrate material with a thin 0.254 mm thickness. The parametric sketch is presented in Figure 1(a), where $a = 1.6$ nm, $r = 1.10$ nm, $w = 0.10$ nm, and $g = 0.50$ nm, respectively. The simulation setup of the designed MTM is presented in Figure 1(b), where a perfect electric conductor (PEC) and perfect magnetic conductor (PMC) are applied along the y and z-axes, respectively. The EM wave is applied along the x-axis in a positive direction. The **simulation** of the developed MTM is done by the CST microwave studio [40].

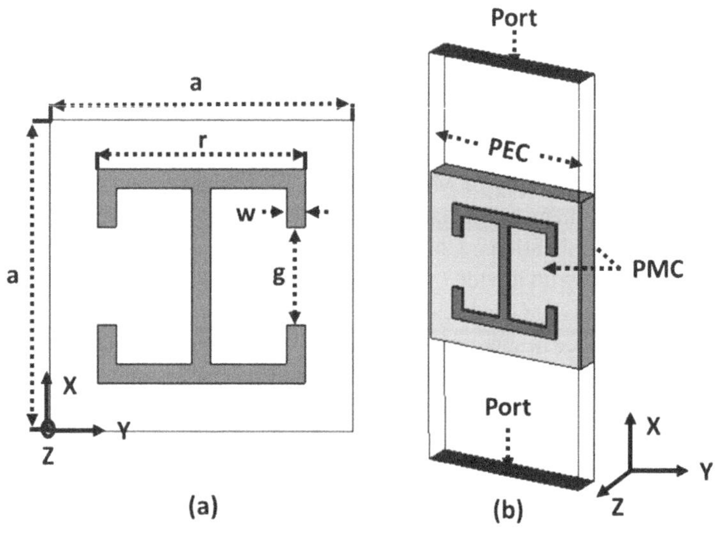

Fig. 1 Horizontal H-shaped metamaterial unit cell (MUC): (a) front view, and (b) simulation setup.

The MTM EM property can be characterized by the Lorentz-Drude model. The Equations (1–3) are used to calculate the relative permittivity (ε_r), relative permeability (μ_r), and effective **refractive index** (η_r), [41–45]. Where, S_{11} and S_{21} are the reflection and transmission coefficients, respectively. d is the propagation distance of the medium, and K_o is the wave number.

$$\varepsilon_r = {}^{2}\!/_{jk_0 d} \times {}^{(1 - S_{11} - S_{21})}\!/_{(1 + S_{11} + S_{21})} \qquad ...(1)$$

$$\mu_r = \frac{2}{jk_0 d} \times \frac{(1 - S_{21} + S_{11})}{(1 + S_{21} - S_{11})} \qquad ...(2)$$

$$\eta = \sqrt{\varepsilon_r \mu_r} \qquad ...(3)$$

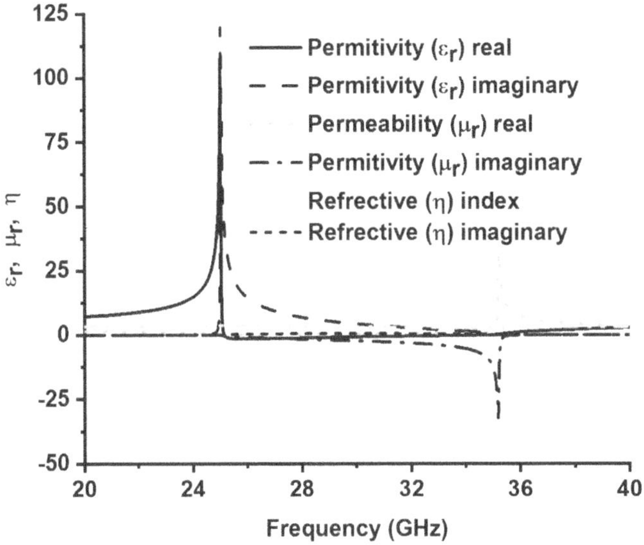

Fig. 2 Metamaterial property of the designed horizontal H- shaped MUC.

Figure 2 illustrates the EM parameter properties of the proposed horizontal-H shaped MTM unit cell (MUC). The results reveal that the designed MUC exhibits a permittivity near zero (ENZ) at the upper-frequency region, while permeability near zero (MNZ) is observed from the lower frequencies up to 35 GHz. These distinctive permittivity and permeability characteristics contribute to the development of a high-refractive-index metamaterial (HRIM). The combination of ENZ and MNZ regions facilitates effective manipulation of EM waves, resulting in a refractive index that is significantly higher than that of the surrounding air medium. The real part of the refractive index, as derived from the EM properties, confirms the HRIM behavior of the MUC, enabling its application in **gain enhancement** for antenna designs. The high refractive index aids in focusing and directing EM wave propagation more effectively, optimizing the antenna's radiation pattern and improving overall performance. These properties underline the potential of the horizontal H-shaped MUC for advanced applications requiring precise EM wave manipulation, particularly in the mm-wave frequency range. This design plays a pivotal role in achieving compact, high-gain antennas for next-generation wireless communication systems. This MTM property is used to enhance the antenna's performance in terms of gain and directivity [46] and provides a unique way to control the direction of the EM radiation based on Snell's law. Figure 3 presents the EM wave propagation from air to HRIM and HRIM to air. According to Snell's law, when the incident angle (θ_i) is fixed, the refracted angle (θ_r) decreases for EM wave propagation transitioning from a low-index medium to a

high-index medium. This indicates that the wave bends toward the normal as it enters the high-refractive-index material. On the other hand, when the EM wave moves from a high-index medium to a low-index medium under the same conditions, θ_r increases, causing the wave to bend away from the normal. This principle is fundamental to the design of MTM lenses. By exploiting the high-refractive-index properties of a MTM, a lens can be created to focus and direct EM waves effectively. The lens aggregates the waves, concentrating their energy along the direction of propagation. This focusing property enables the beam to converge towards the end-fire direction, leading to enhanced radiation characteristics. The ability of a metamaterial lens to focus EM waves has profound implications for antenna performance. It narrows the beam, reduces radiation spread, and minimizes side lobes, resulting in improved gain and directivity. These enhancements are particularly critical for compact and high-frequency systems, such as mm-wave 5G antennas, where efficient energy concentration and precise beam control are essential for addressing propagation losses and achieving reliable communication [47]. The energy transmitted through MTM congregates and is intense in the direction of propagation. This type of MTM can enhance the gain and directivity of the antenna. The EM wave aggregation property can be understood from the MTM's electric (E) field [48], where a spherical wave converts to a planar wave and decreases the half-power beamwidth by the MTM. This process was implemented by various MTM structures on planar antennas [48, 49].

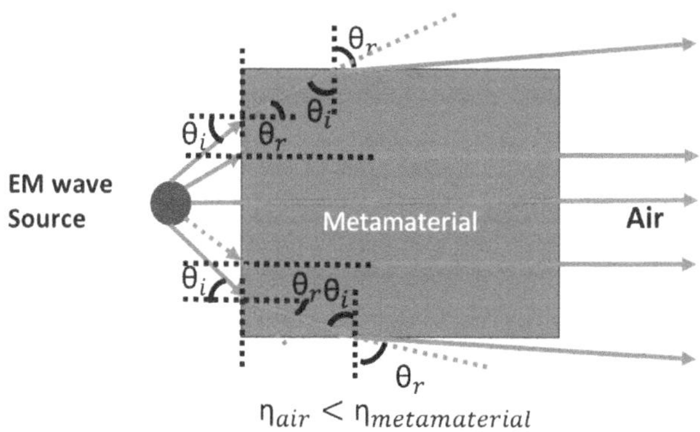

Fig. 3 EM wave propagation through designed HRIM.

The characteristics of the MTM are analyzed by surface current allocation, E-field, and H-field distribution, shown in Figure 4. The relationship between these three field distributions can be understood by analyzing Maxwell's equations [50, 51], where the **electric** ($\vec{D}$) and **magnetic field** induction ($\vec{B}$) can be described by Equations (4) and (5). The ε_0 and μ_0 represent the permittivity and permeability of free space ε_r, and μ_r represents relative permittivity and permeability.

$$\vec{D} = \varepsilon\vec{E} = \varepsilon_0\varepsilon_r\vec{E} \qquad ...(4)$$

$$\vec{B} = \mu\vec{H} = \mu_0\mu_r\vec{H} \qquad ...(5)$$

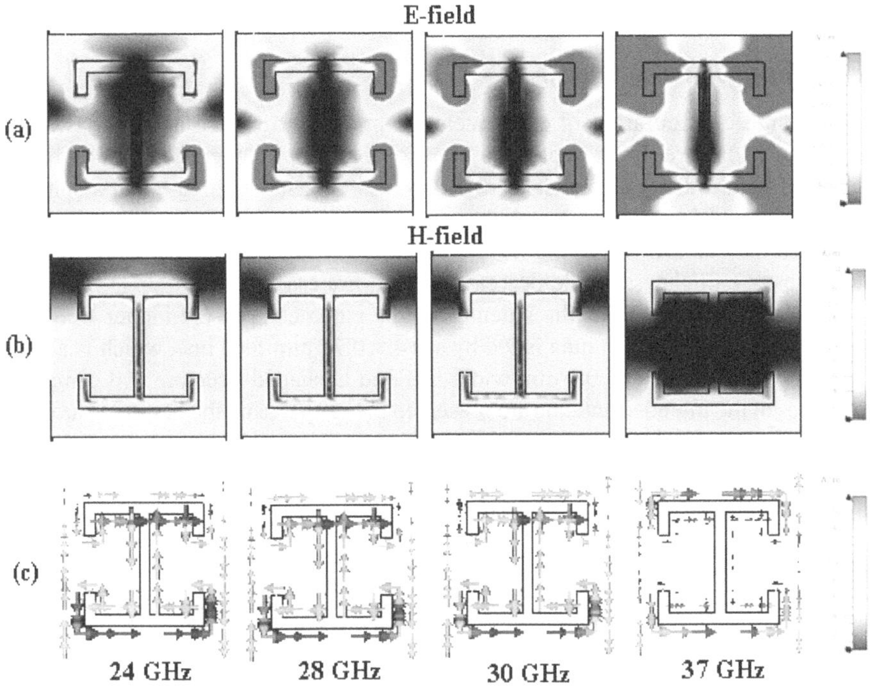

Fig. 4 (a) E-field. (b) H-field, and (c) surface current distribution of the designed MUC.

Figure 4 illustrates the electric field (E-field) distributions of the proposed MUC at four frequencies: 24 GHz, 28 GHz, 30 GHz, and 37 GHz. The results clearly indicate that the E-field density is concentrated at the left and right edges of the MUC, with the field intensity progressively increasing as the frequency rises. This high E-field density effectively aggregates the incident EM waves, directing them along the propagation path, which enhances the overall radiation performance. The magnetic field (H-field) distribution of the MUC, also shown in Figure 4. reveals a frequency-dependent behavior. At lower frequencies, the H-field intensity is more evenly distributed, but as the frequency increases, the intensity becomes more pronounced on the upper side of the MUC, while it diminishes on the lower side. This uneven distribution highlights the metamaterial's ability to manipulate **EM waves** differently across the frequency spectrum. Additionally, the **surface current** distribution of the MUC is presented in Figure 4. The results demonstrate anti-parallel current flows at the splits of the MUC, which play a critical role in the resonant behavior of the structure. As the frequency increases, the surface current intensity gradually decreases. This frequency-dependent surface current behavior contributes to the MUC's ability to aggregate EM waves more effectively at higher frequencies compared to lower ones. Together, these properties—high E-field concentration, distinct H-field behavior, and anti-parallel surface currents—enable the designed MUC to exhibit superior EM wave aggregation capabilities, particularly at higher frequency ranges. These characteristics are pivotal for applications requiring high-frequency performance, such as mm-wave 5G communication systems.

3. Broadband Dipole Antenna Design and Metamaterial Incorporation

3.1 Single Antenna Design

The design configuration of the developed broadband dipole antenna (BDA) is depicted in Figure 5. The antennas were designed on Rogers RT5880 substrate material with a dielectric constant of 2.2, a loss tangent of 0.0009, and a substrate thickness of 0.254 mm. The low dielectric constant is chosen for its ability to reduce electrical power loss at the higher frequency application [52]. On the other hand, low substrate thickness would decrease the end-fire cross-polarisation [13, 53]. The substrate feature improves the antenna results parameters at the higher frequency regions. The designed antenna is fed by a $W4 = 0.71$ mm feed line, which is a series with $W3 = 0.5$ mm and a 0.3 mm width inverted U-shaped structure. All gematrical sketches of the intended antenna are presented in Figure 5, and the design's parameter values are listed in Table 1.

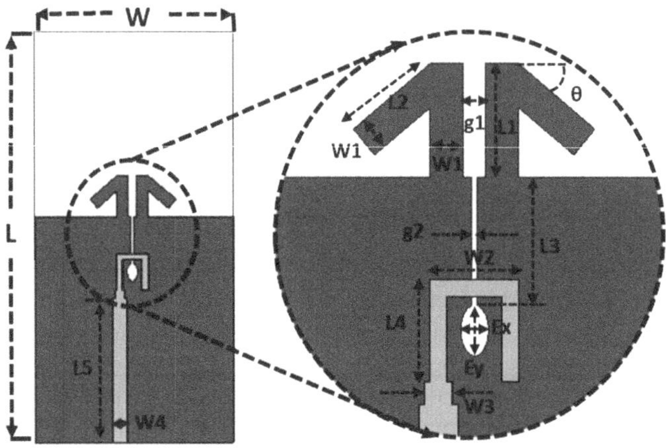

Fig. 5 Front view of a designed BDA with design parametric sketch.

Table 1 Design parameters.

Parameters	*Value (mm)*	*Parameters*	*Value (mm)*	*Parameters*	*Value (mm)*
W	10	W4	0.71	L5	7
L	18	L1	2	g1	0.4
W1	0.6	L2	1.8	g2	0.10
W2	1.6	L3	2.26	Ex	0.25
W3	0.5	L4	1.8	Ey	0.45

The return loss (S_{11}), **efficiency**, and **gain** plot of the designed antenna are presented in Figure 6. This demonstrates the realized gain of the designed antenna between the range of 4.7–6 dBi at the resonant **bandwidth** (24.35–38.7 GHz). The maximum 6 dBi gain appeared at the 34 GHz frequency. Conversely, the efficiency shows around 95% at the resonant bandwidth. The intended antenna radiation pattern at 26 GHz and 28 GHz frequencies in the E-plane and H-plane is presented in Figure 7.

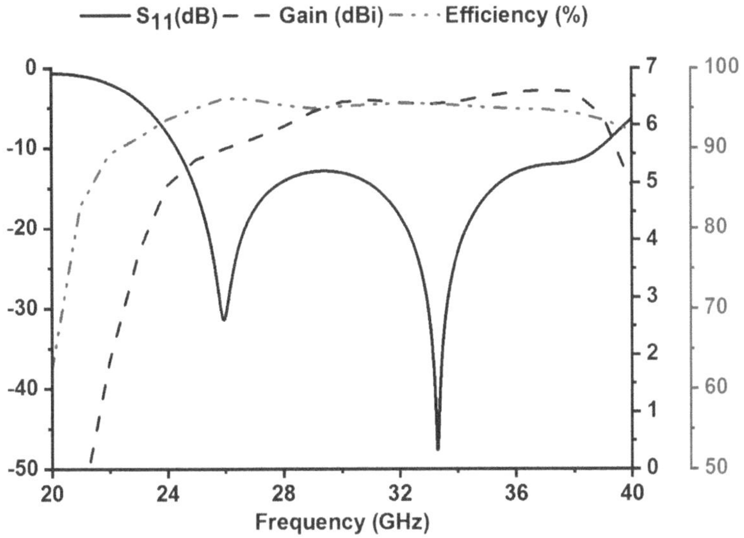

Fig. 6 **Efficiency** and **gain** plot of the designed BDA.

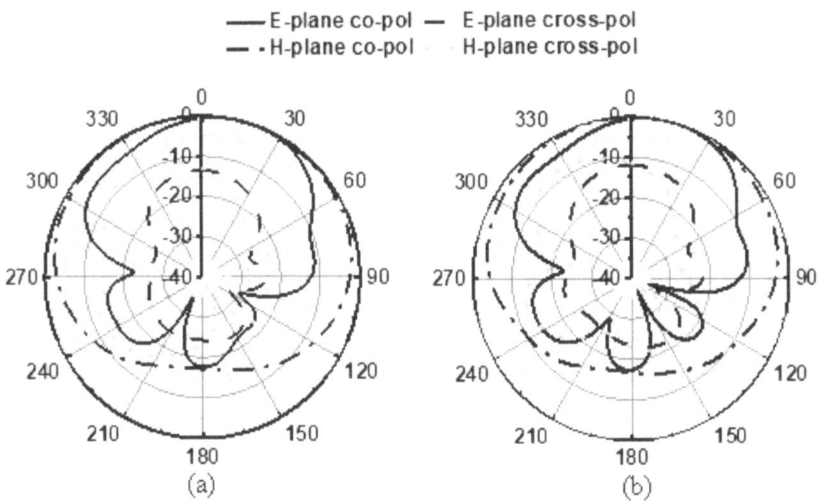

Fig. 7 Simulated radiation pattern of the designed BDA.

The pattern shows good directivity of the designed BDA. The cross-polarization is lower than 8 dB, and the front to the back ratio of co-polarization is more than 15 dB for the entire operating frequency. Figure 8 presents the designed antenna's

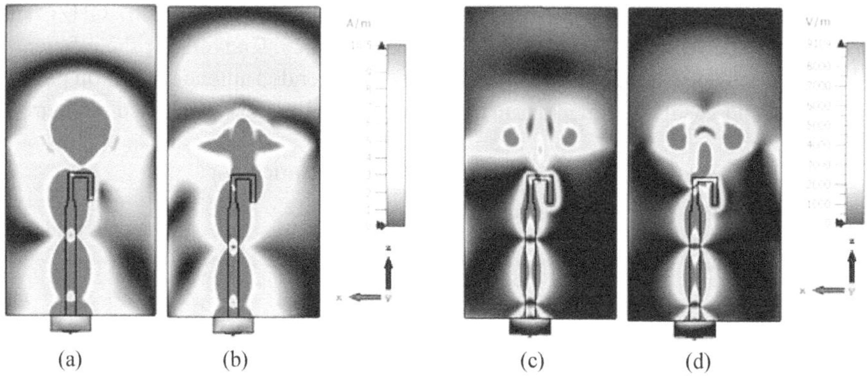

Fig. 8 Magnetic (H)-field at (a) 26 GHz, (b) 28 GHz and electric (E) field (c) 26 GHz, (d) 28 GHz of the designed BDA.

electric (E) and magnetic (H) field distribution. The E-field intensity of the dipole arm is higher at 28 GHz frequency than at 26 GHz, which indicates the higher **directivity** and gain at 28 GHz frequency. Figure 9 shows the **bandwidth** (S_{11}) plot of the different designs. In design one, the BDA does not show any resonance because of the incompetence of the feeding. The feeding on the dipole is completed in design

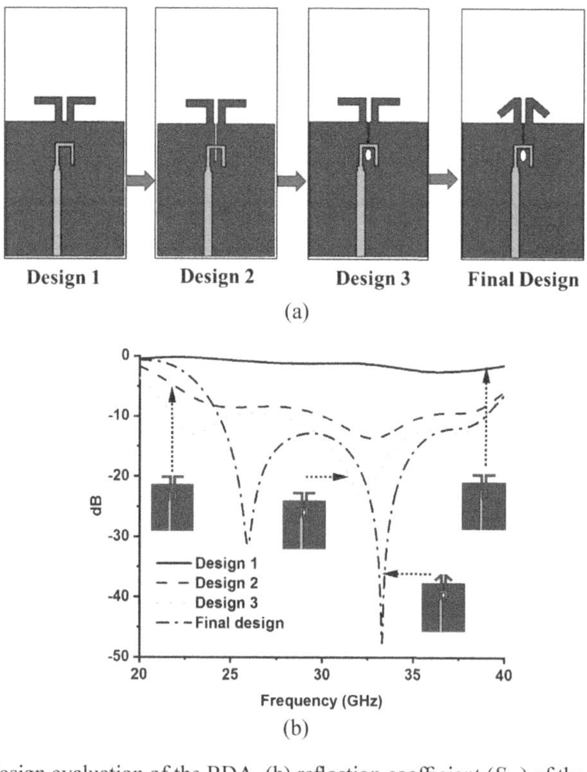

Fig. 9 (a) design evaluation of the BDA, (b) reflection coefficient (S_{11}) of the different steps of antenna designs.

two by making a slot line from the middle of the dipole to the inverted U shape. This design shows 5.54 GHz (– 10 dB) bandwidth from 30–35.54 GHz. The antenna's bandwidth is further enhanced in design 3 by making an elliptical slot at the center of the inverted U-shaped line, which increases the – 10 dB resonance bandwidth (10.8 GHz) from 28–38.8 GHz. Finally, a **broadband** 14.35 GHz (– 10 dB) bandwidth is achieved from 24.35–38.7 GHz frequency by declining the dipole arm by 40-degree.

4. Metamaterial Incorporated Antenna

The BDA's gain is increased significantly by incorporating the MTM structure into the propagation direction of the EM wave. Figure 10 shows the gradual incorporation of HRIM. The S_{11} of the different HIRM stages is outlined in Figure 11(a), which shows a stable – 10 dB impedance bandwidth. The operational bandwidth of the final 5 × 6 HRIM appeared from 23–38 GHz frequency. The HRIM's effect on the BDA's gain is illustrated in Figure 11(b). The BDA's gain increases gradually with the increase of the HRIM unit cell series, and for the final 5×6 HRIM base antenna, the gain is achieved from 6–9.5 dBi. The highest gain of the antenna reached 9.5 dBi at 35 GHz resonant frequency. The equivalent circuit (EC) of the developed HRIMDA is presented in Figure 4.11(c) based on [54–56], where LC represent the feed line, $R_1L_1C_1$ and $R_2L_2C_2$ represent dipole antenna and $R_3L_3C_3$ represent HIRM. The EC is simulated by Advanced Design System (ADS) software, which agrees well with CST simulation shown in Figure 11(d). The EM **wave aggregation** property of the developed HRIMDA can be understood from the antenna's electric, magnetic, and **power flow**. Figure 12 shows the developed electric and magnetic fields at 26 and 28 GHz frequencies. The HRIM unit cell aggregates the EM wave in the direction of propagation. This EM wave aggregation engineering can be more clarified by the power flow of the HRIMDA, which is depicted in Figure 13. The normal and HRIMDA power flow is presented to understand this behavior. The circled region in

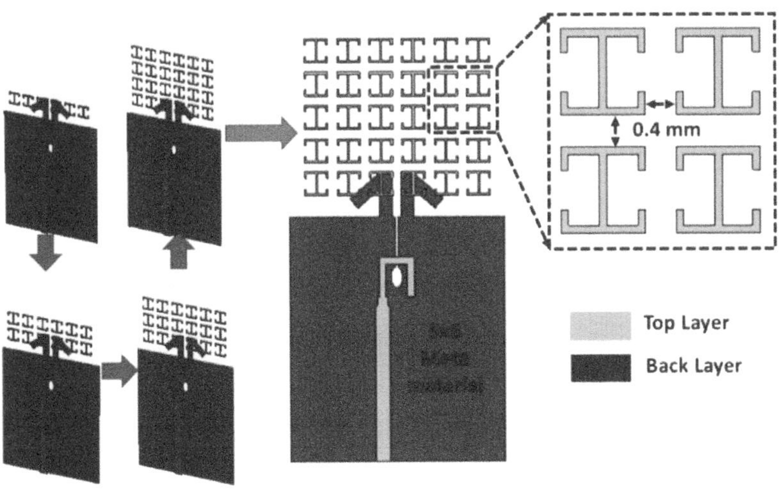

Fig. 10 Incorporation of the HRIM series on the BDA.

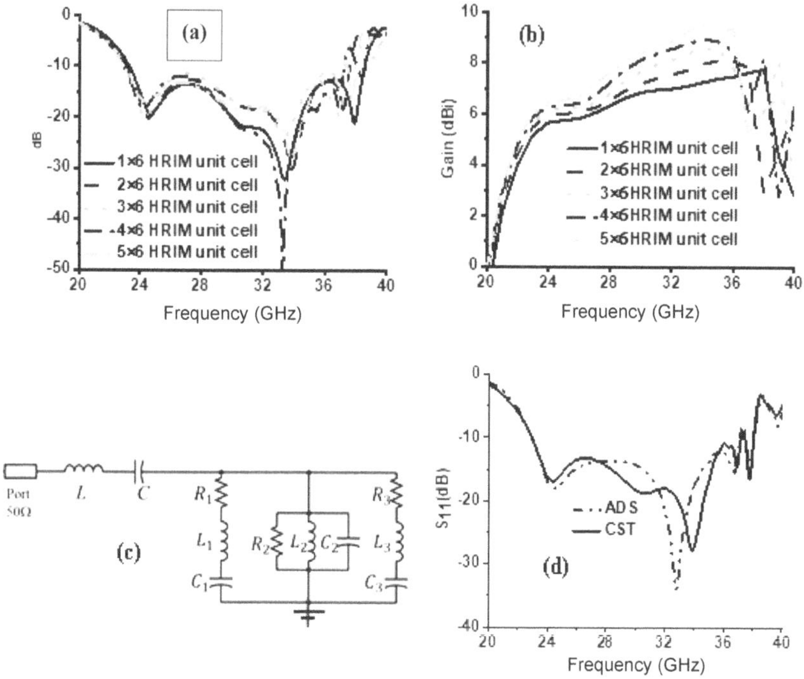

Fig. 11 (a) S_{11}, and (b) gain plot of the intended HRIMDA at a different stage, (c) EC model (L = 1.029 nH C = 0.0259 pF, R_1 = 3.493 Ω, L_1 = 1.814 nH, C_1 = 0.00872 pF, R_2 = 95.011 Ω, L_2 = 0.660 nH, C_2 = 0.032 pF, R_3 = 3.27 Ω, L_3 = 4.550 nH, C_3 = 0.010 pF and (d) S_{11} of ADS and CST simulation.

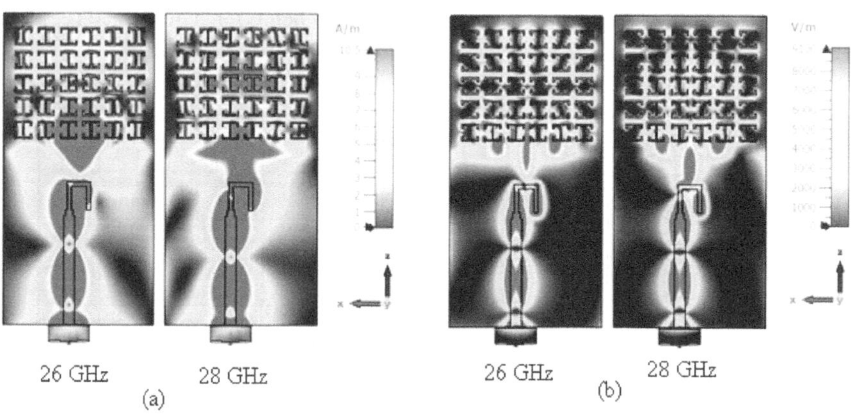

Fig. 12 (a) H-field distribution, and (b) E-field distribution of the developed HRIMDA.

Figure 13 shows the power flow at the side edge of the antenna significantly reduced by the incorporation of HRIM on the antenna. The power flow significantly boosts the direction of the EM wave propagation.

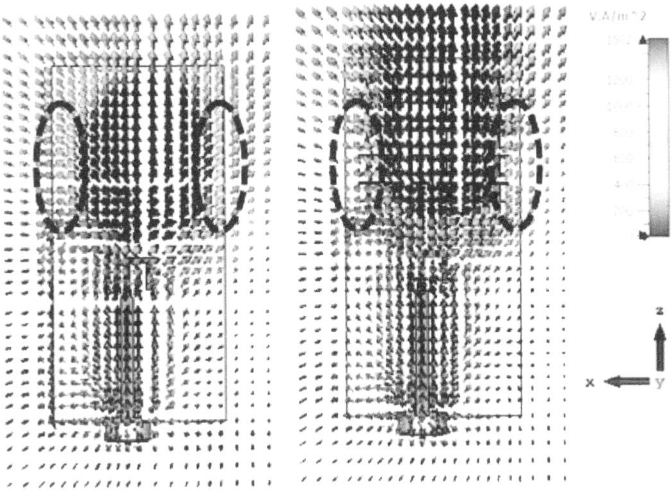

(a) (b)

Fig. 13 Power flow at 28 GHz frequency: (a) normal BDA, and (b) developed HRIMDA.

5. Fabrications and Experimental Results

To verify the intended HRIMDA's gain enhancement, two antennas (typical BDA and HRIMDA) were **fabricated** using 2.54 mm thin Rogers RT5880 ($\varepsilon_r = 2.2$) substrate material. The size of the antenna is 10×18 mm^2. Figure 14 shows the front and rear sides of the fabricated typical and HRIMDA. The vector network analyzer and **far-field** anechoic chamber measurement systems measured the fabricated antennas, as shown in Figure 15. The simulated and measured S_{11} of the typical and HRIMDA is plotted in Figure 16(a), and the S_{11} of the fabricated antenna agrees well with the simulated one. The measured bandwidth of the HRIMDA is 23–38 GHz frequency. The measured gin of the normal and HRIMD is presented in Figure 16 (b), which shows good agreement with the simulation results

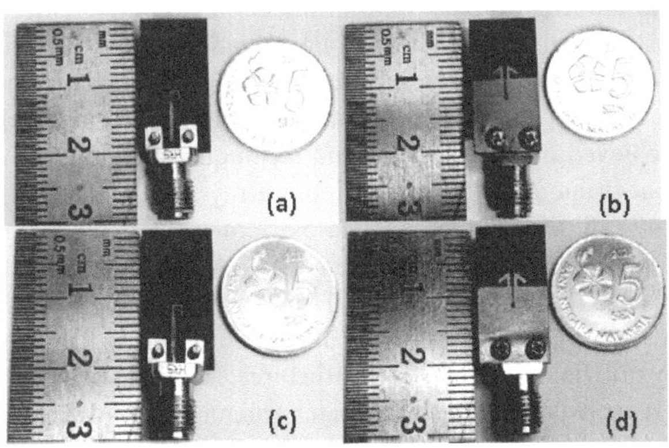

Fig. 14 **Fabricated** typical BDA (a) front view (b) rearview; and Fabricated HRIMDA (c) front view (d) rearview.

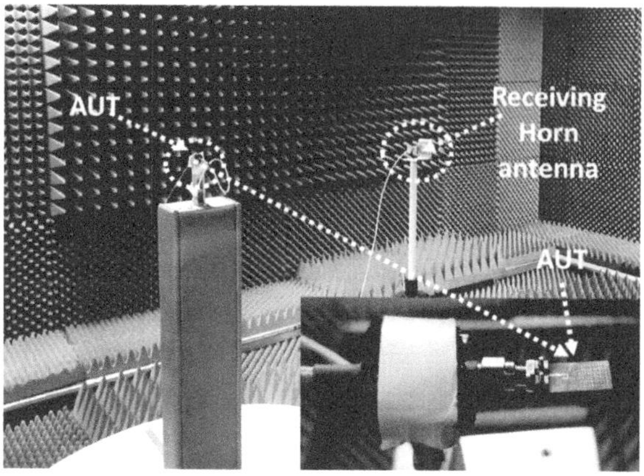

Fig. 15 Anechoic chamber measurement system.

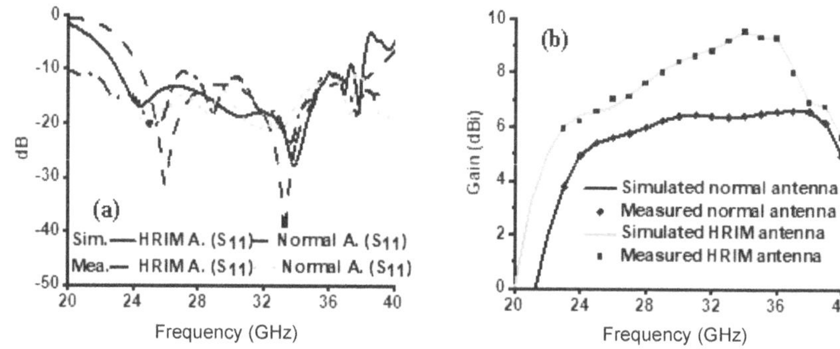

Fig. 16 (a) Simulated and measured S_{11}, and (b) Simulated and measured gain of the typical dipole and HRIMDA.

The simulated and **measured** E-plane and H-plane radiation patterns at 26 GHz and 28 GHz of the developed HRIMDA are presented in Figure 17. The measured radiation pattern agrees well with the simulated one. The cross-polariz ation is less than 10 dB, and the front to back ratio of the co-polariz ation is less than 18 dB at the entire operating frequency, indicating the radiation beam's directivity. A detailed comparison of the developed **antenna** with the existing broadband mm-wave antennas is presented in Table 2. The defective ground antenna is presented in [57, 58], where wideband is achieved by defective ground technique but shows low antenna gain. Different planar-printed log-periodic dipole array (PLPDA) antenna types were presented in [47, 48, 59]. For broadband 5G mm-wave applications, the different types of MTM structures showed significant gains. A planar patch antenna [7], Shared aperture antenna [53], Bow-tie antenna [15], and Antipodal Vivaldi antenna [60] with meta-surface were presented for 5G applications, where the gain was improved significantly by utilizing different MTM structures. Multiple-layer microstrip antenna [25] and dual-layer two-element slotted patch antenna [61] are also presented, where MTM was placed on top of the antenna with an air gap. The MIMO parameters analyses were performed only in [7, 25, 61]. Finally, the developed planar HRIMDA achieved a significant percentage bandwidth enhancement (BW%) compared to the other listed antenna. Also, HRIM enhances gain by 3.5 dBi, which is higher than the existing antennas. The investigation of the MIMO parameters analysis, compact antenna size, larger BW%, and higher gain prioritize the developed antenna over the present antennas.

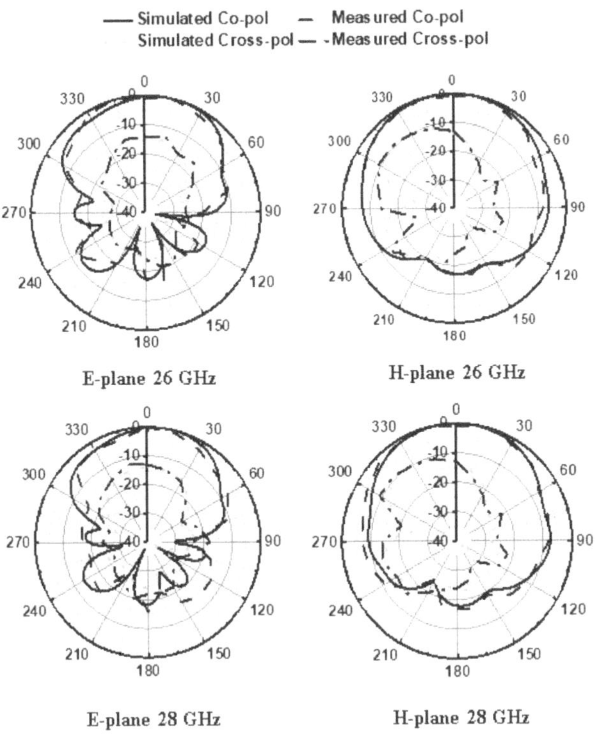

Fig. 17 Simulated and **measured** normalized **radiation pattern** of the intended HRIMDA.

6. MMO Configurations

Figure 18 shows a four-element **MIMO antenna** arrangement of the intended HRIMDA. The designed MIMO configuration is fabricated, and isolation of the two adjacent MIMO elements (antenna 2 and 3) is measured. The front and back views of the fabricated antenna are depicted in Figure 19. Figure 20(a) shows the S_{11} of the simulated and measured MIMO elements. The simulated and measured isolation of the MIMO elements is depicted in Figure 20 (b, c), where the measurement results agree well with the simulated results.

The performance analysis of the designed MIMO antenna is done by investigating the envelope correlation coefficient (ECC) and diversity gain (DG). Equation (6) calculates the ECC value, which indicates the correlation of the MIMO antenna element. Equation (7) calculates the DG value, which indicates the impact of the antenna's radiated power diversity scheme.

Table 2 Comparison with the existing broadband mm-wave antenna.

Ref	*Size (mm^3)*	*Antenna Design*	*Technique and MTM Structure*	*Layer*	*Air Gap*	*Operation BW (GHz)*	*BW %*	*Gain Enhance dBi*	*MIMO*
[57]	$8 \times 8 \times 0.79$ ($0.7\,\lambda_0 \times 0.7\,\lambda_0 \times 0.07\,\lambda_0$)	Umbrella-shaped patch antenna	No MTM is used	1	–	27.5–28.25 36.5–39.25 57–64	23.6 18.4 12.1	–	No
[58]	$16 \times 16 \times 0.79$ ($1.3\,\lambda_0 \times 1.3\,\lambda_0 \times 0.065\,\lambda_0$)	Defected ground antenna	No MTM is used	1	–	25–50	66.6	–	Yes
[59]	$13 \times 22 \times 0.508$ ($1.12\,\lambda_0 \times 1.89\,\lambda_0 \times 0.043\,\lambda_0$)	Landstorfer PLPDA	Semi-circular splits ring	1	No	25.7–40	43.5	1.9	No
[48]	$15 \times 22 \times 0.508$ ($1.3\,\lambda_0 \times 1.91\,\lambda_0 \times 0.044\,\lambda_0$)	Clamped mode PLPDA	Semi ring patch	1	No	26–40	42.4	1.6 dBi	No
[47]	$10 \times 17.5 \times 0.508$ ($1.34\,\lambda_0 \times 2.37\,\lambda_0 \times 0.07\,\lambda_0$)	PLPDA	Bow tie directors	1	No	40–50	22.2	0.73	No
[7]	$12.4 \times 12.4 \times 0.5$ ($1.0\,\lambda_0 \times 1.0\,\lambda_0 \times 0.041\,\lambda_0$)	Patch antenna	Square patch Meta-surface	1	No	24.5–31	23.5	4	Yes
[25]	$10 \times 10 \times \approx 5.12$ ($0.94\,\lambda_0 \times 0.94\,\lambda_0 \times 0.48\,\lambda_0$)	Microstrip antenna	Cross shape	4	Yes	28–34	19.3	N/A	Yes
[53]	$31 \times 31 \times 2$ ($2.71\,\lambda_0 \times 2.71\,\lambda_0 \times 0.181\,\lambda_0$)	Shared aperture antenna	Splits ring resonators	1	No	27–30	10.5	1.6	No
[15]	$30 \times 30.5 \times 0.508$ ($2.46\,\lambda_0 \times 2.5\,\lambda_0 \times 0.041\,\lambda_0$)	Bow-tie antenna	Meander and triple I shape	1	No	24.5–27.5	11.5	2.9	No
[60]	$40 \times 24 \times 1.6$ ($3.3\,\lambda_0 \times 1.99\,\lambda_0 \times 0.13\,\lambda_0$)	Antipodal Vivaldi antenna	Inverted V shape	1	No	24.8–34.52	32.7	1.78	No
[61]	$17 \times 11.7 \times \approx 7.6$ ($1.39\,\lambda_0 \times 0.96\,\lambda_0 \times 0.62\,\lambda_0$)	Two-element slotted patch antenna	Splits ring resonators	2	Yes (b)	24.5–26.5	7.8	2.2	Yes
This work	$10 \times 18 \times 0.254$ ($0.77\,\lambda_0 \times 1.38\,\lambda_0 \times 0.0195\,\lambda_0$)	Dipole antenna	Inverted dual C shape	1	No	23–38	49.2	3.5	Yes

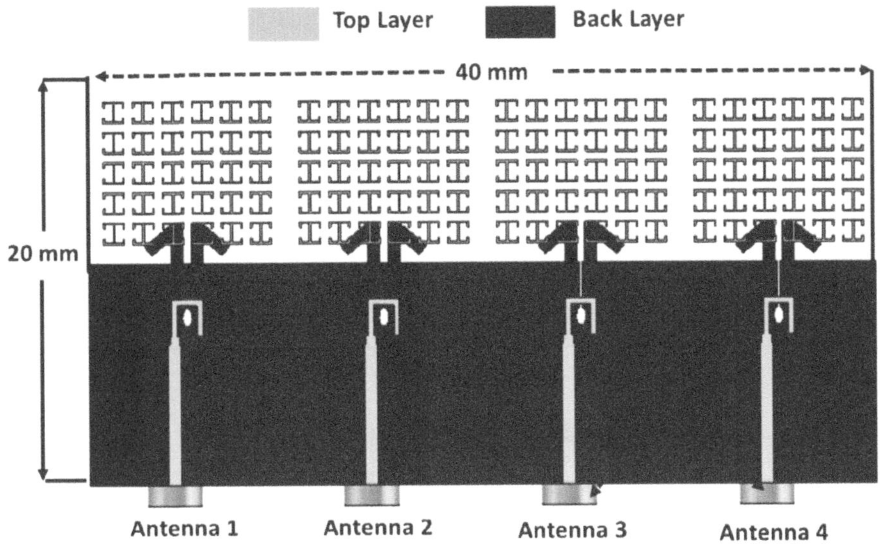

Fig. 18 Four-element MIMO antenna system.

Fig. 19 **Fabricated** MIMO antenna: (a) front view, and (b) back view.

$$\rho_{ec}(i,j) = \frac{|S_{ii} \times S_{ij} + S_{ji} \times S_{jj}|^2}{(1 - |S_{ii}|^2 - S_{ij}^2)(1 - |S_{ji}|^2 - S_{jj}^2)} \quad ...(6)$$

$$DG = 10 \times \sqrt{1 - \rho_{ec}} \quad ...(7)$$

In Equation (6) S_{ii} and S_{jj} are the reflection coefficients; S_{ij} and S_{ji} $(i \neq j)$ are the **transmission coefficients** of MIMO elements. The ECC value of the MIMO antenna should be less than 0.1, and a lower ECC value improves the MIMO performance. The

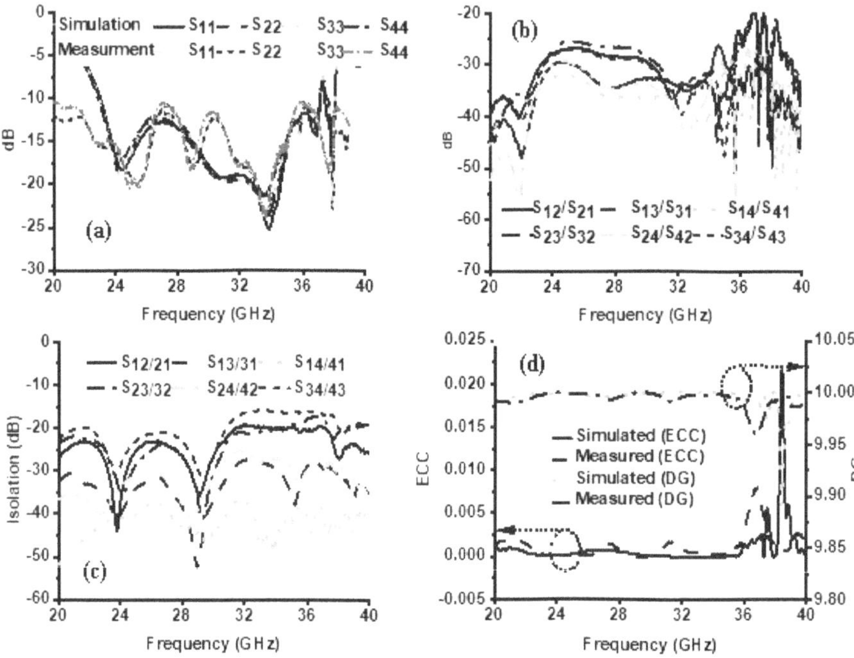

Fig. 20 Simulated and measured: (a) S_{11} of the MIMO element, (b) Simulated **isolation** between the MIMO element, (c) Measured isolation between the MIMO element, and (d) MIMO parameters of two conjugate antennas (antenna 2 and 3).

simulated and measured ECC value of two conjugate antenna elements presented in Figure 20(d) shows an ECC of less than $<1 \times 10^{-3}$, which is the nearly ideal value. The simulated and measured DG value is achieved above 9.9, where the ideal value is 10. A comparison of the developed four-element MIMO and existing MIMO configurations is presented in Table 3, where isolation, ECC, and DG value is compared. The intended MIMO antenna shows near the ideal value of both parameters compared to other existing MIMO antennas. This excellent feature makes the intended **MIMO** antenna a compatible candidate for mm-wave 5G applications.

Table 3 Comparison of intended MIMO antenna and existing MIMO antenna.

Ref.	*Elements*	*Frequency*	*Isolation*	*ECC*	*DG*
[62]	4	25– 39	< – 26	<0.05	N/A
[63]	4	23.5– 29.4	< – 22	<0.004	N/A
[64]	4	27.5– 40	< – 16	<0.001	N/A
[65]	2	25– 29, 37– 41	< – 24	<0.001	>9.9
[66]	4	27– 30	< – 29	<0.16	N/A
Developed	4	23– 38	< – 20	$<1\times10^{-3}$	>9.9

7. Summary

An HRIMDA for mm-wave 5G applications is presented and comprehensively analyzed in this chapter. A horizontal H-shape (HRIM) unit cell is modeled with unique MTM properties to achieve significant performance enhancements. The metamaterial structure effectively increases the gain of the antenna while maintaining compactness and simplicity. The designed HRIMDA is fabricated and verified through experimental measurements, demonstrating remarkable alignment between simulated and measured results. The fabricated antenna features a small size of $0.77\,\lambda_0 \times 1.38\,\lambda_0 \times 0.0195\,\lambda_0$, making it ideal for compact and portable 5G devices.

The developed HRIMDA achieves an impressive – 10 dB impedance bandwidth of 49.2% (23–38 GHz frequency range). This wideband performance, coupled with a notable gain improvement of up to 3.5 dBi, addresses the challenges associated with high propagation losses in the mm-wave spectrum. By incorporating the HRIM structure into the conventional dipole antenna design, the gain enhancement is achieved without compromising efficiency or compactness, providing a practical and scalable solution for advanced 5G applications.

The MIMO performance of the HRIMDA is also thoroughly investigated. The developed **antenna** exhibits exceptional isolation levels (< -30 dB) between adjacent MIMO elements, ensuring minimal interference and superior link reliability. The envelope correlation coefficient (ECC) remains below $<1 \times 10^{-3}$, while the diversity gain (DG) exceeds 9.9, both of which are near-ideal values. These results make the HRIMDA an excellent candidate for high-performance MIMO systems, enabling robust and reliable connectivity in dense 5G environments.

The potential applications of the HRIMDA extend to a wide range of advanced technologies that demand high data rates, low latency, and consistent connectivity. These include smart cities, autonomous vehicles, telemedicine, enhanced mobile broadband, and virtual reality systems. The high gain and wide operational bandwidth of the HRIMDA ensure reliable performance in these use cases, overcoming the challenges of high path loss and atmospheric attenuation inherent in mm-wave frequencies.

Additionally, the compact size and straightforward design of the HRIMDA make it highly versatile, allowing seamless integration into both portable and fixed **wireless devices**. This adaptability ensures that the antenna can support a diverse range of applications, from small-scale IoT devices to large-scale base stations.

Finally, the HRIMDA represents a significant advancement in antenna technology for mm-wave 5G applications. By combining **high gain**, wide bandwidth, and near-ideal MIMO performance metrics, the developed antenna addresses critical challenges in modern wireless communication systems. Its practical design and exceptional performance pave the way for more efficient, reliable, and scalable 5G networks, meeting the ever-growing demand for high-speed connectivity in an increasingly connected world. This work contributes to the ongoing evolution of 5G technologies, offering a robust and innovative solution for the future of wireless communication.

References

[1] FCC. "America's 5G Future." FCC. https://www.fcc.gov/5G (accessed 1 March, 2023).

[2] N.K. Mallat, M. Ishtiaq, A. Ur Rehman and A. Iqbal. "Millimeter-wave in the face of 5G communication potential applications," *IETE Journal of Research*, pp. 1–9, 2020.

[3] S. Wang, X. Fu, R. Ruby and Z. Li. "Pilot spoofing detection for massive MIMO mmWave communication systems with a cooperative relay," *Computer Communications*, vol. 202, pp. 33–41, 2023.

[4] I. Baştürk. "Green communication for OFDMA cellular networks with multiple antennas," *Computer Communications*, vol. 168, pp. 93–101, 2021.

[5] M.L. Hakim, T. Alam and M.T. Islam. "Polarization-Insensitive and Oblique Incident Angle Stable Miniaturized Conformal FSS for 28/38 GHz mm-Wave Band 5G EMI Shielding Applications," *IEEE Antennas and Wireless Propagation Letters*, vol. 22, no. 11, pp. 2644–2648, 2023.

[6] M.L. Hakim, M.T. Islam and T. Alam. "Ultra-Miniaturized Conformal Polarization Insensitive and Incident Angle Stable FSS for n257 Band 5G EMI Shielding Applications," *IEEE Transactions on Antennas and Propagation*, 2024.

[7] N. Hussain, M.-J. Jeong, A. Abbas and N. Kim. "Metasurface-based single-layer wideband circularly polarized MIMO antenna for 5G millimeter-wave systems," *IEEE Access*, vol. 8, pp. 130293–130304, 2020.

[8] M.L. Hakim, M.T. Islam and T. Alam. "Incident Angle Stable Broadband Conformal mmWave FSS for 5G (n257, n258, n260, and n261) band EMI Shielding Application," *IEEE Antennas and Wireless Propagation Letters*, 2023.

[9] A. Hanif, M.L. Hakim, T. Alam, M.T. Islam, H. Alsaif and M.S. Soliman. "Polarization insensitive oblique incident angle stable ultra-thin nano ring resonator-based metamaterial absorber for visible and near-infrared region applications," *Optics & Laser Technology*, vol. 164, p. 109494, 2023.

[10] K. Sultan, M. Ikram and N. Nguyen-Trong. "A Multiband Multibeam Antenna for Sub-6 GHz and mm-Wave 5G Applications," *IEEE Antennas and Wireless Propagation Letters*, vol. 21, no. 6, pp. 1278–1282, 2022.

[11] Y. Soerbakti, R.F. Syahputra, M.D.H. Gamal, D. Irawan, E.H. Putra and R.S. Darwis. "Improvement of low-profile microstrip antenna performance by hexagonal-shaped SRR structure with DNG metamaterial characteristic as UWB application," *Alexandria Engineering Journal*, vol. 61, no. 6, pp. 4241–4252, 2022.

[12] M.S. Islam, S.K. Azam, A.Z. Hossain, M.I. Ibrahimy and S. Motakabber. "A low-profile flexible planar monopole antenna for biomedical applications," *Engineering Science and Technology, An International Journal*, vol. 35, p. 101112, 2022.

[13] M.L. Hakim et al. "Interconnected square splits ring resonator based single negative metamaterial for 5G (N258, N257, N260 and N259) band sensor/EMI shielding/and antenna applications," *Alexandria Engineering Journal*, vol. 81, pp. 419–436, 2023.

[14] M.L. Hakim et al. "Triple-band square split-ring resonator metamaterial absorber design with high effective medium ratio for 5G sub-6 GHz applications," *Nanomaterials*, vol. 13, no. 2, p. 222, 2023.

[15] H. Jiang, L.-M. Si, W. Hu and X. Lv. "A symmetrical dual-beam bowtie antenna with gain enhancement using metamaterial for 5G MIMO applications," *IEEE Photonics Journal*, vol. 11, no. 1, pp. 1–9, 2019.

[16] M. Hussain, U. Rafique, P. Dalal, S.M. Abbas and Y. Zhu. "A Compact and Wide Band Antenna for Millimeter Wave Applications," in *2024 IEEE Wireless Antenna and Microwave Symposium (WAMS)*, 2024: IEEE, pp. 1–4.

[17] B.A. Esmail, S. Koziel and S. Szczepanski. "Overview of planar antenna loading metamaterials for gain performance enhancement: The two decades of progress," *IEEE Access*, vol. 10, pp. 27381–27403, 2022.

[18] C. Milias et al. "Metamaterial-inspired antennas: A review of the state of the art and future design challenges," *IEEE Access*, 2021.

[19] N. Misran, S.H. Yusop, M.T. Islam and M.Y. Ismail. "Analysis of parameterization substrate thickness and permittivity for concentric split ring square reflectarray element," *Jurnal Kejuruteraan (Journal of Engineering)*, vol. 23, pp. 11–16, 2012.

[20] W.-J. Wu and G. Wang. "A Modified AMC-Based Antenna Sensor for Contactless Measurement of Complex Permittivity," *Measurement,* p. 112261, 2022.

[21] M. Hussain, W.A. Awan, M.S. Alzaidi, N. Hussain, E.M. Ali and F. Falcone. "Metamaterials and their application in the performance enhancement of reconfigurable antennas: A review," *Micromachines,* vol. 14, no. 2, p. 349, 2023.

[22] R. Sonak, M. Ameen and R.K. Chaudhary. "High gain dual-band open-ended metamaterial antenna utilizing CRR with broadside radiation characteristics based on left-handed AMC and PEC," *Materials Research Express,* vol. 6, no. 5, p. 055811, 2019.

[23] M.J. Jeong, N. Hussain, J.W. Park, S.G. Park, S.Y. Rhee and N. Kim, "Millimeter-wave microstrip patch antenna using vertically coupled split ring metaplate for gain enhancement," *Microwave and Optical Technology Letters,* vol. 61, no. 10, pp. 2360–2365, 2019.

[24] T. Shaw, D. Bhattacharjee and D. Mitra. "Gain enhancement of slot antenna using zero-index metamaterial superstrate," *International Journal of RF and Microwave Computer-Aided Engineering,* vol. 27, no. 4, p. e21078, 2017.

[25] M. Akbari, H.A. Ghalyon, M. Farahani, A.-R. Sebak and T.A. Denidni, "Spatially decoupling of CP antennas based on FSS for 30-GHz MIMO systems," *IEEE Access,* vol. 5, pp. 6527–6537, 2017.

[26] F. Khajeh-Khalili, M. Amin Honarvar, and A. Dadgarpour. "High-gain bow-tie antenna using an array of two-sided planar metamaterial loading for H-band applications," *International Journal of RF and Microwave Computer-Aided Engineering,* vol. 28, no. 7, p. e21476, 2018.

[27] N.N. Yoon, N. Ha-Van and C. Seo. "High-gain and wideband aperture coupled feed patch antenna using four split ring resonators," *Microwave and Optical Technology Letters,* vol. 60, no. 8, pp. 1997–2001, 2018.

[28] R. Singha, and D. Vakula. "Gain enhancement of the ultra-wideband tapered slot antenna using broadband gradient refractive index metamaterial," *International Journal of RF and Microwave Computer-Aided Engineering,* vol. 28, no. 2, p. e21191, 2018.

[29] H. Ullah and F.A. Tahir. "A high gain and wideband narrow-beam antenna for 5G millimeter-wave applications," *IEEE Access,* vol. 8, pp. 29430–29434, 2020.

[30] M. Pant and L. Malviya. "Design, developments, and applications of 5G antennas: A review," *International Journal of Microwave and Wireless Technologies,* pp. 1–27, 2022.

[31] P.B. Nikam, J. Kumar, V. Sivanagaraju and A. Baidya. "Dual-band reconfigurable EBG loaded circular patch MIMO antenna using defected ground structure (DGS) and PIN diode integrated branch-lines (BLs)," *Measurement,* vol. 195, p. 111127, 2022.

[32] A. Abbas et al. "Highly selective multiple-notched UWB-MIMO antenna with low correlation using an innovative parasitic decoupling structure," *Engineering Science and Technology, An International Journal,* vol. 43, p. 101440, 2023.

[33] G.N. Reddy, C. Ravikumar and A. Rajesh. "Literature review and research direction towards channel estimation and hybrid pre-coding in mmWave massive MIMO communication systems," *Journal of Reliable Intelligent Environments,* pp. 1–20, 2022.

[34] M.L. Bouknia et al. "Effect analysis of the general complex reciprocal gyro-bianisotropic metamaterial medium on the input impedance of a printed dipole antenna," *Alexandria Engineering Journal,* vol. 61, no. 5, pp. 3691–3696, 2022.

[35] C. Zebiri et al. "Gyrotropic reciprocal bianisotropic metamaterial printed dipole antenna: Investigation of negative input impedance and mutual coupling," *Alexandria Engineering Journal,* vol. 61, no. 12, pp. 11677–11685, 2022.

[36] A.K. Dwivedi, A. Sharma, A.K. Singh and V. Singh. "Metamaterial inspired dielectric resonator MIMO antenna for isolation enhancement and linear to circular polarization of waves," *Measurement,* vol. 182, p. 109681, 2021.

[37] M.L. Hakim, T. Alam, M.T. Islam, M.H. Baharuddin, A. Alzamil and M.S. Islam. "Quad-band polarization-insensitive square split-ring resonator (SSRR) with an inner Jerusalem cross

metamaterial absorber for Ku-and K-Band sensing applications," *Sensors,* vol. 22, no. 12, p. 4489, 2022.

[38] A. Musa et al. "Polarization independent metamaterial absorber with anti-reflection coating nanoarchitectonics for visible and infrared window applications," *Materials,* vol. 15, no. 10, p. 3733, 2022.

[39] M.L. Hakim et al. "Ultrawideband polarization-independent nanoarchitectonics: A perfect metamaterial absorber for visible and infrared optical window applications," *Nanomaterials,* vol. 12, no. 16, p. 2849, 2022.

[40] "CST Studio Suit." https://www.3ds.com/products-services/simulia/products/cst-studio-suite/ (accessed 22 February, 2022).

[41] M.L. Hakim, T. Alam, A.F. Almutairi, M.F. Mansor and M.T. Islam. "Polarization insensitivity characterization of dual-band perfect metamaterial absorber for K band sensing applications," *Scientific Reports,* vol. 11, no. 1, pp. 1–14, 2021.

[42] M.L. Hakim et al. "Polarization insensitive symmetrical structured double negative (DNG) metamaterial absorber for Ku-band sensing applications," *Scientific Reports,* vol. 12, no. 1, pp. 1–18, 2022.

[43] A. Musa et al. "Broadband plasmonic metamaterial optical absorber for the visible to near-infrared region," *Nanomaterials,* vol. 13, no. 4, p. 626, 2023.

[44] M.L. Hakim, T. Alam, M.T. Islam, H. Alsaif and M.S. Soliman. "Polarization-independent fractal square splits ring resonator (FSSRR) multiband metamaterial absorber/artificial magnetic conductor/sensor for Ku/K/Ka/5G (mm-Wave) band applications," *Measurement,* vol. 210, p. 112545, 2023.

[45] M.L. Hakim, M.T. Islam, M.S.J. Singh, T. Alam, A.M. Alenezi and M.S. Soliman. "Intensified electric field based highly sensitive backbone-shaped C and X band microwave metamaterial sensor for permittivity characterization applications," *Engineering Science and Technology, An International Journal,* vol. 60, p. 101877, 2024.

[46] L. Chen, Z. Lei, R. Yang, J. Fan and X. Shi. "A broadband artificial material for gain enhancement of antipodal tapered slot antenna," *IEEE Transactions on Antennas and Propagation,* vol. 63, no. 1, pp. 395–400, 2014.

[47] Q.-X. Chu, X.-R. Li and M. Ye. "High-gain printed log-periodic dipole array antenna with parasitic cell for 5G communication," *IEEE Transactions on Antennas and Propagation,* vol. 65, no. 12, pp. 6338–6344, 2017.

[48] G. Zhai et al. "Gain-enhanced planar log-periodic dipole array antenna using non-resonant metamaterial," *IEEE Transactions on Antennas and Propagation,* vol. 67, no. 9, pp. 6193–6198, 2019.

[49] Y. Sun, Z. Du, J. Du, Y. Liu and M.A. Basit. "Enhanced gain and broadband of endfire antenna by using I-shaped resonator structures," *International Journal of RF and Microwave Computer-Aided Engineering,* vol. 28, no. 7, p. e21519, 2018.

[50] M.M. Bait-Suwailam. "Electromagnetic field interaction with metamaterials," *Electromagnetic Fields and Waves,* pp. 1–19, 2019.

[51] M.L. Hakim et al. "Wide-Oblique-Incident-Angle Stable Polarization-Insensitive Ultra-Wideband Metamaterial Perfect Absorber for Visible Optical Wavelength Applications," *Materials,* vol. 15, no. 6, p. 2201, 2022.

[52] E. Campo. "4—Electrical Properties of Polymeric Materials," *Selection of Polymeric Materials,* pp. 141–173, 2008.

[53] K.G. Sadananda, M.P. Abegaonkar and S.K. Koul. "Gain equalized shared-aperture antenna using dual-polarized ZIM for mmWave 5G base stations," *IEEE Antennas and Wireless Propagation Letters,* vol. 18, no. 6, pp. 1100–1104, 2019.

[54] S.A. Khaleel, E.K. Hamad, N.O. Parchin and M.B. Saleh. "Programmable beam-steering capabilities based on graphene plasmonic THz MIMO antenna via reconfigurable intelligent surfaces (RIS) for IoT applications," *Electronics,* vol. 12, no. 1, p. 164, 2022.

[55] S. Hamdan, E.K. Hamad, H.A. Mohamed and S.A. Khaleel. "High-performance MTM inspired two-port MIMO antenna structure for 5G/IoT applications," *Journal of Electrical Engineering,* vol. 75, no. 3, pp. 214–223, 2024.

[56] S.A. Khaleel, E.K. Hamad, N.O. Parchin and M.B. Saleh. "MTM-inspired graphene-based THz MIMO antenna configurations using characteristic mode analysis for 6G/IoT applications," *Electronics,* vol. 11, no. 14, p. 2152, 2022.
[57] M. Hussain et al. "Design and fabrication of a printed tri-band antenna for 5G applications operating across Ka-, and V-band spectrums," *Electronics,* vol. 10, no. 21, p. 2674, 2021.
[58] M. Hussain, Q. Abbas, S.H.H. Gardzi, M. Alibakhshikenari, F. Falcone and E. Limiti. "Ultra-wideband MIMO antenna realization for indoor Ka-band applications," in *2022 United States National Committee of URSI National Radio Science Meeting* (*USNC-URSI NRSM*), 2022: IEEE, pp. 108–109.
[59] Y. Chen et al. "Landstorfer Printed Log-Periodic Dipole Array Antenna With Enhanced Stable High Gain for 5G Communication," *IEEE Transactions on Antennas and Propagation,* vol. 69, no. 12, pp. 8407–8414, 2021.
[60] A.S. Dixit and S. Kumar. "The enhanced gain and cost-effective antipodal Vivaldi antenna for 5G communication applications," *Microwave and Optical Technology Letters,* vol. 62, no. 6, pp. 2365–2374, 2020.
[61] S. Tariq, S.I. Naqvi, N. Hussain and Y. Amin. "A metasurface-based MIMO antenna for 5G millimeter-wave applications," *IEEE Access,* vol. 9, pp. 51805–51817, 2021.
[62] M.A. Khan et al. "mmWave Four-Element MIMO Antenna for Future 5G Systems," *Applied Sciences,* vol. 12, no. 9, p. 4280, 2022.
[63] D.A. Sehrai et al. "Metasurface-based wideband MIMO antenna for 5G millimeter-wave systems," *IEEE Access,* vol. 9, pp. 125348–125357, 2021.
[64] E. Al Abbas, M. Ikram, A.T. Mobashsher and A. Abbosh. "MIMO antenna system for multi-band millimeter-wave 5G and wideband 4G mobile communications," *IEEE Access,* vol. 7, pp. 181916–181923, 2019.
[65] W. Ali, S. Das, H. Medkour and S. Lakri. "Planar dual-band 27/39 GHz millimeter-wave MIMO antenna for 5G applications," *Microsystem Technologies,* vol. 27, no. 1, pp. 283–292, 2021.
[66] M.M. Kamal et al. "Infinity shell-shaped MIMO antenna array for mm-wave 5G applications," *Electronics,* vol. 10, no. 2, p. 165, 2021.

CHAPTER 5

Near Zero Index Metamaterial for Gain Enhancement of Multiband Patch Antenna

Mohammad Tariqul Islam[1] and *Md. Samsuzzaman*[2]

1. Introduction

In the field of modern wireless communication systems, the design and optimization of antennas play a critical role in ensuring efficient transmission and reception of signals across various frequency bands. Patch antennas, known for their simplicity, low profile, and ease of integration, are a popular choice for a wide range of applications. However, traditional patch antennas often face limitations in terms of gain and bandwidth performance. To address these challenges, researchers have turned to innovative materials such as **near zero index metamaterials** (NZIMs). These engineered materials possess unique electromagnetic properties that enable unprecedented control over wave propagation, leading to the enhancement of antenna gain and directivity. Metamaterials are a type of engineered material that possess extraordinary electromagnetic properties not found in natural materials. Metamaterials are commonly referred to as artificially composite materials. The remarkable breakthrough in modern science has captivated researchers. In recent decades, research on **electromagnetic** (EM) metamaterials for the microwave spectrum has generated a significant and widespread interest among the research community [1]. The metamaterial can exist in various forms, such as single negative, double negative, and double positive. When either epsilon (ε) or mu (μ) is negative, the feature is called single negative (SNG). If both are negative, it is called double negative (DNG). If both are positive, it is called double positive (DPS) metamaterial. Single negative

[1] Department of Electrical, Electronic, and Systems Engineering, Universiti Kebangsaan Malaysia, Bangi, Selangor, Malaysia.

[2] Department of Computer and Communication Engineering, Patuakhali Science and Technology University, Patuakhali, Bangladesh.

* Corresponding author: tariqul@ukm.edu.my[1]; sobuz@pstu.ac.bd[2]

metamaterial can be classified into two distinct types. If the permittivity is negative, it is referred to as epsilon negative (ENG), and if the permeability is negative, it is referred to as mu negative (MNG). In 1968, Victor Veselago made the initial discovery of a substance that exhibited negative permeability (μ) and permittivity (ε), which had distinct properties different from those of ordinary natural materials [2]. However, this term was not widely used until 1999, as there was a scarcity of such natural materials. In 2000, Smith et al. successfully discovered a new composite substance with the enigmatic property [2], i.e., both permeability (μ) and permittivity (ε) are negative, which is normal to the left hand as a metamaterial. The metamaterials can be applied to a number of applications, such as **antenna performance enhancement** [3–18, 19], SAR reduction [20], super lenses [21], filters [22], invisibility cloaking [23], EM absorber [24], sensors [25, 26], etc.

Metamaterials are intentionally created to cater to specific applications in various frequency bands. **Multiple applications** are present in each frequency band. The frequency ranges for the S-band, C-band, and X-band are 2–4 GHz, 4–8 GHz, and 8–12 GHz, respectively. S-band is utilized in target indication radar, weather radar, and surface ship radar. C-band is commonly used in satellite TV, radio communications, GEO satellites, and weather radar. Conversely, X-band is utilized for radar applications specifically for the purpose of weather monitoring. For the improvement of gain of the antenna, the metamaterial approach was proposed by Y. Liu et al. [27]. Pendry et al. demonstrated MTM in the 1990s [28], ZIMs or NZIMs were also highly discovered and used to improve antenna gain. Enoch et al. [29] first executed extreme-directive radiation by the use of a monopole source implanted in ZIM. Yao Guo et al. illustrated the patch antenna [30], which is broadband omnidirectional and the horizontal gain improved by NZIM cover. Varieties of metamaterial shapes according to the applications are suggested. E. Ekmekci illustrated an **SNG metamaterial** only for X-band and its effective medium ratio is 7.45 [31]. For Ku-band Hayet Benosman designed a metamaterial and its EMR is low, only 5.25 [32]. S.S. Islam discussed an ENG metamaterial but it was applicable only for S-band and the size of the unit cell was 28 × 28 mm^2 and EMR is 3.06 [33]. H. Zhou recommended double-Z-shape metamaterial, but the size of the unit cell is 8.5 × 8.5 mm^2 and its EMR was 4.80. M. M. Hasan proposed negative index metamaterial which was applicable for multiband but its dimension and EMR were 10 × 10 mm^2 and 8.90 [34]. An E-LC resonator was presented for gaining ENG response and cover only S-band. The EMR of this design is 10.2 and its size is 14 × 14 mm^2 [35]. A compact triple band metamaterial of negative permittivity has been presented for C-, and X-band applications. Authors claimed triple-band but showed two bands C- and X-. EMR of this work is 11.5 at lower resonance frequency [36]. For the imaging purpose, an antenna was presented which is a microwave absorber. Authors propose structure of the microwave imaging with GSM, ISM, Wi-Fi, and WiMax where operating frequency at 1.8 GHz, 2.45 GHz, and 5.80 GHz [37]. In energy harvesting, metamaterial-based frequency bands are used for satellite communication and GSM at the resonance frequencies 900 MHz, 1.37 GHz, 1.61 GHz, and 2.55 GHz [38]. A 10 × 10 mm^2 unit cell has been designed for C-, X-, and Ku- bands and its EMR is 3.90 [39]. A 5.5 × 5.5 mm^2 unit cell was deliberate for C-band only and the EMR of this configuration is 7.33 [40]. Islam et al. presented a unit cell for S-, C-, and X- bands and its size and EMR were 8 × 8 mm^2

and 13.16, respectively [41]. A left- handed dual band metamaterial presented for the application of satellites, its size and EMR are 10 × 10 mm^2 and 6.47, respectively [42]. A miniaturized Greek-key pattern **multiband metamaterial** has been published in 2012. The EMR and numerical size of the pattern was 12.5 and 10 × 10 mm^2 [43]. A dual band gap coupled hexagonal split-ring resonator (SRR)-based metamaterial is presented for microwave applications, its dimension and EMR are 10 × 10 mm^2 and 8.4, respectively [44]. Rezapour et al. designed a single negative metamaterial which reduced the mutual coupling effect of the array antenna [45]. A double negative modified Z-shaped metamaterial has been designed for the application of wideband operation, but the optimized size and EMR were 10 × 10 mm^2 and 4, respectively [46]. A multi-split ring resonator based metamaterial has been discussed for L-, and S-band applications in [47]. DNG metamaterial with modified H-shaped has been designed for microwave application, but its size is 9 × 9 mm^2 and EMR is 3 [48]. A metamaterial presented for energy harvester to improve the image quality in [49]. To find out the subwavelength feasibility image detection of an incident wave by MTM-based energy harvester. For C- and K-band radar applications, natural and artificial chiral nihility were studied in [50]. In this article, transmission theory with conversions of polarization asymmetric was exhibited. Asymmetric **triple-band** propagation of line polarization of distorted S-shape bilayer chiral MTM has been designed for only X-band and the physical length, EMR of this cell are 8.5 × 8.5 mm^2 and 2.09 [51]. A butterfly pattern metamaterial has been discussed whose electrical size and EMR at lower frequency were $0.16\lambda \times 0.21\lambda$ and 4.7, respectively [52]. The application of X-band imaging by the devices of resistive where the tunable EM band gap characterization material with a disordered resonator of cavity [53]. For X-band communication, parametric studies of composite on split S-shaped meta atoms have been designed in which physical dimensions and EMR are 10 × 10 mm^2 and 2.4 [54]. A U-ring nested resonator for multiband metamaterial was demonstrated for X- and Ku-bands and the EMR of this resonator is 5.6 at minimum resonance frequency [55]. Modified **circular-based** metamaterial for electric resonators was designed which is left-handed, its size and EMR are 5 × 5 mm^2 and 8.45, respectively [56]. Permittivity negative properties can be used in the communication sector for several purposes such as antenna bandwidth enhancement [57], design of microwave filter, etc. Conversely, the NZIM property has been used for various purposes in the communication section, such as to improve antenna gain [58] and directivity [29].

The potential advantages of integrating NZIMs with multiband patch antennas are vast, offering novel solutions to longstanding challenges in antenna design. By leveraging the unique properties of NZIMs, such as their ability to manipulate EM waves with exceptional precision, it is possible to achieve enhanced gain, directivity, and bandwidth for antennas operating across multiple frequency bands. This chapter delves into the theoretical foundations, practical design considerations, and empirical findings related to the application of NZIMs in multiband patch antennas. Through a combination of rigorous analysis and experimental validation, this chapter aims to provide a comprehensive understanding of how NZIMs can revolutionize antenna technology and pave the way for more efficient and versatile wireless communication systems.

2. Concept of Metamaterial and Antenna Gain Enhancement

Metamaterials are engineered materials with unique properties not commonly found in nature, derived from their carefully designed structures rather than their chemical composition. These materials can manipulate EM waves in unconventional ways, such as bending light or sound waves in reverse, creating negative refraction, and altering the phase velocity of waves. Such abilities make metamaterials particularly valuable for antenna design, where they can significantly enhance performance by improving **gain, directivity**, and bandwidth [59]. The concept of gain enhancement using metamaterials involves incorporating these materials into antenna designs to achieve superior radiation patterns and higher efficiency. Metamaterials, due to their tailored EM properties, enable antennas to focus energy more precisely in desired directions, thus increasing gain and directivity [60]. This targeted energy distribution not only strengthens the signal in specified areas but also reduces interference and energy wastage in undesired directions. One area of particular interest is the use of NZIMs, which exhibit near-zero permittivity and permeability. These materials allow for the creation of innovative antenna designs with unprecedented control over wave propagation, enhancing gain and efficiency while maintaining compactness and low weight [61]. Such properties are critical for modern wireless communication systems, including 5G and Internet of things (IoT) applications, where high-performance antennas are required in constrained spaces. Metamaterials also open avenues for advanced features such as beam steering, **polarization control**, and adaptive radiation patterns, enabling antennas to dynamically adjust to changing communication needs [59, 62]. These capabilities contribute to improved signal quality, stability, and reliability, even in challenging environments with obstacles and interference [63].

Moreover, by optimizing radiation patterns and energy distribution, metamaterials enhance the overall efficiency of antennas, resulting in reduced power consumption and longer device lifespans. This is crucial for the sustainability of wireless communication networks, especially as demand for connectivity continues to grow exponentially. Recent research has explored the integration of metamaterials in various antenna designs, including patch, dipole, and phased array antennas, with promising outcomes such as increased gain, better directivity, and improved bandwidth [64]. Such advances underscore the potential of metamaterials to revolutionize antenna technology and drive innovation in wireless communications. In summary, the concept of metamaterial-based gain enhancement in antennas presents a transformative approach to meeting the evolving demands of modern connectivity. By leveraging the unique electromagnetic properties of metamaterials, antenna designs can achieve superior performance across various applications, from **wearable devices** and IoT to aerospace and 5G networks [62]. As research and development in this field continue to progress, metamaterial-enhanced antennas are poised to play a pivotal role in shaping the future of wireless communication systems.

3. Importance of Near Zero Index Metamaterial for Gain Enhancement

NZIMs represent a significant advancement in antenna technology, offering transformative potential for enhancing gain and performance in multiband patch antennas. Characterized by their near-zero permittivity and permeability, NZIMs enable precise control over electromagnetic wave propagation, leading to notable improvements in antenna gain, directivity, and bandwidth across multiple frequency bands [65]. This capability is particularly important in modern wireless communication systems, where antennas must support diverse standards and protocols across various applications. The integration of NZIMs with multiband patch antennas allows for the development of compact and lightweight designs that maintain high performance, making them suitable for space-constrained applications such as wearable technology, (IoT devices, and aerospace systems. Moreover, NZIMs enable **advanced antenna features** such as beam steering, polarization control, and adaptive radiation patterns, which are critical for the successful deployment of next-generation communication networks like 5G and beyond [66].

By harnessing NZIMs' unique properties, antennas can achieve more stable and reliable communication links even in complex environments, which is especially beneficial for critical applications such as telemedicine, autonomous vehicles, and smart city infrastructure [67]. Additionally, the manipulation of EM waves facilitated by NZIMs contributes to higher efficiency and reduced energy loss in antenna systems, supporting sustainable and power-efficient wireless networks. Recent research continues to explore the integration of NZIMs in multiband patch antennas, yielding promising results such as tailored radiation patterns and enhanced gain across a wide range of frequencies [68]. These advancements highlight the adaptability of NZIM-based antennas, which can be customized to meet specific application requirements and operational environments. In summary, the use of NZIMs for gain enhancement in multiband patch antennas represents a significant leap forward in antenna technology. The benefits of NZIMs extend beyond improved gain and directivity to include advanced capabilities such as beam steering and polarization control, increased efficiency, and enhanced signal quality. As wireless communication systems continue to evolve, **NZIM-based antennas** will play a pivotal role in meeting the growing demands of modern connectivity, ushering in a new era of high-performance, energy-efficient communication solutions.

The use of NZIMs for gain enhancement in multiband patch antennas is of great importance for several specific reasons as follows:

- **Improved Gain and Directivity:** NZIMs exhibit distinct EM characteristics that enable them to manipulate the propagation of waves in innovative manners, resulting in substantial improvements in antenna gain and directivity. This enhanced performance is crucial for applications where signal strength and quality are of utmost importance.
- **Enhanced Bandwidth:** The integration of patch antennas with NZIMs allows for broader bandwidths, enabling them to function efficiently across various frequency bands. The ability to adapt to different protocols and standards is very beneficial for contemporary wireless communication systems.
- **Size Reduction:** NZIMs enable the shrinking of antennas by enabling the creation of small structures without compromising performance. This is crucial for applications that have limited space, such as wearable devices and IoT applications.

- **Improved Efficiency:** NZIMs possess the capability to efficiently modify electromagnetic waves, resulting in enhanced radiation patterns and decreased energy losses. This improves the overall efficiency of the antenna system, leading to enhanced power management.
- **Enhanced Signal Quality:** NZIMs enhance signal quality and minimize interference, ensuring dependable communication links in intricate situations by enhancing gain and directivity.
- **Advanced Applications:** The distinctive qualities of NZIMs allow for the development of inventive applications, including beam steering, polarization control, and adaptive antennas. These sophisticated functionalities are essential for upcoming communication technologies like 5G and beyond.

4. Metamaterial Structure Design and Analysis for Gain Enhancement

4.1 Unit Cell Geometric Layout

Figure 1(a and b) displays the top view and perspective view of the optimized square enclosed circle SRR (**SEC-SRR**) design imported from the CST software. The unit cell in question consists of a square and a circular split ring resonator, both constructed from copper with a thickness of 0.035 mm. The substrate material utilized is Rogers RO4350B, with copper serving as the resonator positioned on the substrate. The back side of the substrate lacks copper. The substrate employed has a thickness of 1.524 mm, a dielectric constant of 3.66, and a loss tangent of 0.0031. The unit cell has been refined to a size of 8 mm × 8 mm, whereas the square ring has a size of 7 mm × 7 mm. The diameter of the outer ring and inner ring of the circular resonator are 5 mm and 4 mm, respectively. The gap between the square and **circular resonator** is uniform, measuring 0.25 mm. The metal strip connecting the square and circular resonator has dimensions of 0.5 mm × 0.25 mm (length and breadth). When the copper-structured ring resonator is placed on the substrate, it generates electromagnetic radiation and related impedance. The requirements for the SEC-SRR

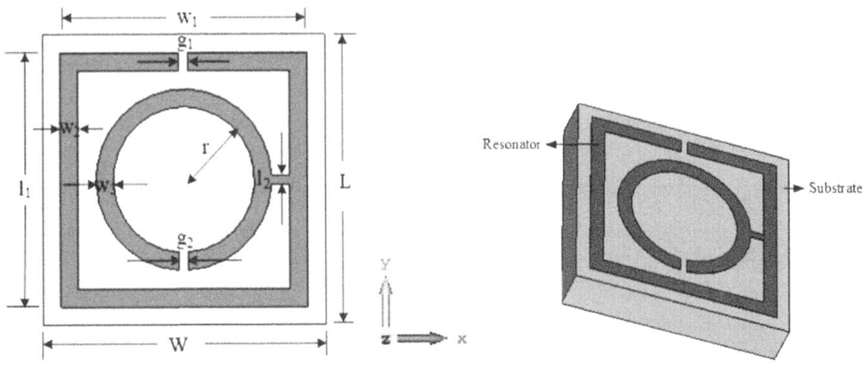

(a) (b)

Fig. 1 SEC-SRR design (a) Top view (b) Perspective view.

unit cell are as follows: the width (W) is 8 mm, the length (L) is 8 mm, the widths of the three arms are $w1 = 7$ mm, $w2 = 0.5$ mm, and $w3 = 0.5$ mm, the lengths of the two arms are $l1 = 7$ mm and $l2 = 0.25$ mm, and the gaps between the arms are $g1 = 0.25$ mm and $g3 = 0.25$ mm. Additionally, the radius (r) of the unit cell is 2 mm.

4.2 Simulation Setup with Metamaterial Theory Analysis

The simulation configuration shown in Figure 2, with T representing transmission and *R* representing reflection, can be seen using the schematic design in Figure 3. An EM wave propagates along the *z*-axis and generates electric and magnetic fields. The electric field is oriented along the *x*-axis, whereas the magnetic field is oriented along the *y*-axis. The **incident wave** propagates through the unoccupied space of region 1 on a metamaterial slab. A portion of the wave is present in area 2, where it is transmitted (ahead wave) and reflected back to region 1 as a portion (reverse wave). Area 2 transfers the wave to area 3, which is also a vacant space, and the wave continues to propagate in this medium. The 3D EM CST Microwave Studio, analyzed using the **Finite Integration Technique** (FIT), has been utilized to evaluate the effective parameters. The specified unit cell has been placed between the positive and negative *z*-axis of the two waveguide ports. The boundary conditions were set to be fixed in both the *x* and *y* directions in order to achieve the desired electric field (PEC) and magnetic field (PMC). The *Z*-axis was selected for the propulsion of free space maneuvering, as depicted in Figure 2. To analyze the unit cell and array design in the frequency range of 2 – 12 GHz, a tetrahedral mesh obtained from a frequency-domain solver was used. To comprehend the unusual outcomes of the EM phenomena, an analysis is conducted on the reflection response (*S*11) and the transmission response (*S*21).

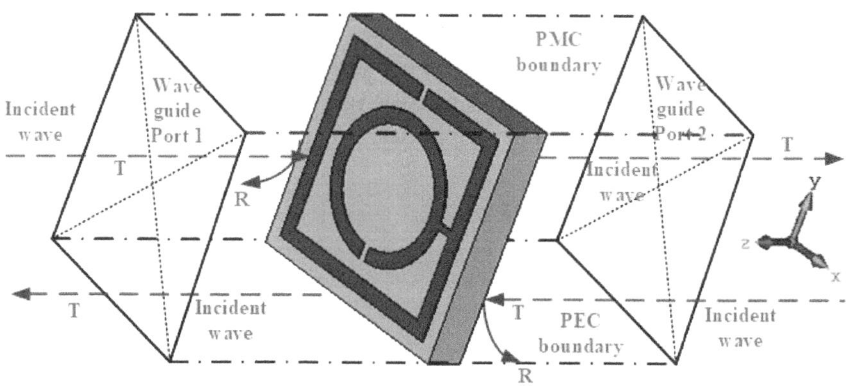

Fig. 2 The boundary complaint of the suggested metamaterial unit cell with the purpose of incidence, reflection, and transmission.

The electric and magnetic fields can be described for these three regions using the following equations as implemented in [69]. The phasor of an electric and magnetic field for region 1 is stated as,

$$E_0 = \left(E_0^t e^{-ik_0 z} + E_0^r e^{-ik_0 z}\right) a_x \qquad ...(1)$$

$$H_0 = \left(H_0^t e^{-ik_0 z} + H_0^r e^{-ik_0 z}\right) a_y \qquad ...(2)$$

Here, E_0^t, E_0^r, $H_0^t = E_0^t/z_0$, $H_0^r = -E_0^r/z_0$ are the forward (transmitted) and backward (reflected) electric and magnetic field amplitudes, respectively. $k_0 = \omega\sqrt{\mu_0 \varepsilon_0}$ is the wavenumber of the incident wave and $z_0 = \omega\sqrt{\mu_0/\varepsilon_0}$ is the wave impedance in region 1 where μ_0 and ε_0 are free space permeability and permittivity. For region 2, the phasor form can be articulated by the Equations:

$$E_1 = \left(E_1^t e^{-ik_1 z} + E_1^r e^{-ik_1 z}\right) a_x \qquad ...(3)$$

$$H_1 = \left(H_1^t e^{-ik_1 z} + H_1^r e^{-ik_1 z}\right) a_y \qquad ...(4)$$

Here, E_1^t, E_1^r, $H_1^t = E_1^t/z_1$, $H_1^r = -E_1^r/z_1$ are the forward (transmitted) and backward (reflected) electric and magnetic field amplitudes, respectively. $k_1 = \omega\sqrt{\mu_1 \varepsilon_1}$ is the wave number and $z_1 = \omega\sqrt{\mu_1/\varepsilon_1}$ is the wave impedance in region 2 where μ_1, ε_1 are the metamaterial permeability and permittivity. Similarly, for region 3, phasor form of the electric and magnetic field will be:

$$E_2 = E_2^t e^{-ik_0 z} a_x \qquad ...(5)$$

$$H_2 = H_2^t e^{-ik_0 z} a_y \qquad ...(6)$$

E_2^t, $H_2^t = E_2^t/z_0$ are the forward (transmitted) electric and magnetic field amplitude in region 3.

Along the z-direction, $z = 0$ and $z = t$ and using normalized wave impedance at metamaterial, $Z = Z_1/Z_0 = \sqrt{\mu_r/\varepsilon_r}$. The different **transmitted and reflected** wave related with the forward wave of region 1 can be achieved as in [69].

$$E_0^r = \frac{i(Z^2 - 1)\sin(k_1 t)}{2Z_1 Z_0 \cos(k_1 t) + i(Z^2 + 1)\sin(k_1 t)} E_0^t \qquad ...(7)$$

$$E_1^t = \frac{(Z^2 + Z)e^{-ik_1 t}}{2Z\cos(k_1 t) + i(Z^2 + 1)\sin(k_1 t)} E_0^t \qquad ...(8)$$

$$E_1^r = \frac{z_1(Z - Z^2)e^{-ik_1 t}}{2Z\cos(k_1 t) + i(Z^2 + 1)\sin(k_1 t)} E_0^t \qquad ...(9)$$

$$E_2^t = \frac{2Ze^{-ik_0 t}}{2Z\cos(k_1 t) + i(Z^2 + 1)\sin(k_1 t)} E_0^t \qquad ...(10)$$

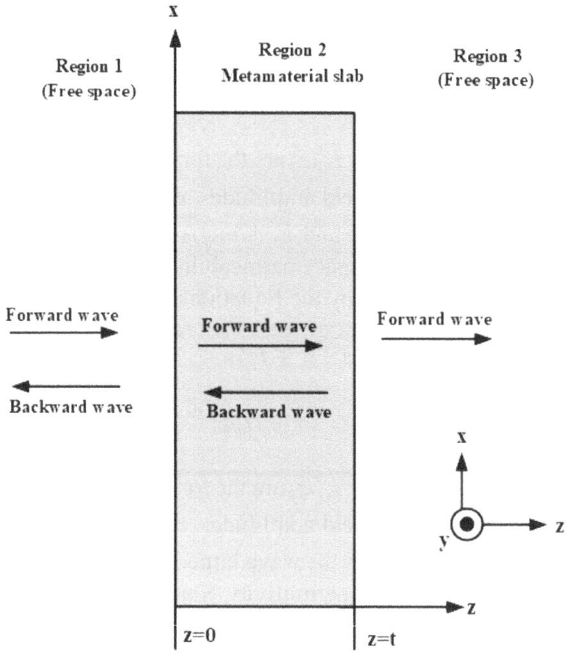

Fig. 3 The normal incidence of EM waves on the thickness of the metamaterial, *t*.

$$V_1 = |S_{11}| + |S_{21}| \quad ...(11)$$

$$V_2 = |S_{21}| - |S_{11}| \quad ...(12)$$

If,

$$X = \frac{1 - V_1 V_2}{V_1 - V_2}$$

Now the coefficient of reflection of the **incident wave**,

$$\Gamma = X \pm \sqrt{X^2 - 1} \quad ...(13)$$

The coefficient of transmission,

$$X = \frac{V_1 - \Gamma}{1 - V_1 \Gamma} \quad ...(14)$$

The relative index, n and impedance, z can be derived from these equations. The relative permittivity, ε_r, and relative permeability, μ_r can be obtained from Equation (9) and (10) [70].

$$\varepsilon_r \sim \frac{2}{jk_0 d} \times \frac{(1 - V_1)}{(1 + V_1)} \quad ...(15)$$

$$\mu_r \sim \frac{2}{jk_0 d} \times \frac{(1 - V_2)}{(1 + V_2)} \quad ...(16)$$

where, $k_0 = \frac{2\pi f}{c}$, c is light speed, and d is substrate thickness.

n_r is symbolized in equation (17)

$$n_r = \sqrt{\mu_r \varepsilon_r} \quad ...(17)$$

Normalized impedance, $$Z = \sqrt{\mu_r / \varepsilon_r} \quad ...(18)$$

Figure 4 indicates the frequency versus response of the reflection and transmission of the SEC-SRR unit cell, where the resonance frequencies are observed at 2.61, 6.32, and 9.29 GHz, respectively.

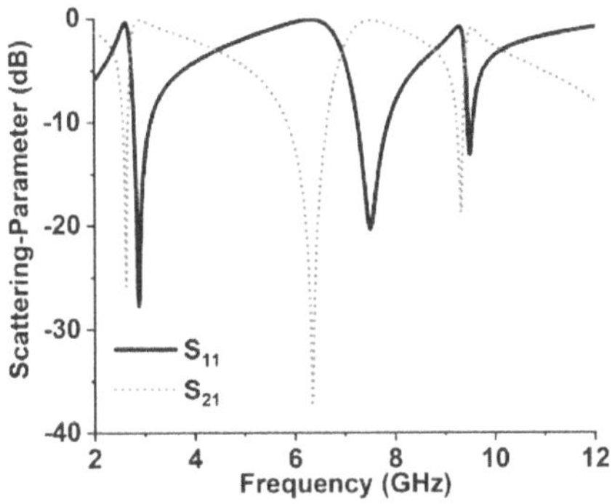

Fig. 4 Frequency versus S_{11} and S_{21} of the SEC-SRR unit cell.

4.3 Surface Current, Current Density, Electric and Magnetic Field Analysis

Figure 5 illustrates the distribution of **surface currents** in the proposed unit cell at three specific resonance frequencies. A distinct surface current is observed in the bottom section of the outer ring at the minimum resonance frequency of 2.61 GHz. A moderate stream passes through the upper portion of this ring. The upper half of the circular ring exhibits a **substantial current flow**, with a larger concentration of current observed near the inner edge of the ring and the **edge of the split gap**. The current pattern undergoes reversal at a frequency of 6.32 GHz. Currently, there is a greater concentration of electric current along the inner edge of the outer ring, with a particularly high current in the upper half of the circular ring. At the resonant frequency of 9.29 GHz, there is a significant presence of high current on the right side of the unit cell. However, the current distribution throughout the entire unit cell is not uniform. Figure 6 displays the current density of the unit cell. The current density undergoes a shift from the right side of the unit cell to the left. The lower left region of the outer ring has a low density of current. Amperes' rule establishes a correlation between the magnetic field and the electric current. The equation represents the magnetic **flux density** at a wire at a distance r.

$$B = \frac{\mu I}{2\pi r} \quad ...(19)$$

Where, μ is the free space permeability constant. From the H field structure introduced in Figure 7(b), it is evident that there is an intensified H field pattern present in the region with high current density. The **magnitude of the *H* field** diminishes gradually as the distance from the ring increases, in accordance with the established laws and equations. Figure 7(a) depicts the electric field distribution of the proposed unit cell, which satisfies Maxwell's equation as stated in Equation (20), which states that any alteration in the magnetic field results in the formation of a curl in the electric field. This law is clearly demonstrated by the image depicted in Figure 7(a) of the E file. When comparing the E field with the H field, it is evident that the E field is close to zero when the H field is constant, and the presence of the E field is observed when there is variation in the H field. A greater disparity in the H field corresponds to a higher magnitude of the E field.

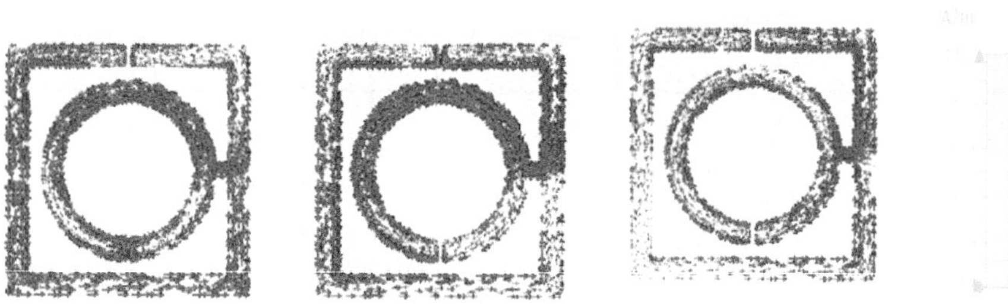

(a) (b) (c)

Fig. 5 Surface current distribution at: (a) 2.61, (b) 6.32, and (c) 9.29 GHz.

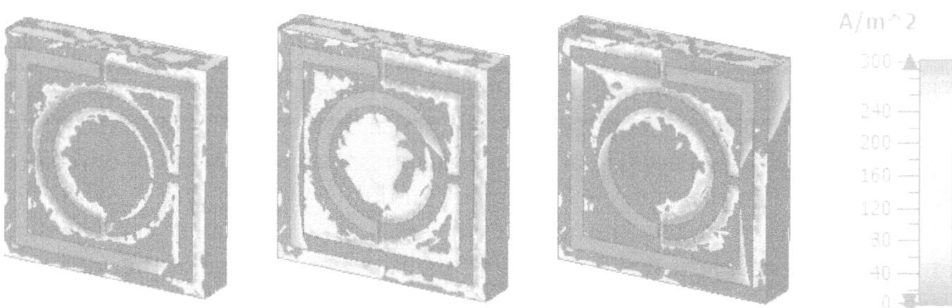

(a) (b) (c)

Fig. 6 Current density plot (J-surface current) at: (a) 2.61, (b) 6.32, and (c) 9.29 GHz.

The permittivity (ε) and permeability (μ) are frequency-dependent in a lossless **isotropic medium**.

$$D = \varepsilon E \qquad \text{...(20)}$$

$$B = \mu H \qquad \text{...(21)}$$

Furthermore, the two constitutive relationships, which define the medium's response to the fields applied, μ_0

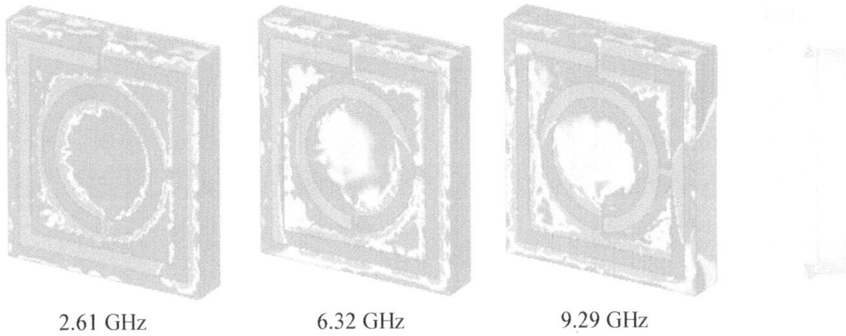

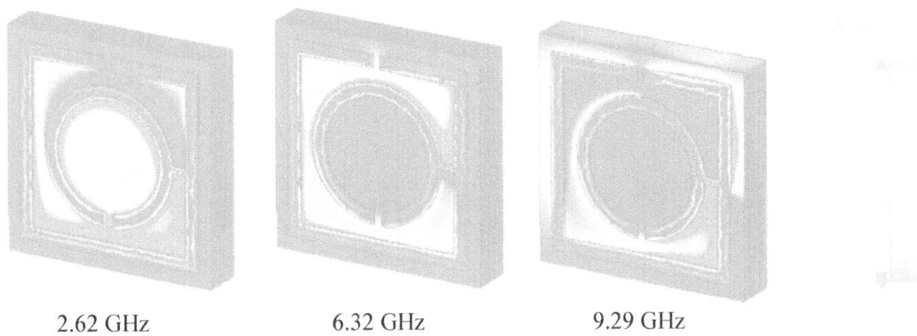

Fig. 7 Optimized metamaterial: (a) **electric field,** (b) magnetic field distribution at 2.61, 6.32 GH, and 9.29 GHz, respectively.

$$D = \varepsilon_0 \varepsilon E \qquad \text{...(22)}$$

$$B = \mu_0 \mu E \qquad \text{...(23)}$$

The **DPS-ENG-DPS** system generates unconventional EM phenomena, such as reflection and refraction waves, within the DPS medium. For the transmission of EM waves, it is most advantageous to have a complex configuration for both SNG and DPS media. An artificially created lossy ENG slab is placed next to a naturally occurring DPS slab made of air, with the same size and shape. Figure 8 depicts the DPS-ENG-DPS slab that was utilized for EM wave propagation tests. The intensity distribution of the electric and magnetic fields for the ENG medium is measured and compared with the reference DPS slabs. Figure 9 and 10 depict the intensity of the electric field and magnetic field at frequencies of 2.61 GHz, 6.32 GHz, and 9.29 GHz, respectively. Observation revealed the presence of negative refracted waves in the ENG slabs and the occurrence of reverse wave propagation. Given, $\varepsilon < 0$ and $\mu > 0$, the Drude function can be defined as the energy dissipation factor (Γ) divided by the frequency (ω).

$$\varepsilon(\omega) = 1 - \frac{\omega_p^2}{\omega(\omega + j\Gamma)} \qquad \text{...(24)}$$

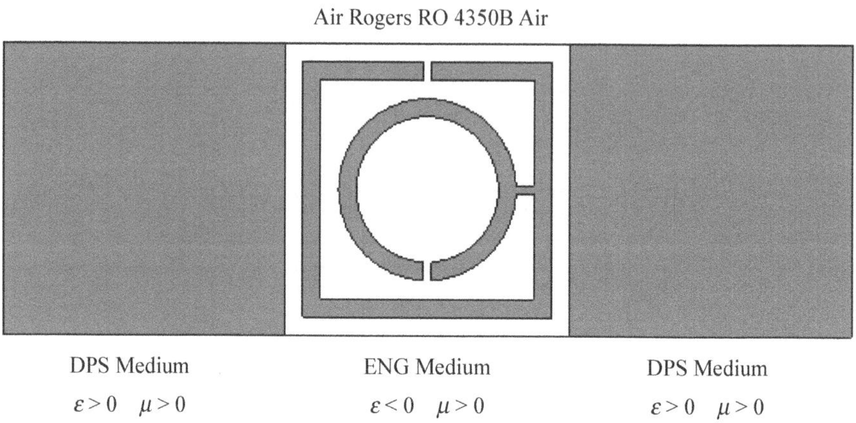

Fig. 8 DPS-ENG-DPS slab.

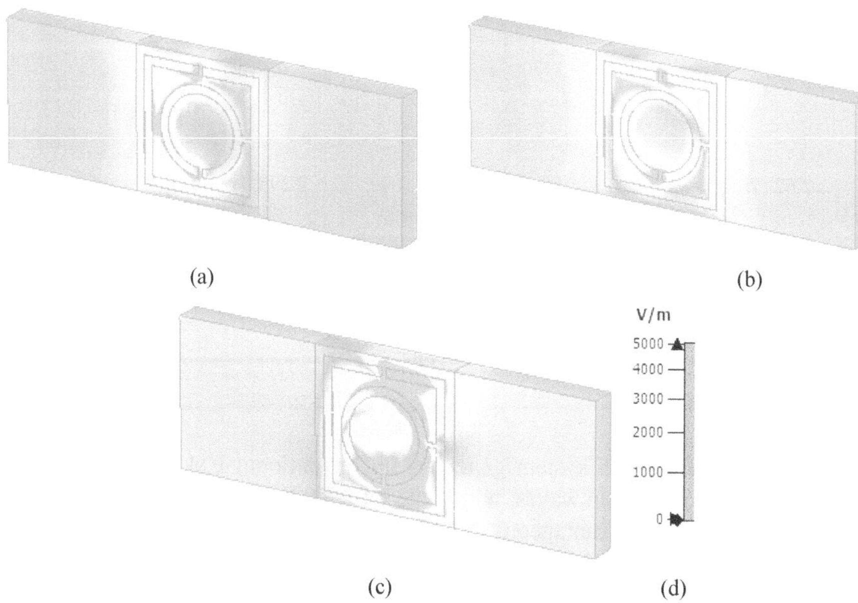

Fig. 9 Interaction of electromagnetic waves on electric field: (a) at 2.61, (b) at 6.32, and (c) at 9.29 GHz, (d) Scale.

4.4 Unit Cell Design with Circuit Model ing and Performance Analysis

Since the mentioned form involves passive components, i.e., **inductive-capacitive** (*LC*) elements, the resonance frequency (f) is given below [71].

$$f = \frac{1}{2\pi\sqrt{LC}} \quad ...(25)$$

L represents the total inductance, while *C* represents the total capacitance. In the specified arrangement, inductance is created by the metal strip, while capacitance is

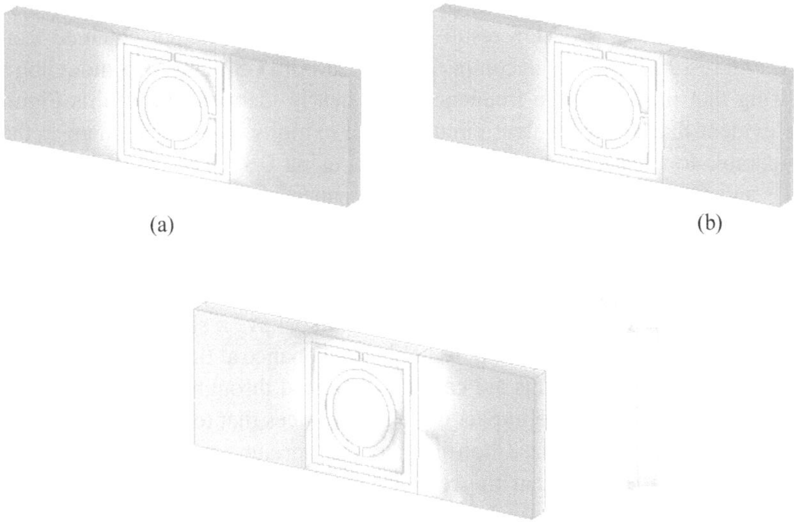

Fig. 10 Interaction of electromagnetic waves on magnetic field: (a) at 2.61, (b) at 6.32, and (c) at 9.29 GHz.

created by the splits. The electrical resonances are generated through the coupling between the gaps and electric fields. Furthermore, the magnetic fields are coupled with the loops to form magnetic resonances. The capacitance

$$C = \varepsilon_0 \varepsilon_r \frac{A}{d} \ (F) \qquad ...(26)$$

where, $\varepsilon_0 = 8.854 \times 10^{-12}$ F/m, ε_r = relative permittivity, A and d are the region and distance of the split, respectively.

In addition, the whole inductance (L) can be concluded by [72]:

$$L = 0.01 \times \mu_0 \left\{ \frac{2(d+g+h)^2}{(2w+g+h)^2} + \frac{\sqrt{(2w+g+h)^2 + l^2}}{(d+g+h)} \right\} t \qquad ...(27)$$

where, $\mu_0 = 4\pi \times 10^{-7}$ H/m.

Thus, the whole capacitance (C) can be concluded by

$$C = \varepsilon_0 \left[\frac{2w+g+h}{2\pi(d+h)^2} \ln \left\{ \frac{2(d+g+h)}{(a-l)} \right\} \right] t \qquad ...(28)$$

Figure 11 displays an estimated equivalent circuit of the suggested unit cell. The circuit includes **microstrip lines** represented by inductors $L1$, $L3$, $L4$, and capacitors $C1$, $C2$, and $C3$, which are associated with the split gaps. $L2$ represents the inductor resulting from the coupling of two rings, whereas $C2$ represents the coupling capacitor. The values of these elements are modified by the utilization of advanced design software (ADS). By adjusting the parameter values, the transmission coefficient is found to ensure that the ADS result closely matches the output obtained from the CST. Figure 12 illustrates the comparison of *S*21 both CST and ADS modeling tools, showing that the resonance frequencies are nearly identical. From this Figure 12, it is evident that the equivalent circuit model exhibits tri-band resonances that are comparable to those observed in the unit cell using CST. The parallel combination of *L*3 and *C*3 in the circuit generates a resonance circuit known as a parallel LC circuit, which symbolizes the equivalency of the inner ring. The resonance at 9.29 GHz is triggered by *L*3 and *C*3. Optimizing the resonance frequency can be achieved by altering the values of *L*3 and *C*3. Upon careful examination of the analogous circuit in **ADS,** it becomes apparent that *L*1 and *C*1 have a significant impact on the resonances occurring at 6.32 GHz. To achieve resonance at this frequency, the values of the *L*1 and *C*1 components are carefully adjusted through tuning. The pairing of *L*2 and *C*2 demonstrates the impact on the resonances that took place at frequencies of 2.61 GHz and 9.29 GHz. Choosing the appropriate parameter values for these two components ensures that the resonances occur at these specific frequencies. The **inductor-capacitor pair** *L*4 and *C*4 demonstrate their impact on the cutoff frequencies of the resulting bands. Therefore, it is necessary to accurately change these numbers in order to choose the appropriate frequency bands.

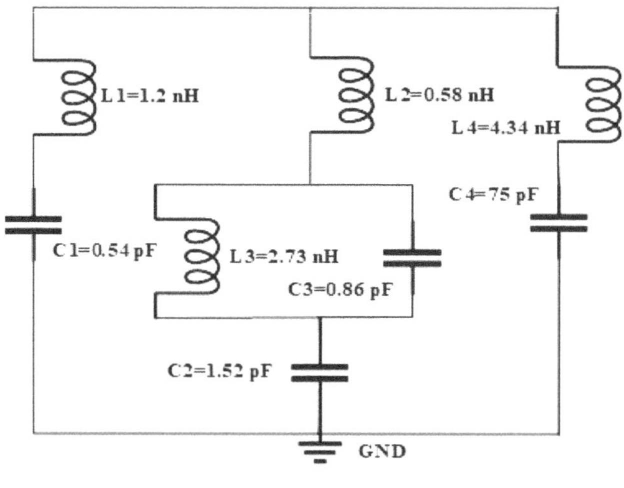

Fig. 11 Equivalent circuit of the designed unit cell.

4.5 *Parametric Study of the Designed Unit Cell*

a. Effect of changing of the model

To choose the design, we have created various models and analyzed the responses of *S*11 and *S*21. Model 1 comprises a square ring and a circular ring, model 2 consists of a split square ring and a circular ring, and model 3 contains a square ring and a split circular ring. Furthermore, model 4 comprised of a divided square ring and a divided circular ring, whereas the suggested model also includes a divided square ring and a

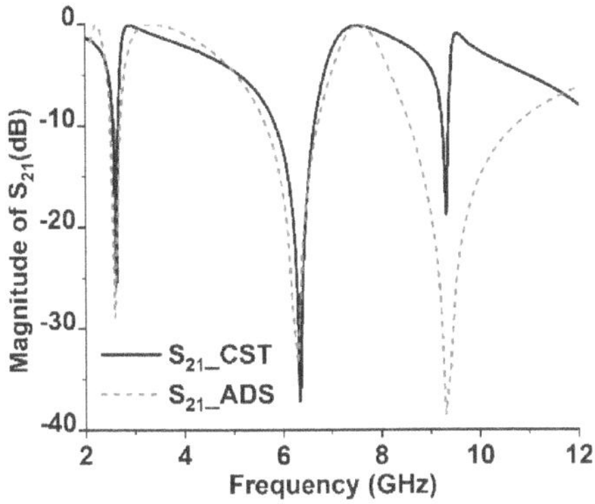

Fig. 12 Frequency versus magnitude of S_{21} for CST and ADS analysis.

divided circular ring, joined by a metal strip line. Figure 13(a, b) reveals that models 1, 2, and 3 have a single frequency band and resonance frequency. Model 4 exhibits the presence of two frequency bands and two resonance frequencies. The model we offer is compatible with **triple frequency bands** and three resonance frequencies. The chosen model is characterized by a straightforward design, consisting of a split square ring and a split circular ring joined by a **metal strip line**. Despite its simplicity, this model is capable of covering a wide range of frequencies. Figure 13(a, b) reveals that the suggested model outperforms another model in terms of reflection and transmission properties. Table 1 displays various parameters for different models..

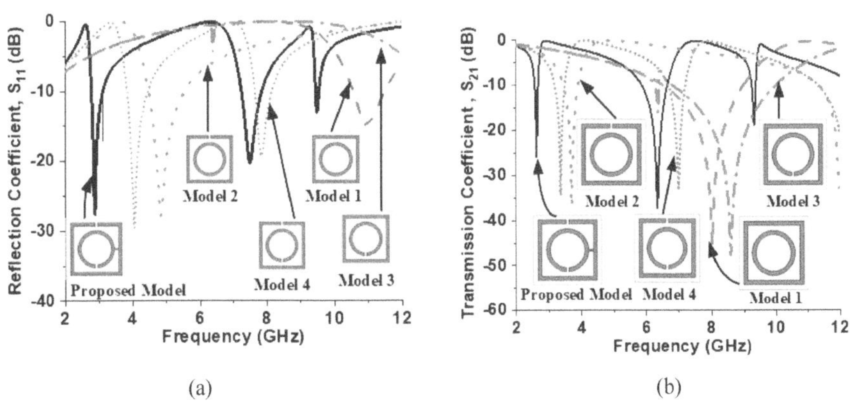

Fig. 13 For various models: (a) Change of S_{11} (b) Change of S_{21}.

b. Effect of changing of the unit cell size

The initial dimension has been determined using the general **subwavelength guidelines** of lambda/10. Our objective was to create a multiband metamaterial that could be used for *S*-band (2–4 GHz), *C*-band (4–8 GHz), and *X*-band (8–12 GHz) applications. According to the lambda/10 connection, the initial dimension of the metamaterial unit cell is 15 mm. To achieve downsizing of the unit cell dimension, we

Table 1 The values of analyzed resonance frequency f_0 (GHz), resonance peak S_{21} (dB) and covering bands for different models.

Model	*Resonance Frequency f_0 (GHz)*	*Resonance Peak S21 (dB)*	*Covering Bands*
1	7.99	– 44.37	*C*-band
2	3.66	– 36.28	*S*-band
3	8.53	– 47.43	*X*-band
4	3.49, 7.11	– 34.54, – 32.93	*S*-, *C*-bands
Proposed	2.61, 6.32, 9.29	– 26.32, – 37.51, – 19.13	*S*-, *C*-, *X*-bands

have adjusted the dimension (either increased or decreased) in CST using optimization techniques. Additionally, we have incorporated **circular and rectangular split gaps** into the unit cell structure. We have successfully developed an optimized square enclosed circle split ring resonator (SEC-SRR) with dimensions of 7 mm in length and breadth of the main unit cell. These dimensions were chosen to ensure the desired resonance for *S*-, *C*-, and *X*-band applications. The resonance frequency value is inversely proportional to the diameter of the unit cell. To achieve the lowest resonance frequency, maximize the coverage of bands, and obtain the largest value of **effective medium ratio (EMR)**, we have considered four different sizes for the unit cell while also focusing on miniaturization. The analysis in Figure 14 includes four different sizes of the unit cell: 5 mm, 6 mm, 7 mm, and 8 mm. The transmission response of the unit cell of various sizes is illustrated in Figure 15. Table 2 presents the resonance frequency, bandwidth, coverage band, number of bands, and maximum EM radiation for the various sizes of unit cells. The unit cell has dimensions of 5 millimeters by 5 millimeters squared. The resonance frequency of the system is 5.89 GHz, and its bandwidth ranges from 5.67–6.05 GHz. The unit cell encompasses the C-band and possesses an EMR value of 10.19. The unit cell has dimensions of 6 millimeters by 6 millimeters squared. The resonance frequencies are 4.18 GHz and 10.24 GHz, and they cover the *C*- and *X*-bands. The measured EMR value is 11.77. The unit cell, which has dimensions of 7×7 mm^2, exhibits resonance frequencies of 3.48 GHz, 8.23 GHz, and 11.76 GHz. The frequencies mentioned are associated with the *S*- and *X*-bands, and the measurement of **effective medium ratio** is 12.32. The resonant frequencies for a unit cell with dimensions of 8×8 mm^2 are 2.61, 6.32, and 9.29 GHz. The frequencies encompass the *S*-, *C*-, and *X*-bands, with an EMR value of 14.37.

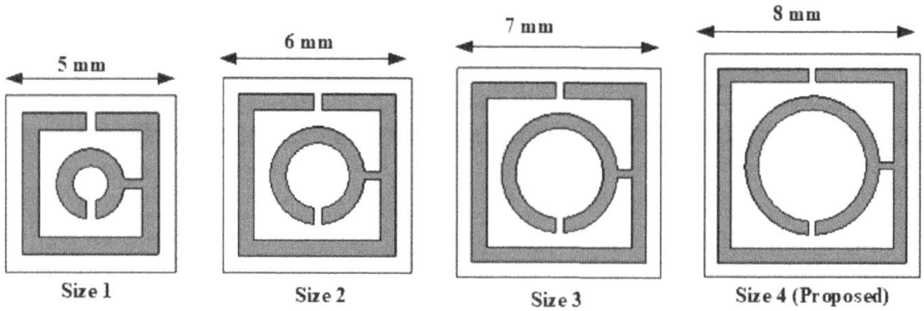

Fig. 14 Different size of the unit cell.

The unit cell exhibits its highest resonance frequency, widest frequency range, and maximum EMR when its physical dimensions are 8 × 8 mm^2 and its electrical size is 0.070λ × 0.070λ at a frequency of 2.61 GHz. The unit cell has dimensions of 5 × 5 mm^2. The resonance frequency of the system is 5.89 GHz, and its bandwidth ranges from 5.67–6.05 GHz. The unit cell encompasses the *C*-band and possesses an EMR value of 10.19. The unit cell has dimensions of 6 × 6 mm^2 The resonance frequencies are 4.18 GHz and 10.24 GHz, and they cover the *C*- and *X*-bands. The EMR value is 11.77. The unit cell, which has dimensions of 7 × 7 mm^2, exhibits resonance frequencies of 3.48 GHz, 8.23 GHz, and 11.76 GHz. The frequencies mentioned are associated with the *S*- and *X*-bands, and the measurement of EMR is 12.32. The resonant frequencies for a unit cell with dimensions of 8 × 8 mm^2 are 2.61, 6.32, and 9.29 GHz. The frequencies encompass the *S*-, *C*-, and *X*-bands, with an EMR value of 14.37. The unit cell exhibits its highest resonance frequency, widest frequency range, and maximum EMR when its physical dimensions are 8 × 8 mm^2 and its **electrical size** is 0.070λ × 0.070λ at a frequency of 2.61 GHz.

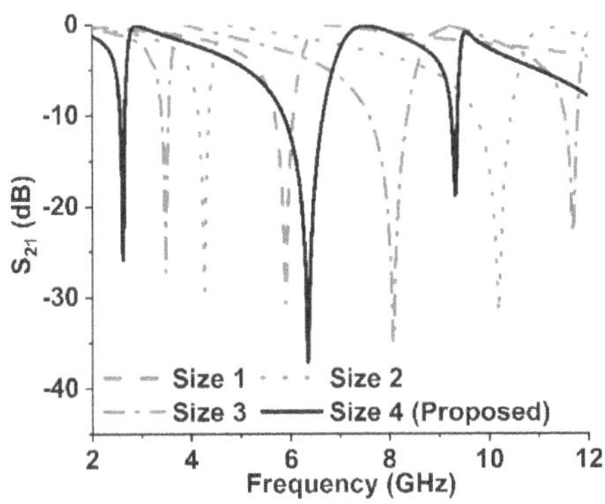

Fig. 15 Frequency versus S_{21} of the different size of the unit cell.

Table 2 An analysis of the **resonance frequency, bandwidth, coverage band**, number of bands, and maximum electromagnetic radiation (EMR) of unit cells of various sizes.

Different Size	*Resonance Frequency (GHz)*	*Bandwidth (GHz)*	*Covering Band*	*Number of Bands*	*Maximum EMR*
Size 1	5.89	5.67–6.05	C-	1	10.19
Size 2	4.18, 10.24	4.15–4.35, 9.74–10.44	C-, X-	2	11.77
Size 3	3.48, 8.23, 11.76	3.41–3.54, 8.01–8.63, 11.72–11.87	S-, X-	2	12.32
Size 4 (Proposed)	2.61, 6.32. 9.29	2.55–2.66, 5.84–6.63, 9.21–9.36	S-, C-, X-	3	14.37

c. Effect of different port signals

An in-depth analysis has been conducted on the different **boundary conditions** and port signals depicted in Figure 16. Figure 16(a, b) depicts the unit cell structure of the SEC-SRR with a port signal oriented along the *X*-axis and *Y*-axis, respectively. Figure

16(c) shows the X-axis port signal for the reflection and transmission coefficients. Figure 16(d) depicts the signal at the Y-axis port, which corresponds to the reflection and transmission coefficients. There is a distinct presence of robust transmission frequency bands in contrast to the frequency bands of the reflection coefficient.

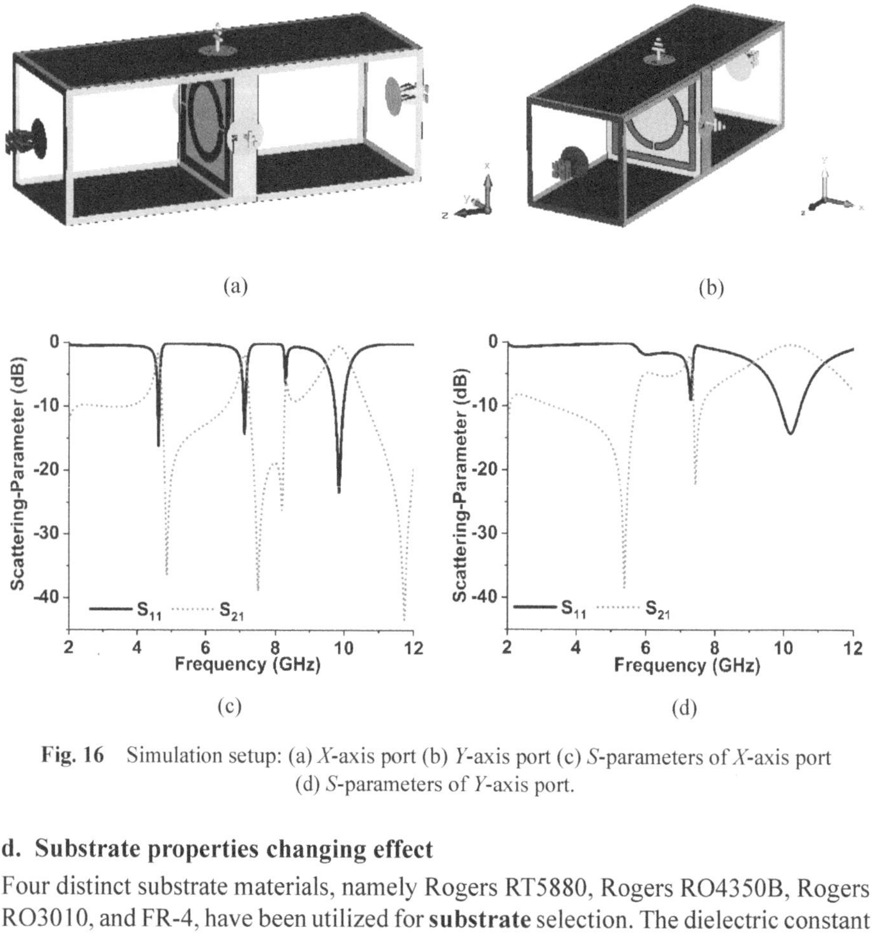

Fig. 16 Simulation setup: (a) X-axis port (b) Y-axis port (c) S-parameters of X-axis port (d) S-parameters of Y-axis port.

d. Substrate properties changing effect

Four distinct substrate materials, namely Rogers RT5880, Rogers RO4350B, Rogers RO3010, and FR-4, have been utilized for **substrate** selection. The dielectric constant (DC), loss tangent (LT), and thickness of each substrate material vary. The dielectric constant (DC), loss tangent (LT), and thickness of Rogers RT5880 are 2.2, 0.0009, and 1.575 mm, respectively. The DC for Rogers RO4350B is 3.66, the LT is 0.0031, and the thickness is 1.524 mm. The DC for Rogers RO3010 is 10.2, the LT is 0.0035, and the thickness is 1.27 mm. The DC for FR-4 is 4.4, the LT is 0.025, and the thickness is 1.5 mm. For this investigation, the dimensions of the substrate material and the SEC-SRR design remained constant. Figure 17(a, b) reveals that both reflection and transmission responses exhibit three bands for each substrate material. However, the substrate material **Rogers RO4350B** outperforms the other substrate materials. The Rogers RO4350B substrate material was selected exclusively due to its suitability for the purpose of this investigation. Table 3 displays various characteristics for different substrates.

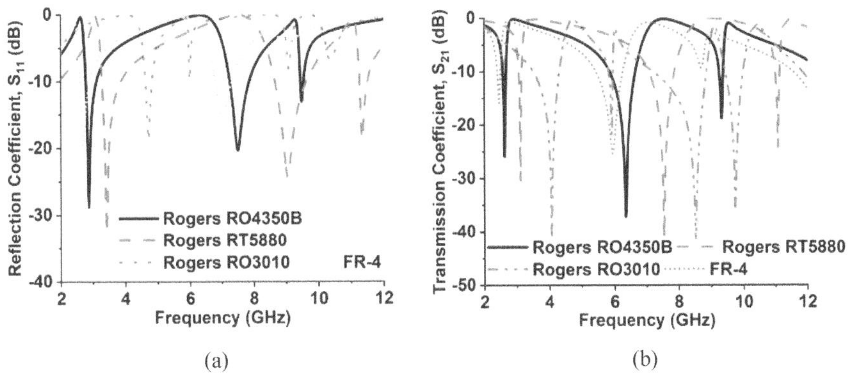

Fig. 17 Effect of change of the substrate on (a) S_{11} (b) S_{21}.

Table 3 Outcome of substrate belongings on analyzed resonant frequency f_0 (GHz), S_{21} (dB) and EMR $\left(\frac{\lambda_0}{L}\right)$.

Substrate	ε_r	*tan*	f_0*(GHz)*	S_{21} *(dB)*	$\left(\frac{\lambda_0}{L}\right)$
Rogers RT5880	2.2	0.0009	3.12,7.72, 10.98	– 30.34, – 40.67, – 25.17	12.02, 4.86, 3.42
Rogers RO3010	10.2	0.0035	4.07, 8.51, 9.74	– 40.23, – 39.94. – 36.12	9.21, 4.41, 3.85
FR-4	4.4	0.0025	2.57,6.15	– 15,32, – 25.52	14.59, 6.09
Rogers RO4350B	3.66	0.0031	2.61, 6.32, 9.29	– 26.32, – 37.5, – 19.13	14.37, 5.93, 4.04

e. Effect of the changing of strip width

We have utilized various strip widths to analyze the relationship between frequency and *S*11, as well as frequency and *S*21. By increasing the **strip width**, the inductance decreases, which in turn increases the resonance frequency. This is because the resonance frequency is inversely proportional to the inductance. The strip width has been measured at four different sizes: 0.25, 0.40, 0.50, and 0.60 mm. Upon analyzing the strip width, it is shown that **the lower resonance** frequency remains nearly constant for varied strip widths. However, the middle resonance frequency shows a tiny increase, and the last resonance frequency displays an increase. The analysis utilized a strip width of 0.25 mm, represented by a black solid line. Figure 18(a) illustrates the impact of strip width on *S*11, whereas Figure 18(b) demonstrates the influence of strip width on *S*21.

f. Effect of the changing of split gap

We have utilized various **split gap** sizes to observe the relationship between frequency and *S*11, as well as the relationship between frequency and *S*21. By raising the split gap, the capacitance decreases, resulting in an increase in the resonance frequency. This is because the resonance frequency is inversely proportional to the capacitance. The split gap has been divided into four sizes: 0.25, 0.30, 0.40, and 0.50 mm. Upon analyzing the split gap, it is evident that the lower resonance frequency remains nearly constant for varied gap sizes. However, the middle resonance frequency shows a

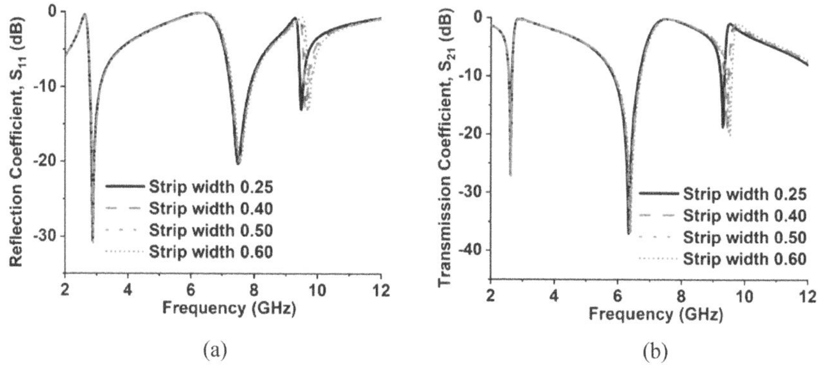

Fig. 18 Effect of the distinction of strip width: (a) frequency versus S_{11} (b) frequency versus S_{21}.

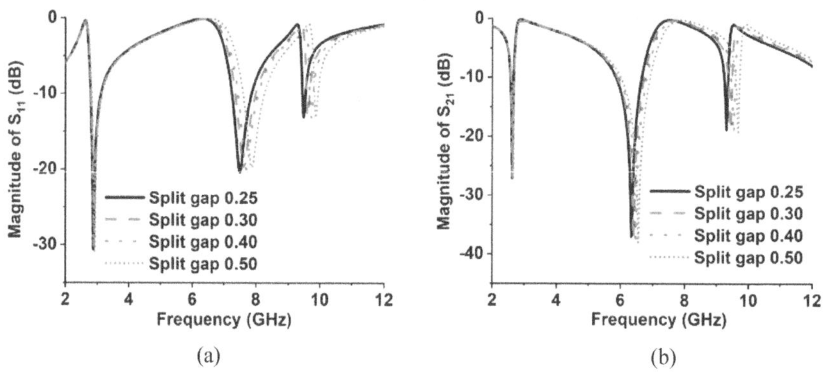

Fig. 19 Effect of the distinction of split gap: (a) frequency versus S_{11} (b) frequency versus S_{21}.

tiny increase, and the last resonance frequency experiences a greater increase. In this investigation, a split gap of 0.25 mm has been used and is represented by a black solid line. Figure 19(a) demonstrates the impact of the split gap on *S*11, whereas Figure 19(b) illustrates the effect of the split gap on *S*21.

4.6 Characterization of Metamaterial Property

Figure 20(a, b) displays the analysis of **scattering parameters** and the spectra of the phase of *S*11 and *S*21. The graph in Figure 21(a–d) represents the permeability (μ), permittivity (ε), refractive index (η), and impedance (z). Figure 20(a) illustrates that *S*21 exhibits three passbands where its value is below – 10 dB. These passbands occur within the frequency ranges of 2.55–2.67 GHz, 5.84–6.63 GHz, and 9.21–9.36 GHz, which correspond to the *S*-, *C*-, and *X*-bands. The minimum transmission coefficients of the cell are –26.32, –37.51, and – 19.13 for the resonance frequencies of 2.61 GHz, 6.32 GHz, and 9.29 GHz, respectively. The values of the reflection coefficient *S*11 are –27.53, –21.23, and – 14.27 at frequencies of 2.89 GHz, 7.57 GHz, and 9.63 GHz, respectively. Various methods, including the Lossy Dude Method, TR-Method, Nicolson Ross Weir (NRW) Method, and Robust method, are employed to extract

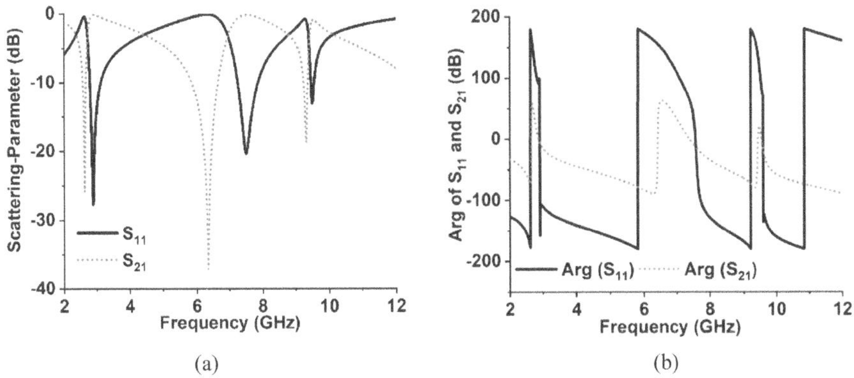

Fig. 20 (a) Reflection response (S_{11}), transmission response (S_{21}) (b) Phase of S_{11} and S_{21}.

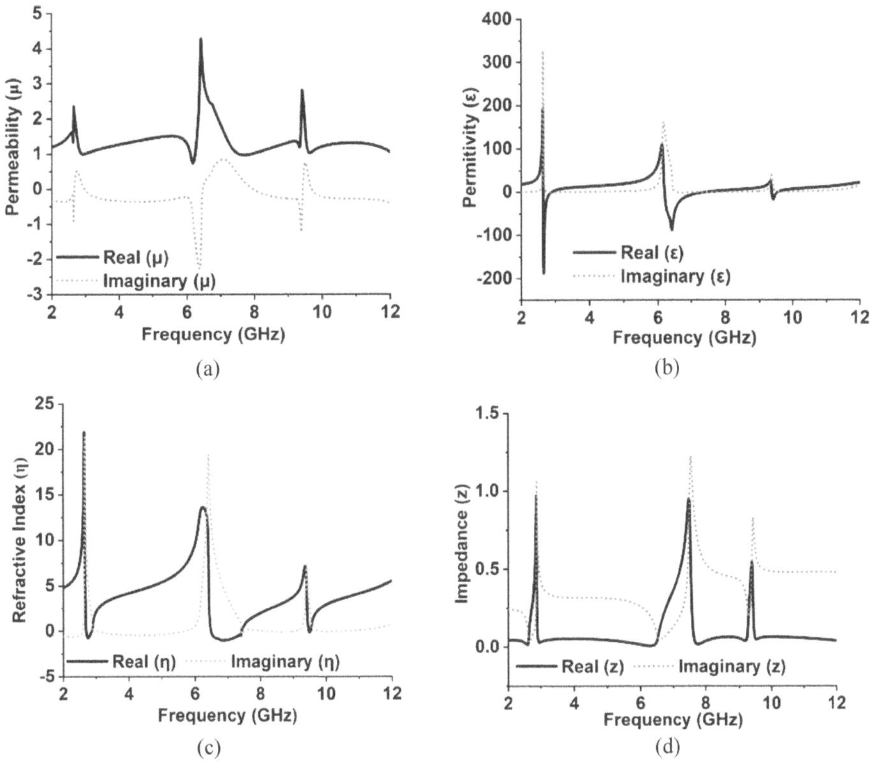

Fig. 21 Real and imaginary graph: (a) Permeability, (b) Permittivity, (c) Near zero refractive index, (d) Impedance.

the effective parameter of metamaterial. The **NRW Method** has been employed to extract the values of permeability and permittivity based on the complex *S*11 and *S*21 characteristics. At the resonance frequencies of 2.61 GHz, 6.32 GHz, and 9.29 GHz, the permittivity values are – 186.8, – 87.84, and – 17.18, respectively, as shown in Figure 21(b). The permeability values at these frequencies are 2.32, 4.28, and 2.8, respectively, as depicted in Figure 21(a). Hence, this particular design is classified as an epsilon negative (ENG), often known as a single negative metamaterial. Figure 21(c) displays a curve depicting the relationship between refractive index and frequency. At resonance frequencies of –0.73, –0.8, and –0.15, the refractive index exhibits a negative value close to zero. Figure 21(a–d) displays graphs of permittivity, permeability, refractive index, and impedance using the NRW technique in MATLAB. Table 4 provides evidence of the frequency range, **negative permittivity** magnitude, near-zero negative **refractive index**, and *S*21.

Table 4 Range of frequency region and magnitude of negative permittivity, negative refractive index, and bandwidth.

Effective Parameters	*Range of Frequency Region in GHz*	*Magnitude*
Negative permittivity	2.63–2.87, 6.39–7.45 and 9.37–9.52	– 186.8, – 87.84, – 17.18
Negative refractive index	2.66–2.74, 6.37–7.27 and 9.41–9.63	– 0.73, – 0.8, – 0.15
Transmission coefficient $\|S_{21}\|$	2.55–2.67, 5.84–6.63 and 9.21–9.36	– 26.32, – 37.51, – 19.13

4.7 Metamaterial Array Design and Performance Analysis

Figure 22(a) indicates the transmission characteristics of different configurations of the **unit cell**, including a 1 × 2 (8 × 16 mm^2) array, a 2 × 2 (16 × 16 mm^2) array, and a 4 × 4 (32 × 32 mm^2) array. A 1 × 2 array is formed by two-unit cells connected horizontally, whereas a 2 × 2 array is formed by four-unit cells connected

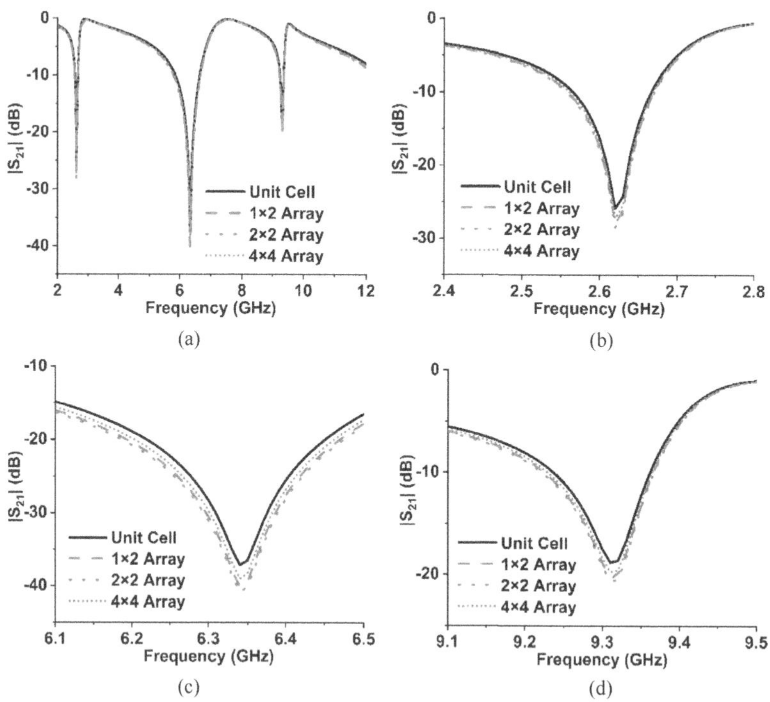

Fig. 22 Comparison among unit cell and different array configuration based on S_{21}: (a) different array, (b) 1st resonance frequency for different array, (c) 2nd resonance frequency for different array, (d) 3rd resonance frequency for different array.

both horizontally and vertically. Similarly, a 4 × 4 (32 × 32 mm^2) array consists of sixteen-unit cells connected both horizontally and vertically. Figure 22(b–d) illustrates the changes in each resonance frequency with varying **array arrangement**. The resonance frequency, transmission bandwidth, and properties of these arrays closely resemble the results obtained from the unit cell, as indicated in Table 5. The array result has been somewhat displaced due to the mutual coupling effect on the unit cell result.

Table 5 Comparison of different parameters among unit cells, 1 × 2, 2 × 2 and 4 × 4 array of the unit cell.

Parameter	*Unit Cell (GHz)*	*1×2 Array (GHz)*	*2×2 Array (GHz)*	*4×4 Array (GHz)*	*Maximum Shift of Resonance (%)*
1st Resonance and bandwidth	2.61 0.11	2.60 0.12	2.62 0.12	2.61 0.11	0.38
2nd Resonance and bandwidth	6.29 0.79	6.30 0.80	6.29 0.80	6.30 0.81	0.15
3rd Resonance and bandwidth	9.31 0.15	9.32 0.16	9.31 0.16	9.30 0.16	0.22

4.8 Experimental Results and Discussions of the Unit Cell and Array

The **experimental setup** was created by constructing the unit cell and a 35 × 20 array to examine the accuracy of the simulation outcome. Figures 23(a) and 24 depict the construction prototype and experimental arrangement of the unit cell. The unit cell experiment is performed by positioning the unit cell at the midpoint between two waveguide ports. The waveguide ports were linked to the vector network analyzer Agilent N5227A (PNA) using a coaxial cable for connection. For this study, we utilized three sets of wave guides to quantify the various resonance frequencies. The three sets of waveguides are as follows: 340WCAS with a frequency range of 2.20–3.30 GHz and a resonance frequency of 2.61 GHz, 137 WCAS with a frequency range of 5.85–8.20 GHz and a resonance frequency of 6.32 GHz, and 112WCAS with a frequency range of 7.05–10 GHz and a resonance frequency of 9.29 GHz. The simulated and experimental results of the unit cell, as shown in Figure 25, exhibit a close correspondence. Due to the measurement procedure and fabrication errors, a slight mismatch may have occurred. Figure 23(b) and Figure 25 display the constructed array prototype and the organization of the measurement procedure. The dimension of the array prototype is 280 × 160 mm^2. Two horn antennas have been placed for transmitting and receiving functions at the distance of a far field. The array of the unit cell was measured by the SAS-571 of the A.H systems which are double ridge guided horn antennas. The size of each antenna is 27.9 × 14.2 × 24.4 cm and range of the operating frequency is 700 MHz to 18 GHz. Far field distance $\geq \frac{2D^2}{\lambda}$ for an antenna, here D is the antenna dimension and is the radiation wavelength. Since 700 MHz is the lower cut off frequency and 27.9 cm is the highest dimension, the distance of the far field for the horn antenna will be greater than equal to 36.32 cm.

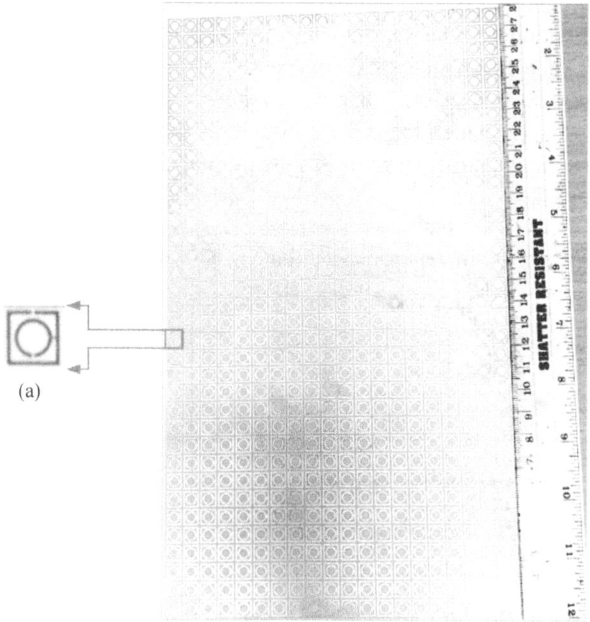

(a)

(b)

Fig. 23 (a) Fabricated metamaterial unit cell, and (b) 35 × 20 array metamaterial.

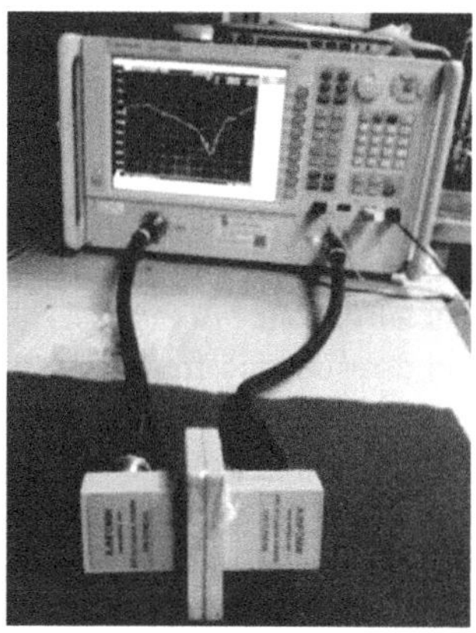

Fig. 24 Measurement setup for unit cell through the vector network analyzer and waveguide.

Because D = 27.9 cm and = 42.86 cm then $\frac{2D^2}{\lambda}$ = 36.32 cm. Consequently, two antennas located 0.40 meters apart validate the far field distance separating them. The metamaterial array is centrally located between the horn antennas, with a separation distance of 0.20 m or 20 cm from each antenna. The electromagnetic wave is being transmitted from one horn antenna on the prototype of an array and received by another horn antenna in the array. The **array prototype** has been designed to match the size of the horn antenna in order to efficiently transmit the total electromagnetic wave on the array prototype. To minimize interference from external noise, horn antennas are placed inside an **anechoic chamber** and connected to the Vector Network Analyzer (VNA) using a lengthy cable with high signal loss. The resonance frequencies of the measured unit cell are 2.63, 6.30, and 9.27 GHz, corresponding to the *S*-, *C*-, and *X*-bands, respectively. These frequencies are illustrated in Figure 26. Figure 27 illustrates the transmission response of the array measurement. The resonance frequencies have undergone slight shifts, resulting in the presence of certain harmonics. Due to a minor fabrication error and the effect of mutual coupling, there is a slight discrepancy between the measured result and the simulated result. The graph of the array measurement displays a small portion of the noise that may occur due to the use of a lossy cable connecting the horn antenna and **VNA**. Finally, the desired *S*-, *C*-, and *X*-bands were successfully achieved for the prototype array.

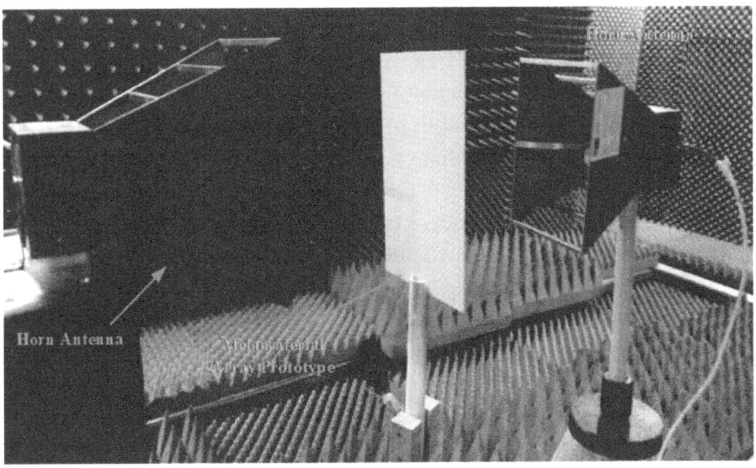

Fig. 25 Array measurement setup with two-horn antenna.

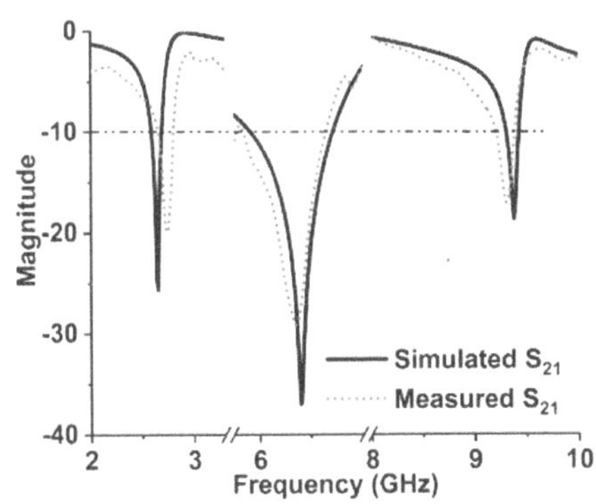

Fig. 26 Unit cell transmission coefficient of the simulated and measured results.

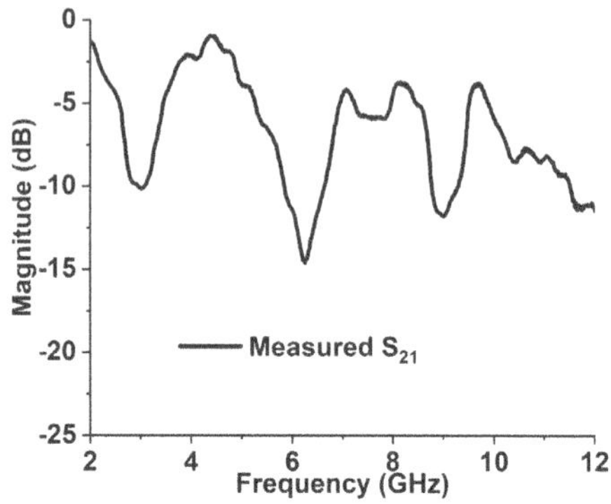

Fig. 27 Measured transmission coefficient of the metamaterial array.

4.9 EMR and Comparison Study

At present, the EMR is a crucial area of study for the metamaterials that enable its **compactness.** The formula for calculating the EMR is as follows: where it represents the wavelength and denotes the length of the optimized unit cell. The EMR values of the recommended SEC-SRR metamaterial are 14.37, 5.93, and 4.04 at the resonance frequencies of 2.61, 6.32, and 9.29 GHz, respectively. The EMR value of 14.37 enhances its uniformity and reduces its electrical size, without being constrained by any manufacturing limitations. If the EMR exceeds a value of 4, the proposed design exhibits negative permittivity and/or permeability. This increase in gain is achieved by minimizing the cost of connecting to the external field, as evidenced by the lowest transmission value [44, 73]. A performance has been compared in Table 6 where it is seen that the references [39, 41, 41, 46, 46] have the electrical dimension and resonance frequencies are high and EMR are low compared to our proposed work. From references [46, 44] it is realized that the resonance frequencies are high and EMR are lower than the mentioned SEC-SRR design though the unit cell size is equal. It is also seen that the unit cell size of the references [44, 42, [35, 54, 56] are low compared to the suggested unit cell but the lower resonance frequencies are high and the EMR are low compared to our designed unit cell. So, it can be said that the performance of the proposed unit cell design is better than the references mentioned in Table 6.

5. Integration of Designed Metamaterial with Multiband Antenna for Gain Enhancement

To apply the designed NZI epsilon negative multiband metamaterial (SEC-SRR), a printed log periodic dipole array (LPDA) of *S*-, *C*-, *X*-bands antenna is designed using CST simulation software. The design is based on existing log periodic research [76, 77] and is tailored to the required operating band. The multiband antenna is constructed using an inexpensive FR4 substrate with a dielectric constant of 4.3, as shown in Figure 28(a). The LPDA is designed with a copper top and bottom view and is excited using a 50-ohm microstrip edge feed technique. The dimensions of the **LPDA antenna** are 75.60 mm by 49 mm. The proposed MTM is now utilized

Table 6 Comparison between mentioned unit cell and some published unit cell.

References	*Year*	*Metamateri al Shape*	*Dimensions Physical (mm^2) Electrical ($\lambda\times\lambda$)*	*Resonance Frequencies (GHz)*	*Frequency Band*	*EMR*
[41]	2020	Modified CSRR	8×8 $0.076\lambda\times0.076\lambda$	2.85, 4.83, 8.39, 11.42	S-, C-, and X-bands	13.6, 7.76, 4.47, 3.28
[39]	2019	Inverse Double L	10×10 $0.26\lambda\times0.26\lambda$	7.69, 8.46, 13.12, 14.03	C-, X-, and Ku-bands	3.9, 3.55, 2.89, 2.14
[40]	2019	CSRR	5.5×5.5 $0.14\lambda\times0.14\lambda$	7.44	C-band	7.33
[74]	2019	RLC	10×8 $0.29\lambda\times0.24\lambda$	8.84	X-band	3.39
[42]	2020	CSR	8×8 $0.15\lambda\times0.15\lambda$	5.80, 16.61	C-, Ku-bands	6.47,
[44]	2020	Hexagonal	10×10 $0.12\lambda\times0.182\lambda$	3.56, 11.67	S-, X-bands	8.4, 2.57
[75]	2018	Ring	20×20 $0.76\lambda\times0.76\lambda$	11.5, 13.5	X-, Ku-bands	1.3, 1.11
[48]	2015	Modified H	9×9 $0.29\lambda\times0.29\lambda$	9.52, 12.44, 14.94	X-, Ku-bands	3.5, 2.68, 2.23
[32]	2012	Double S	3.65×3.65 $0.29\lambda\times0.29\lambda$	15.68	Ku-band	5.25
[52]	2012	Butterfly pattern resonator	6×8 $0.16\lambda\times0.21\lambda$	7.97, 10.31, 12.08	C-, X-, Ku-bands	4.7, 3.63, 3.10
[55]	2012	Nested U-ring resonator	6.5×6.5 $0.17\lambda\times0.17\lambda$	8.24, 9.86, 12.42	X-, Ku-bands	5.6, 4.68, 3.71
[43]	2015	Greek-key pattern	10×10 $0.08\lambda\times0.08\lambda$	2.4, 3.5, 4	S-, C-bands	12.5, 8.57, 7.5
Propos-ed	2020	SEC-SRR	8×8 $0.070\lambda\times0.070\lambda$	2.61, 6.32, 9.29	S-, C-, X-bands	14.37, 5.93, 4.04

to optimize the gain parameter in the application of the proposed SEC-SRR MTM, thereby improving its performance. Due to its smaller size, the designed MTM unit cell is unable to fully encompass the LPDA antenna. A 9 × 6 metamaterial array unit cell is specifically designed to completely cover the LPDA antenna. The 9 × 6 MTM array is positioned above the LPDA antenna, as depicted in Figure 28(b), with an optimized distance of 20 mm, as illustrated in Figure 28(c). Upon finishing the design in CST software, a simulation was conducted to evaluate the performance of the antenna parameter with and without the implementation of the metamaterial array. Figure 29(a) displays the reflection coefficient outcome of the LPDA antenna with and without metamaterial. The graph exhibits multiband operating characteristics with a –10 dB magnitude. However, Figure 29(b) illustrates the actual increase in performance of the LPDA antenna when combined with the SEC-SRR 9 × 6 metamaterial array cover. The observed gain array has experienced a substantial increase, particularly in the higher *C* and *X* band resonances, when compared to using only the LPDA antenna. At a frequency of 2.61 GHz, the gain has experienced a slight increase of 6.30 dB compared to the previous value of 6.15 dB. However, at frequencies of 6.32 GHz and 9.29 GHz, the realized gain has increased by 1.3 dB (from 3.25–4.60 dB) and

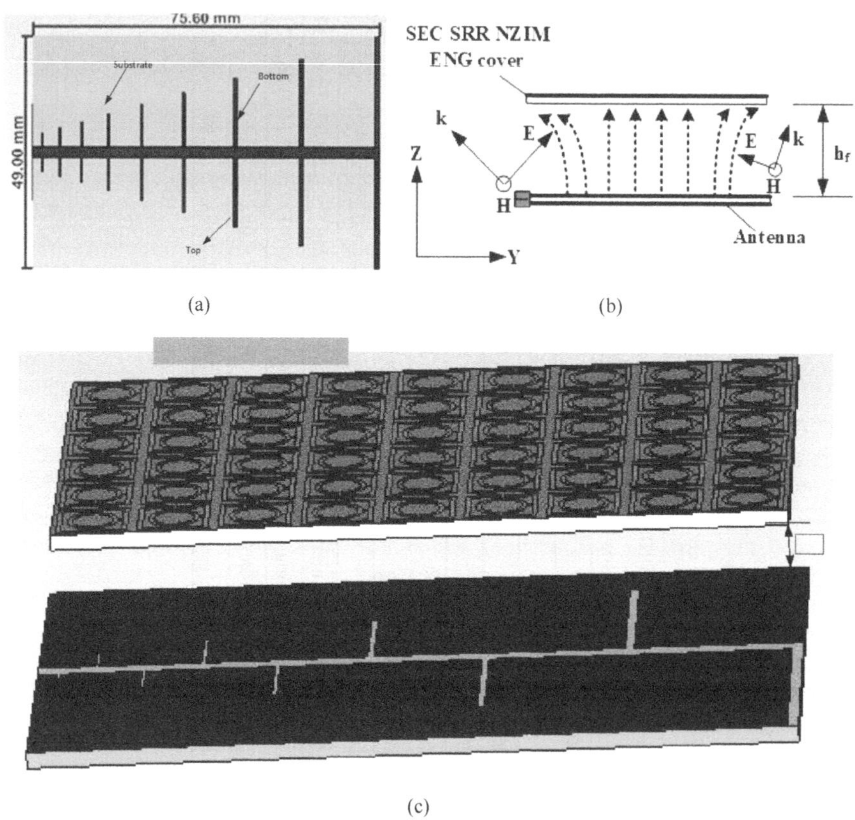

Fig. 28 Geometric layout of the (a) LPDA antenna, and (b) Schematic structure of antenna with SEC-SRR Metamaterial, (c) LPDA antenna with proposed metamaterial array.

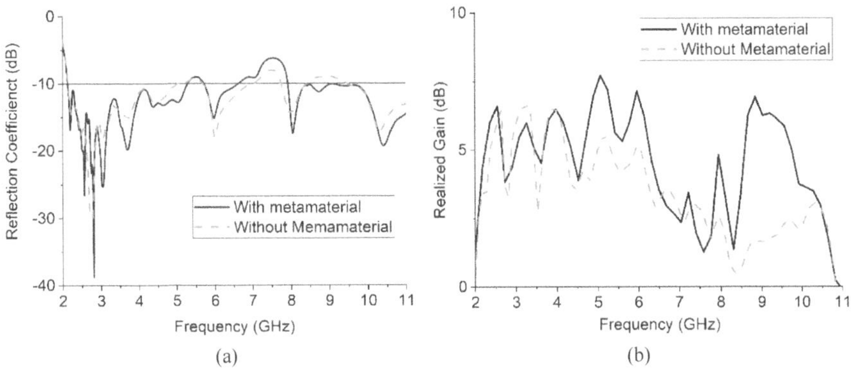

Fig. 29 (a) Reflection coefficient, (b) realized gain of the LPDA antenna and with metamaterial array.

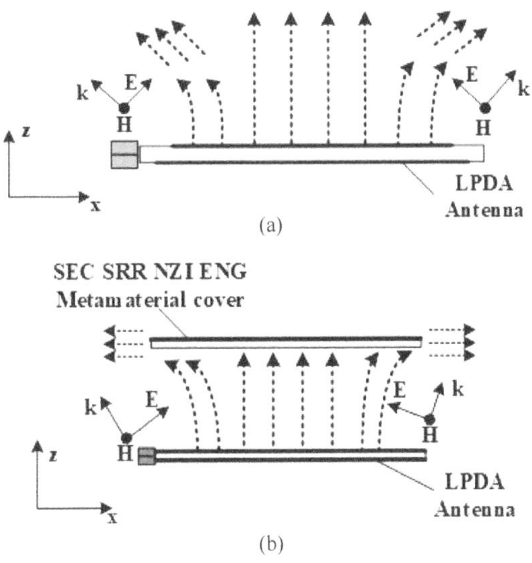

Fig. 30 Radiation schematic diagram of: (a) LPDA antenna only, (b) LPDA Antenna with NZI SNG metamaterial cover.

4.47 dB (from 1.84–6.31 dB), respectively. The method by which the LPDA antenna achieves increased gain through the use of NZI ENG metamaterial can be elucidated by referring to Figure 29(a, b). When the sideward EM wave hits the NZI ENG cover, the sideward radiation power has an EM field E component that is perpendicular to the cover (along the Z-axis). The **NZI ENG mechanism** causes the emission of EM waves from the NZI cover in a direction that is aligned with the interfaces. As a result, certain EM waves that pass through the metamaterial cover become parallel to the XY plane. Due to this, the amplified EM waves align with the LPDA's horizontal plane, causing a shift in the direction of the primary beam of the LPDA, resulting in an increase in gain.

6. Summary

This book chapter explores the application of near-zero index metamaterials to amplify the **gain of a multiband patch antenna**. The chapter commences by presenting the notion of metamaterials, which are man-made structures designed to display distinctive electromagnetic characteristics that are absent in natural materials. More specifically, near-zero index metamaterials are emphasized for their capacity to **control electromagnetic waves** in innovative manners. The chapter further explores the design and execution of the suggested metamaterial structure with the aim of amplifying the gain of a multiband patch antenna. The metamaterial structure is deliberately incorporated into the LPDA antenna system to produce an increase in gain across many frequency bands. Thorough study and simulations are performed to optimize the design parameters of the metamaterial and guarantee that it is **compatible with the patch antenna**. In summary, the chapter demonstrates how near-zero index metamaterials have the ability to greatly improve the gain and overall performance of multiband patch antennas. **Metamaterials integration** provides novel opportunities for advancement in antenna design and wireless communication systems.

References

[1] I. Vendik and O. Vendik, "Metamaterials and their application in microwaves: A review," *Technical Physics,* vol. 58, no. 1, pp. 1–24, 2013.

[2] D.R. Smith, W.J. Padilla, D. Vier, S.C. Nemat-Nasser and S. Schultz. "Composite medium with simultaneously negative permeability and permittivity," *Physical Review Letters,* vol. 84, no. 18, p. 4184, 2000.

[3] M. Hussain, W.A. Awan, M.S. Alzaidi, N. Hussain, E.M. Ali and F. Falcone. "Metamaterials and their application in the performance enhancement of reconfigurable antennas: A review," *Micromachines,* vol. 14, no. 2, p. 349, 2023.

[4] K. Zhang, P.J. Soh and S. Yan. "Meta-wearable antennas: A review of metamaterial based antennas in wireless body area networks," *Materials,* vol. 14, no. 1, p. 149, 2020.

[5] T. Shaw and D. Mitra. "Design of highly directive slot antenna using metamaterial structure under optimal condition," *International Journal of RF and Microwave Computer-Aided Engineering,* vol. 31, no. 9, p. e22791, 2021.

[6] Y.-H. Ke, L.-L. Yang, Y.-Y. Zhu, J. Wang and J.-X. Chen. "Filtering quasi-yagi strip-loaded DRR antenna with enhanced gain and selectivity by metamaterial," *IEEE Access,* vol. 9, pp. 31755–31761, 2021.

[7] K. Sumathi, S. Lavadiya, P. Yin, J. Parmar and S.K. Patel. "High gain multiband and frequency reconfigurable metamaterial superstrate microstrip patch antenna for C/X/Ku-band wireless network applications," *Wireless Networks,* vol. 27, pp. 2131–2146, 2021.

[8] M. Moniruzzaman, M.T. Islam, N. Misran, M. Samsuzzaman, T. Alam and M.E. Chowdhury. "Inductively tuned modified split ring resonator based quad band epsilon negative (ENG) with near zero index (NZI) metamaterial for multiband antenna performance enhancement," *Sci. Rep.,* vol. 11, no. 1, p. 11950, 2021.

[9] A.H. Jabire et al. "Metamaterial based design of compact UWB/MIMO monopoles antenna with characteristic mode analysis," *Applied Sciences,* vol. 11, no. 4, p. 1542, 2021.

[10] S. Das, A. Gupta and S. Sahu. "Metamaterial based fractal-ground loaded frequency-reconfigurable monopole-antenna with gain-bandwidth enhancement," *AEU-International Journal of Electronics and Communications,* vol. 132, p. 153593, 2021.

[11] M. Ameen and R. Chaudhary. "Metamaterial circularly polarized antennas: Integrating an epsilon negative transmission line and single split ring-type resonator," *IEEE Antennas and Propagation Magazine,* vol. 63, no. 4, pp. 60–77, 2020.

[12] P. Mishra, S.S. Pattnaik and B.S. Dhaliwal. "Square-shaped fractal antenna under metamaterial loaded condition for bandwidth enhancement," *Progress in Electromagnetics Research C,* vol. 78, pp. 183–192, 2017.

[13] H. Sakli, C. Abdelhamid, C. Essid and N. Sakli. "Metamaterial-based antenna performance enhancement for MIMO system applications," *IEEE Access,* vol. 9, pp. 38546–38556, 2021.

[14] D.A. Sehrai, F. Muhammad, S.H. Kiani, Z.H. Abbas, M. Tufail and S. Kim. "Gain-enhanced metamaterial based antenna for 5G communication standards," *CMC Comput. Mater. Contin,* vol. 64, pp. 1587–1599, 2020.

[15] H. Jeong, Y. Kim, M.M. Tentzeris and S. Lim. "Gain-enhanced metamaterial absorber-loaded monopole antenna for reduced radar cross-section and back radiation," *Materials,* vol. 13, no. 5, p. 1247, 2020.

[16] B. Zarghooni, A. Dadgarpour and T.A. Denidni. "Millimeter-wave antenna using two-sectioned metamaterial medium," *IEEE Antennas and Wireless Propagation Letters,* vol. 15, pp. 960–963, 2015.

[17] K.N. Olan-Nuñez and R.S. Murphy-Arteaga. "A novel metamaterial-based antenna for on-chip applications for the 72.5–81 GHz frequency range," *Scientific Reports,* vol. 12, no. 1, p. 1699, 2022.

[18] F. Khajeh-Khalili, M.A. Honarvar, M. Naser-Moghadasi and M. Dolatshahi. "High-gain, high-isolation, and wideband millimetre-wave closely spaced multiple-input multiple-output antenna with metamaterial wall and metamaterial superstrate for 5G applications," *IET Microwaves, Antennas, & Propagation,* vol. 15, no. 4, pp. 379–388, 2021.

[19] M. Islam, M.T. Islam, M. Samsuzzaman and M.R.I. Faruque. "Compact metamaterial antenna for UWB applications," *Electron. Lett.,* vol. 51, no. 16, pp. 1222–1224, 2015.

[20] S. il Kwak, D.-U. Sim, J.H. Kwon and Y.J. Yoon. "Design of PIFA with metamaterials for body-SAR reduction in wearable applications," *IEEE Transactions on Electromagnetic Compatibility,* vol. 59, no. 1, pp. 297–300, 2016.

[21] M. Chen, H. Jiang, H. Zhang, D. Li and Y. Wang. "Design of an acoustic superlens using single-phase metamaterials with a star-shaped lattice structure," *Scientific Reports,* vol. 8, no. 1, p. 1861, 2018.

[22] Z. Zheng, W. Wang, H.-T. Zhang, Y.-Y. Zheng and X.-L. Liang. "Dual-Band Anti-Interference Slot Antenna Using Metamaterial Structure," in *2019 IEEE International Symposium on Antennas and Propagation and USNC-URSI Radio Science Meeting,* 2019: IEEE, pp. 327–328.

[23] G. Zhao, S. Bi, M. Niu and Y. Cui. "A zero refraction metamaterial and its application in electromagnetic stealth cloak," *Materials Today Communications,* vol. 21, p. 100603, 2019.

[24] M. Baqir and P. Choudhury. "Hyperbolic metamaterial-based UV absorber," *IEEE Photonics Technology Letters,* vol. 29, no. 18, pp. 1548–1551, 2017.

[25] W. Liu, H. Sun and L. Xu. "A microwave method for dielectric characterization measurement of small liquids using a metamaterial-based sensor," *Sensors,* vol. 18, no. 5, p. 1438, 2018.

[26] N. Misran, S.H. Yusop, M.T. Islam and M.Y. Ismail. "Analysis of Parameterization Substrate Thickness and Permittivity for Concentric Split Ring Square Reflectarray Element," *Jurnal Kejuruteraan (Journal of Engineering),* vol. 23, pp. 11–16, 2012.

[27] Y. Liu and X. Zhao. "Perfect absorber metamaterial for designing low-RCS patch antenna," *IEEE Antennas and Wireless Propagation Letters,* vol. 13, pp. 1473–1476, 2014.

[28] J.B. Pendry, A. Holden, W. Stewart and I. Youngs. "Extremely low frequency plasmons in metallic mesostructures," *Physical Review Letters,* vol. 76, no. 25, p. 4773, 1996.

[29] S. Enoch, G. Tayeb, P. Sabouroux, N. Guérin and P. Vincent. "A metamaterial for directive emission," *Physical Review Letters,* vol. 89, no. 21, p. 213902, 2002.

[30] Y. Guo, J. Zhao, Q. Hou and X. Zhao. "Broadband omnidirectional patch antenna with horizontal gain enhanced by near-zero-index metamaterial cover," *IET Microwaves, Antennas, & Propagation,* vol. 14, no. 7, pp. 671–676, 2020.

[31] E. Ekmekçi, and G. Turhan-Sayan. "Investigation of effective permittivity and permeability for a novel v-shaped metamaterial using simulated s-parameters," in *5th International Conference on Electrical and Electronics Engineering*, 2007: sn.

[32] H. Benosman and N.B. Hacene. "Design and simulation of double 's' shaped metamaterial," *International Journal of Computer Science Issues (IJCSI),* vol. 9, no. 2, p. 534, 2012.

[33] S.S. Islam, M.R.I. Faruque and M.T. Islam. "Design of a new ENG metamaterial for S-band microwave applications," *Journal of Electrical and Electronics Engineering,* vol. 7, no. 2, p. 13, 2014.

[34] M.M. Hasan, M.R.I. Faruque, M.M. Islam and M.T. Islam. "Multi band negative index metamaterial based on split square shape resonators," in *2016 19th International Conference on Computer and Information Technology (ICCIT),* 2016: IEEE, pp. 99–102.

[35] W. Withayachumnankul, C. Fumeaux and D. Abbott. "Compact electric-LC resonators for metamaterials," *Optics Express,* vol. 18, no. 25, pp. 25912–25921, 2010.

[36] D. Marathe and K. Kulat. "A compact triple-band negative permittivity metamaterial for C, X-band applications," *International Journal of Antennas and Propagation,* vol. 2017, 2017.

[37] F.O. Alkurt et al. "Antenna-based microwave absorber for imaging in the frequencies of 1.8, 2.45, and 5.8 GHz," *Optical Engineering,* vol. 57, no. 11, p. 113102, 2018.

[38] M. Bakır et al. "Metamaterial-based energy harvesting for GSM and satellite communication frequency bands," *Optical Engineering,* vol. 57, no. 8, p. 087110, 2018.

[39] A.M. Tamim, M.R.I. Faruque, M.J. Alam, S.S. Islam and M.T. Islam. "Split ring resonator loaded horizontally inverse double *L*-shaped metamaterial for *C*-, *X*- and Ku-Band Microwave applications," *Results in Physics,* vol. 12, pp. 2112–2122, 2019.

[40] A.F. Almutairi, M.S. Islam, M. Samsuzzaman, M.T. Islam, N. Misran and M.T. Islam. "A complementary split ring resonator based metamaterial with effective medium ratio for C-band microwave applications," *Results in Physics,* vol. 15, p. 102675, 2019.

[41] M.R. Islam, M. Samsuzzaman, N. Misran, G.K. Beng and M.T. Islam. "A tri-band left-handed meta-atom enabled designed with high effective medium ratio for microwave based applications," *Results in Physics,* p. 103032, 2020.

[42] T. Ramachandran, M.R.I. Faruque and M.T. Islam. "A dual band left-handed metamaterial-enabled design for satellite applications," *Results in Physics,* vol. 16, p. 102942, 2020.

[43] B. Zarghooni, A. Dadgarpour and T.A. Denidni. "Greek-key pattern as a miniaturized multiband metamaterial unit-cell," *IEEE Antennas and Wireless Propagation Letters,* vol. 14, pp. 1254–1257, 2015.

[44] M.S. Islam, M. Samsuzzaman, G.K. Beng, N. Misran, N. Amin and M.T. Islam. "A Gap Coupled Hexagonal Split Ring Resonator Based Metamaterial for S-Band and X-Band Microwave Applications," *IEEE Access,* vol. 8, pp. 68239–68253, 2020.

[45] M. Rezapour, J.A. Rashed-Mohassel, A. Keshtkar and M.-N. Moghadasi. "Suppression of mutual coupling in rectangular dielectric resonator antenna arrays using Epsilon-Negative metamaterials (ENG)," *Journal of Electromagnetic Waves and Applications,* vol. 33, no. 9, pp. 1211–1223, 2019.

[46] M. Hasan, M.R.I. Faruque, S.S. Islam and M.T. Islam. "A new compact double-negative miniaturized metamaterial for wideband operation," *Materials,* vol. 9, no. 10, p. 830, 2016.

[47] T. Ramachandran, M.R.I. Faruque, E. Ahamed and S. Abdullah. "Specific absorption rate reduction of multi split square ring metamaterial for *L*-and *S*-band application," *Results in Physics,* vol. 15, p. 102668, 2019.

[48] T.M. Hossain, M.F. Jamlos, M.A. Jamlos, P.J. Soh, M.I. Islam and R. Khan. "Modified H-shaped DNG metamaterial for multiband microwave application," *Applied Physics A,* vol. 124, no. 2, p. 183, 2018.

[49] F.O. Alkurt et al. "Enhancement of image quality by using metamaterial inspired energy harvester," *Physics Letters A,* vol. 384, no. 1, p. 126041, 2020.

[50] O. Derın, M. Karaaslan, E. Ünal, F. Karadağ, O. Altintaş and O. Akgöl. "Exhibition of polarization conversions with asymmetric transmission theory, natural like chiral, artificial chiral nihility and retrieval studies for *K*-and *C*-band radar applications," *Bulletin of Materials Science,* vol. 42, no. 4, p. 191, 2019.

[51] Z. Zhou and H. Yang/"Triple-band asymmetric transmission of linear polarization with deformed S-shape bilayer chiral metamaterial," *Applied Physics A,* vol. 119, no. 1, pp. 115–119, 2015.
[52] G. Du and C. Liu. "Multiband metamaterial structure: Butterfly-pattern resonator," *Microwave and Optical Technology Letters,* vol. 54, no. 9, pp. 2179–2181, 2012.
[53] F.O. Alkurt and M. Karaaslan. "Characterization of tunable electromagnetic band gap material with disordered cavity resonator for X band imaging applications by resistive devices," *Optical and Quantum Electronics,* vol. 51, no. 8, p. 279, 2019.
[54] M. Hasan, M.R.I. Faruque and M.T. Islam. "Parametric studies on split S-shaped composite meta atom for X-band communication," *Bulletin of the Polish Academy of Sciences. Technical Sciences,* vol. 65, no. 4, 2017.
[55] O. Turkmen, E. Ekmekci and G. Turhan-Sayan. "Nested U-ring resonators: A novel multi-band metamaterial design in microwave region," *IET Microwaves, Antennas, & Propagation,* vol. 6, no. 10, pp. 1102–1108, 2012.
[56] S.-H. Liu, L.-X. Guo and J.-C. Li. "Left-handed metamaterials based on only modified circular electric resonators," *Journal of Modern Optics,* vol. 63, no. 21, pp. 2220–2225, 2016.
[57] P. Dawar and A. De. "Bandwidth enhancement of RMPA using ENG metamaterials at THz," in *2013 4th International Conference on Computer and Communication Technology (ICCCT),* 2013: IEEE, pp. 11–16.
[58] M. Bouzouad, S. Chaker, D. Bensafielddine and E. Laamari. "Gain enhancement with near-zero-index metamaterial superstrate," *Applied Physics A,* vol. 121, no. 3, pp. 1075–1080, 2015.
[59] A. Erentok and R.W. Ziolkowski. "Metamaterial-inspired efficient electrically small antennas," *IEEE Trans. Antennas Propag.,* vol. 56, no. 3, pp. 691–707, 2008.
[60] N.T. Selvi, P.T. Selvan, S. Babu and R. Pandeeswari. "Multiband metamaterial-inspired antenna using split ring resonator," *Computers & Electrical Engineering,* vol. 84, p. 106613, 2020.
[61] M. Moniruzzaman et al. "Gap coupled symmetric split ring resonator based near zero index ENG metamaterial for gain improvement of monopole antenna," *Scientific Reports,* vol. 12, no. 1, p. 7406, 2022.
[62] B.A. Esmail and S. Koziel. "Design and optimization of metamaterial-based highly-isolated MIMO antenna with high gain and beam tilting ability for 5G millimeter wave applications," *Scientific Reports,* vol. 14, no. 1, p. 3203, 2024.
[63] M. El Ghzaoui, J. Belkadid and A. Benbassou. "Near zero index Metamaterial-based SIW antenna for 6G Sub-Terahertz applications," *Results in Optics,* vol. 12, p. 100468, 2023/07/01/ 2023, doi: https://doi.org/10.1016/j.rio.2023.100468.
[64] P. Kumar, T. Ali and M.M. Pai. "Electromagnetic metamaterials: A new paradigm of antenna design," *IEEE Access,* vol. 9, pp. 18722–18751, 2021.
[65] I. Liberal and N. Engheta. "Near-zero-index metastructures," in *Plasmonic Materials and Metastructures*: Elsevier, 2024, pp. 197–226.
[66] R. Giri and R. Payal. "Negative-Index Metamaterials," *Electromagnetic Metamaterials: Properties and Applications,* pp. 205–217, 2023.
[67] S.S. Al-Bawri, M.T. Islam, T. Shabbir, G. Muhammad, M.S. Islam and H.Y. Wong. "Hexagonal shaped near zero index (NZI) metamaterial-based MIMO antenna for millimeter-wave application," *IEEE Access,* vol. 8, pp. 181003–181013, 2020.
[68] B. Farooq, T. Parveen, I. Ali and S. Ullah. "Antenna design for advance wireless systems using metamaterial surfaces," in *2016 13th International Bhurban Conference on Applied Sciences and Technology (IBCAST),* 12–16 Jan. 2016, 2016, pp. 641–646, doi: 10.1109/IBCAST.2016.7429947.
[69] M.M. Bait-Suwailam. "Electromagnetic Field Interaction with Metamaterials," in *Electromagnetic Fields and Waves*: IntechOpen, 2019.
[70] R.W. Ziolkowski. "Design, fabrication, and testing of double negative metamaterials," *IEEE Transactions on Antennas and Propagation,* vol. 51, no. 7, pp. 1516–1529, 2003.
[71] P. Gay-Balmaz and O.J. Martin. "Electromagnetic resonances in individual and coupled split-ring resonators," *J. Appl. Phys.,* vol. 92, no. 5, pp. 2929–2936, 2002.
[72] C.R. Paul. *Inductance: Loop and Partial.* John Wiley & Sons, 2011.

[73] M.J. Hossain, M.R.I. Faruque and M.T. Islam. "Design and analysis of a new composite double negative metamaterial for multi-band communication," *Current Applied Physics,* vol. 17, no. 7, pp. 931–939, 2017.

[74] E. Ahamed, M.R.I. Faruque, M.F.B. Mansor and M.T. Islam. "Polarization-dependent tunneled metamaterial structure with enhanced fields properties for X-band application," *Results in Physics,* vol. 15, p. 102530, 2019.

[75] A. Hoque, M. Tariqul Islam, A.F. Almutairi, T. Alam, M. Jit Singh and N. Amin. "A polarization independent quasi-TEM metamaterial absorber for X and Ku band sensing applications," *Sensors,* vol. 18, no. 12, p. 4209, 2018.

CHAPTER 6

Isolation Enhancement of Wideband MIMO Antenna Using T-structured Metamaterial Shield

Md. Mhedi Hasan[1] and *Mohammad Tariqul Islam*[2]

1. Introduction

Multiple-input multiple-output (MIMO) antennas are crucial for fifth-generation (5G) communications that enhance data transmission rate, spectrum utilization, channel quality, and reliability. MIMO antenna provides ultra-fast data transmission and reception concurrently through multipath propagation than a single antenna. It also mitigates the multiple fading and achieves a high information rate, enhanced link capacity and enhanced coverage without wasting more spectrum and power [1–3]. Due to the fastest-growing demand for compact high-speed wideband 5G MIMO antennas, the design and integration of multiple antennas in a compact MIMO system are very challenging [1, 4–6]. Moreover, the number of MIMO elements and simple antenna geometry are highly desirable to enhance the MIMO antenna elements within the limited space. On the other side, integrating multiple antenna components into a limited space results in significant mutual coupling between antenna components, which degrades overall antenna performance. The mitigation of the coupling effect among the adjacent antenna elements is a crucial challenge in MIMO technology [7]. In this chapter, a new T-structured metamaterial (MM) shield incorporated isolation technique is developed in a 5G MIMO system. A highly isolated MM-integrated compact wideband MIMO antenna is designed for 5G frequency range 1 (FR1) **sub-6 GHz** applications. The design target of the proposed MIMO antenna is to develop a highly isolated **5G sub-6 GHz** compact MM-integrated MIMO antenna. Therefore, a new symmetric T-structured MM shield is designed to enhance isolation between two elements of a MIMO antenna useful for 5G **n79 band** applications. The novelty of the designed MIMO antenna is its unique symmetric T-shaped MM shield-embedded

[1,2] Department of Electrical, Electronic and Systems Engineering, Faculty of Engineering and Built Environment, Universiti Kebangsaan Malaysia (UKM), 43600 UKM Bangi, Selangor, Malaysia.
* Corresponding author: tariqul@ukm.edu.my

compact two-element MIMO structure, very high enhanced isolation of >29 dB, and an enhanced very low ECC. The hanging coupled T-structured MM shield is employed between a mic-slotted two-element MIMO antenna, enhancing isolation and operating spectrum as well as reducing ECC substantially which confirms high isolation and channel capacity. The electric fields are also explored to elucidate the isolation strategy with MM shield and contribution to 5G MIMO antenna technology. The diversity performance of the developed MM-integrated MIMO antenna is investigated and found excellent performance within acceptable limits. The MIMO antenna performance is analyzed by EM simulator software CST microwave studio and verified by a measurement system with good agreement. Finally, the designed MM-integrated MIMO antenna is compared with state-of-the-art designs to prove their suitability for 5G sub-6 GHz applications.

2. Wide Stopband Metamaterial Design

2.1 Design Analysis

In this chapter, a hanging coupled T-shaped symmetric style **MM** with epsilon negative (ENG), and Mu-near-zero (MNZ) characteristics is proposed to minimize the electromagnetic (EM) mutual coupling among MIMO components. The proposed hanging coupled **T-shaped symmetric metamaterial** (CTS-MM) initial design target is to achieve a compact dimension and **wide band-stop** characteristics within the desired 5G frequency range 1 (FR1) sub-6 GHz spectrums. Hence, the initial estimated size of the proposed MM structure is 9 mm, which is opted at the lower 5G spectrum of the n79 band, obeying the sub-wavelength standard. Therefore, the promising characteristics of the developed MM can be reached [8]. The proposed symmetric MM resonator is engineered into four similar unique quartiles with coupled hanging T-structured strips, as depicted in Figure 1(a). Each quartile is composed of a split square ring encased hanging strips. The proposed symmetric MM has a miniaturized dimension of $0.12\lambda_{min} \times 0.12\lambda_{min} \times 0.02\lambda_{min}$ at 4.14 GHz and is designed using inexpensive FR-4 epoxy substrate material ($\varepsilon_r = 4.3$ and height = 1.6 mm). Figure 1(a) also exhibits the complete metamaterial structure design details with initial to proposed configurations.

The proposed symmetric resonator is developed and investigated utilizing an EM simulator, CST. During the simulation, the electric field and magnetic field are applied to the x-axis and y-axis, respectively, whereas the excitation ports are directed to the z-axis. Initially, four split gaps integrated square ring is introduced on the top side of the FR-4 substrate and the back side is non-copper, as depicted in step 1 (Figure 1(a)). In this configuration, no resonance exists in the desired 5G spectrum, which is depicted in Figure 1(b). A plus-shaped connecting strip is incorporated in the middle of the ring in the next stage, which looks like a connected four T-structured geometry, resulting in resonance at 5.06 GHz, in Figure 1(a, b). Finally, two copper strips are hung in the left and upper arms of each quartile of step 2 configuration, forming the targeted hanging coupled T-shaped architecture with a desired performance, in Figure 1(a, b). The incorporation of hanging metal strips shifted the resonance to 4.75 GHz, covering the expected application area of the 5G n79 band. Figure 2(a) shows the designed

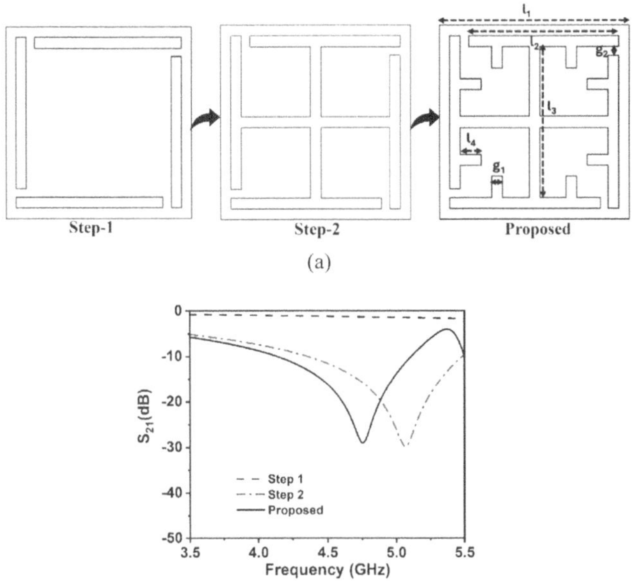

Fig. 1 (a) MM unit cell configuration with design specifications, (b) Simulated responses (S_{21}) for initial to proposed configurations.

1 × 2 **array** configuration and Figure 2(b) depicts the simulated **reflection** (S_{11}) and **transmission** (S_{21}) characteristics of the developed MM unit cell and 1 × 2 array structure. The S_{21} reveals that the designed MM structure has an outstanding wide – 10 dB bandwidth from 4.14–5.1 GHz, resulting in wide reflection characteristics [9, 10]. Moreover, the transmission characteristics indicated in Figure 2(b) confirm excellent wide stop-band properties from 4.14–5.1 GHz (– 10 dB), covering the 5G n79 band. Table 1 shows the CTS-MM design parameters. Therefore, the wide linear high reflection properties (S_{11}) of the developed MM structure can help in reducing EM coupling energy in MIMO antenna technology by applying between MIMO elements [10–13].

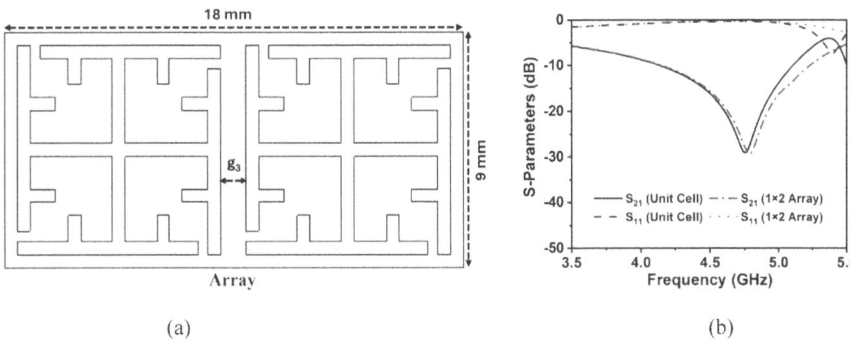

Fig. 2 (a) Array configuration with design specifications, (b) Simulated S_{21} and S_{11} for array and unit cell configurations.

Table 1 Design details of developed MM configuration.

Parameters	*Size (mm)*	*Parameters*	*Size (mm)*
g_1	0.5	l_2	7.1
g_2	0.4	l_3	7.0
g_3	1	l_4	1.0
l_1	9	–	–

2.2 Parametric Analysis

A **parametric investigation** on the outer 4-corner split gaps (g_2) is presented in Figure 3 to optimize the MM architecture for the desired 5G spectrum. The change of split gap controls the MM characteristics by obeying the frequency and capacitance relationships. Figure 3 depicts the effect of corner four split gaps on the scattering performance of the developed MM. The four split gaps change from 0.2 – 1 mm with 0.2 mm steps, which shifts the S_{21} resonance to a higher frequency by increasing the gap width. The final optimized g_2 value is chosen as 0.4 mm for the desired frequency application, which nicely covers the 5G n79 band.

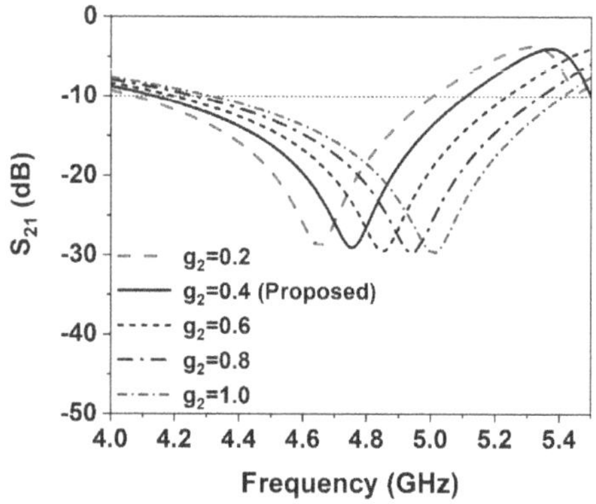

Fig. 3 MM parametric study with four split gap width (g_2) changes.

2.3 Metamaterial Properties

The designed CTS-MM effective parameters (permittivity, permeability, and refractive index) have been investigated to further explore the **MM characteristics** in depth. The simulation platform CST integrated post-processing module can be used to extract the MM effective characteristics from the scattering parameters. The post-processing template uses a robust retrieval technique, and the parameters are determined with the following equations [14].

Reflection coefficient,

$$S_{11} = \frac{R_{01}(1 - e^{i2nk_0 d})}{1 - R_{01}^2 e^{i2nk_0 d}} \quad ...(1)$$

Transmission coefficient,

$$S_{21} = \frac{(1 - R_{01}^2)\, e^{ink_0d}}{1 - R_{01}^2\, e^{i\,2nk_0d}} \qquad ...(2)$$

where, the thickness of the substrate and free space wave number are denoted by d and k_0, respectively.

$$R_{01} = \frac{z - 1}{z + 1} \qquad ...(3)$$

From expressions (1) and (2):

Impedance, $$z = \pm\sqrt{\frac{(1 + S_{11})^2 - S_{21}^2}{(1 - S_{11})^2 - S_{21}^2}} \qquad ...(4)$$

where, $e^{ink_0d} = X \pm i\sqrt{1 - X^2}$ and $X = 1/2S_{21}\,(1 - S_{11}^2 + S_{21}^2)$.

Refractive index,

$$n = \frac{1}{K_0d}\left\{\left[[\ln(e^{ink_0d})]'' + 2m\pi\right] - i\left[\ln(e^{ink_0d})\right]'\right\} \qquad ...(5)$$

where, (.)′ and (.)″ indicate the operator's real and imaginary parts, respectively, and m indicates an integer associated with the branch index of n'. The effective **permittivity**, ε, and **permeability**, μ are determined by Equations (6) and (7), respectively:

Permittivity,

$$\varepsilon = \frac{n}{Z} \qquad ...(6)$$

Permeability,

$$\mu = nz \qquad ...(7)$$

Figure 4 presents the developed CTS-MM unit cell extracted parameters. From the graph depicted in Figure 4, a wide negative permittivity real value of 4.76–5.4 GHz can be noted for the MM structure, confirming ENG characteristics. Moreover, the near-zero imaginary values of ε are seen from 3.5–4.3 GHz frequency spectrum. Interestingly, the unique T-shaped symmetric engineered MM structure shows near-zero permeability in the desired lower 5G band, as presented in Figure 4, proving its applicability in 5G MIMO antenna technology to reduce EM mutual coupling between MIMO components. The near-zero imaginary permeability values are also found in the spectrum of 3.5–5.5 GHz. Maxwell Equations (8) and (9), indicate that **near-zero characteristics** of any material can decouple the electric and magnetic fields [11, 15].

$$\nabla \times E = i\omega\mu H \qquad ...(8)$$

$$\nabla \times H = -i\omega\varepsilon E \qquad ...(9)$$

where, the E-field is presented by ***E*** and H-field by ***H*** for the frequency of ω. Furthermore, the designed MM unit cell and array architecture shows a negative refractive index (NRI) feature, in Figure 4, which indicates that this could significantly enhance the antenna performance [16]. Table 2 illustrates the extracted CTS-MM properties. Therefore, the developed symmetric engineered ENG and μ-near-zero (MNZ) MM with **wide band-stop characteristics** can be a potential approach to improve the antenna performance specially in isolation in a 5G MIMO antenna system [10, 11, 17–19].

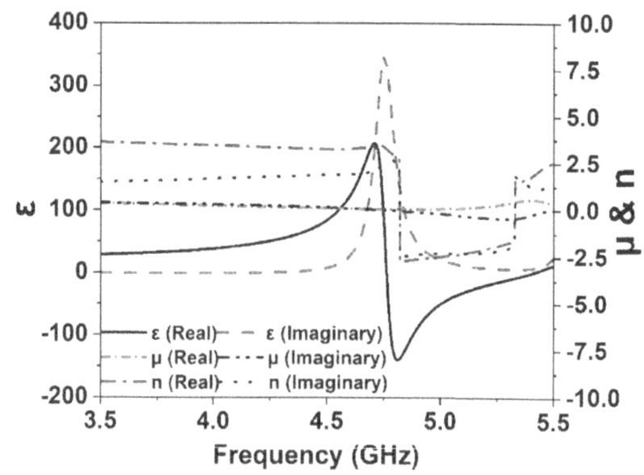

Fig. 4 Retrieved properties of MM unit cell.

Table 2 Retrieved properties of the developed unit cell.

Parameter	*Frequency Range (GHz)*	*Parameter*	*Frequency Range (GHz)*
Negative real permittivity	4.76–5.4	Near-zero permittivity (Imaginary)	3.5–4.3
Negative refractive index	4.8–5.3	Near-zero permeability (Imaginary)	3.5–5.5
Near-zero real permeability	3.5–5.3	–	–

2.4 Shielding Effectiveness

Shielding effectiveness (SE) is the most influential parameter in verifying the shielding performance of MM. The SE of the proposed CTS-MM structures is computed as follows [20, 21]:

$$SE_{dB} = S_{21,\text{ without shield}} - S_{21,\text{ with shield}} \quad ...(10)$$

where, $S_{21,\text{ with shield}}$, and $S_{21,\text{ without shield}}$ are denoted the transmission coefficient with and without metamaterial-based shield. Figure 5 depicts the SE of the developed array and unit cell MM structure, indicating a wide > 10 dB SE within the spectrum of 4.14–5.2 GHz and 4.14–5.1 GHz, respectively. The highest SE value is noted at about 29.4 dB, confirming its applicability for EM wave shielding applications. Therefore, the obtained SE of the designed MM structure proves its applicability at the desired 5G spectrum for EM coupling shielding between MIMO elements.

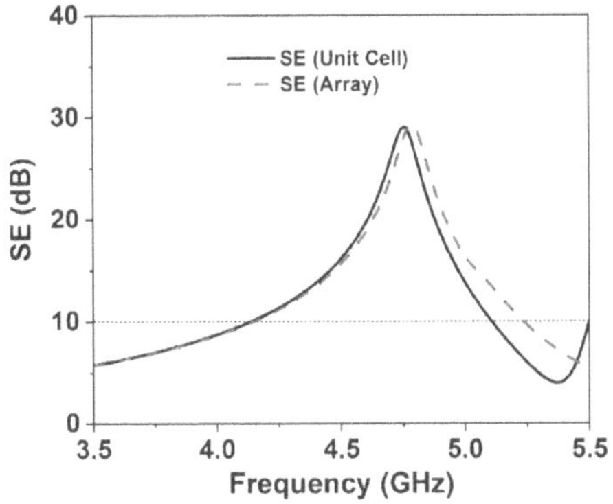

Fig. 5 SE for the developed MM structure.

2.5 T-Structured MM Shield Design

Figure 6 depicts the developed T-structured MM shield and its geometrical details. The developed 1 × 2 array of unit cells with a size of 9 × 18 mm^2 and an MM unit element has been employed to develop a unique **T-structured metamaterial shield**, which is revealed in Figure 6 (a). One 1 × 2 array of MM is joined vertically to another MM array and a unit cell is attached to the tail, resulting in a T-shaped MM shield. This unique **MM shield** can be employed in a MIMO system to reduce EM coupling energy among MIMO components. Figure 6 (b) also depicts the fabricated T-structured MM shield, and its constituent elements.

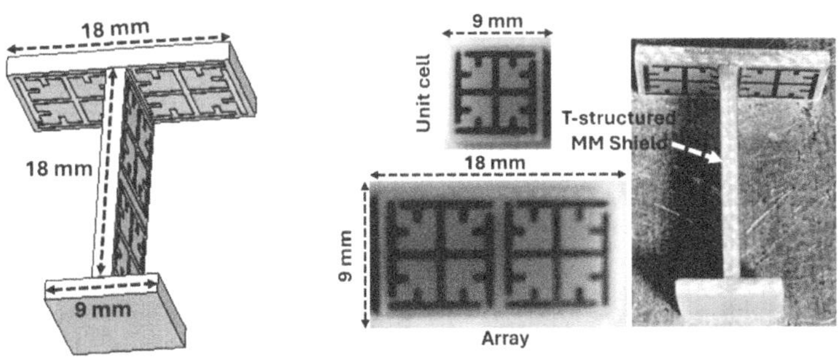

(a) (b)

Fig. 6 (a) T-structured MM geometry, (b) fabricated prototype.

3. MIMO Antenna Configuration with MM Shield

3.1 Design and Result Analysis

Figure 7 reveals the designed two-port MIMO antenna geometrical configuration. The most available inexpensive FR-4 printed substrate material is utilized in this antenna.

The initial two rectangular identical radiators with a mic shape slot are developed on a top layer of the 30 mm × 50 mm substrate material. A microstrip line is employed to excite the proposed antenna radiators. The antennas are positioned side by side and fed coaxially. The H-plane coupled MIMO interelement distance is only about $0.20\lambda_{min}$, resulting in structural compactness. The proximity between individual antenna components directly affects the isolation between MIMO antennas. So, the large interelement distance between MIMO elements can enhance isolation, but the overall antenna dimension will be increased. Conversely, the **mutual coupling** between adjacent antenna elements will enhance while the interelement spacing and overall antenna size are reduced. As the objective of this chapter is to develop a highly isolated compact wideband 5G MIMO antenna, a unique symmetric engineered T-structured MM isolation shield is proposed between MIMO antennas for isolation enhancement. Figure 7 reveals the antenna design evolutions and Figure 8(a,b) represents the corresponding reflection (S_{11}) and mutual coupling (MC) simulation results. In the initial antenna, two symmetric rectangular structure patches with a mic-shaped slot are utilized. The initial antenna patch dimensions are predicted using conventional mathematical equations [22]. As the proposed identical two antenna radiators share the same ground plane, the surface current influences the individual performance of the antenna. Thus, two vertical slots are introduced in this phase to improve the isolation between the two antenna elements. The antenna operates within the frequency spectrum of 4.43–4.99 GHz (11.89%) with a least isolation of only 16.1 dB between two adjacent MIMO antennas. After that, the proposed unique symmetric engineered metamaterial array (1 × 2 array).is introduced between two MIMO elements as an isolation shield in the next stage (stage 2). Moreover, a new **inverse split T-shaped** defected ground structure (DGS) is incorporated during stage 2. The single-layered MM shield with DGS offers impedance bandwidth (IBW) of 11.37% (4.48–5.02 GHz) and minimum isolation of about 16.3 dB over the spectrum of 4.48–5 GHz, as illustrated in Figure 8(a, b). Also, the antenna fractional bandwidth is reduced by 0.52% with the single layer of MM. In stage 3, a back-to-back MM array (two layers of MM) is placed in a two-port antenna for isolation enhancement, but it is still comparatively low, almost 18 dB in the operating band of 4.47–4.99 GHz, as illustrated in Figures 7 and 8. Moreover, the operating spectrum (4.47–4.99 GHz, 10.99%) is reduced by 0.38% compared to stage 2.

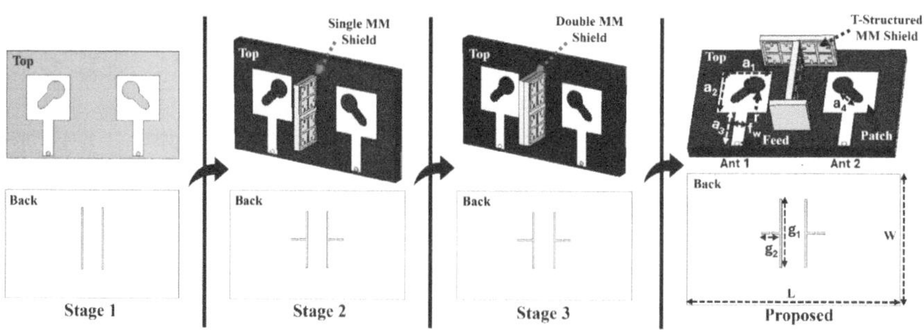

Fig. 7 Design evolutions of T-structured MM-integrated **MIMO antenna**.

Finally, a unique T-structured MM shield is integrated with the developed MIMO antenna in between the antenna elements, resulting in significantly enhanced isolation of greater than 29 dB over the desired spectrum of 4.45–5 GHz, as indicated in Figures 7 and 8. The proposed unique T-structured MM shield can stop the propagation of coupling energy between two adjacent antennas, hence reducing the EM coupling energy between two MIMO elements. Moreover, the designed antenna with T-structured MM provides an excellent wide IBW of 4.45–5.04 GHz (12.4%) with an improvement of 0.51% compared to the initial antenna, which nicely covers the 5G n79 band, in Figure 8 (a). Additionally, the developed MM-shield achieves 1.1% miniaturization. The degree of **miniaturization** is determined by $\rho = \frac{f_1 - f_2}{f_1} \times 100\%$, where f_1 and f_2 represent the resonant frequencies before and after miniaturization, respectively [23]. Figure 8(a, b) clearly reveals the developed MM contribution among the MIMO configurations. A notable improvement in isolation of higher than 29 dB in the desired spectrum is seen after implementing the developed T-shaped MM shield, implying that the near-field EM mutual coupling is greatly reduced by >12.9 dB compared to the initial antenna. The highest isolation for the initial antenna (stage 1) is 26.7 dB, which is enhanced to 48.7 dB (enhancement 22 dB) after loading the designed T-structured MM shield. Table 3 shows the developed MIMO antenna design parameters.

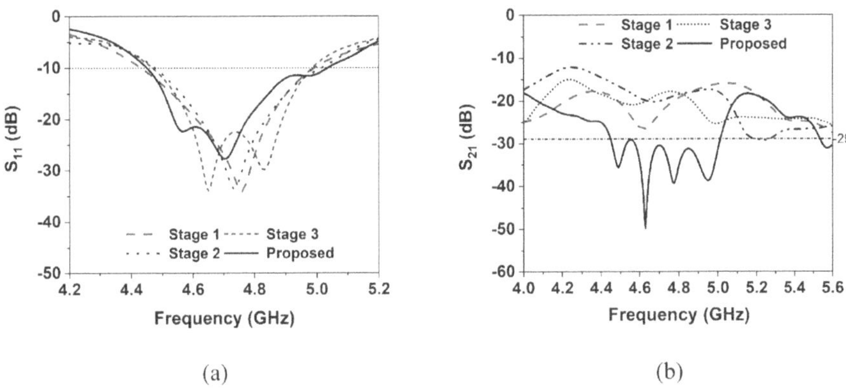

Fig. 8 Simulated *S*-parameter results for different evolution stages of MIMO antenna: (a) S_{11}, (b) S_{21}.

Table 3 Design specifications of the developed T-structured MM-integrated MIMO antenna.

Parameters	*Size (mm)*	*Parameters*	*Size (mm)*
a_1	12.4	g_1	15.5
a_2	12.6	g_2	4.3
a_3	10	r	2.5
a_4	4.03	L	50
f_w	3	W	30

3.2 Antenna Parametric Analysis

The effects of the various thicknesses of the T-structured MM shield on antenna performance are investigated with **parametric analysis** which is demonstrated in Figure 9. The middle MM thickness of the developed T-structured shields, such as single MM (proposed), two-layered MM (2 MM), and three-layered MM (3 MM) has a significant influence on antenna performance specially isolation between MIMO elements, which is presented in Figure 9. It is evident that the final proposed T-structured MM shield achieves **higher isolation** and wide operating bandwidth than others. The final optimized antenna geometry is chosen for the desired maximum performance in 5G n79 applications.

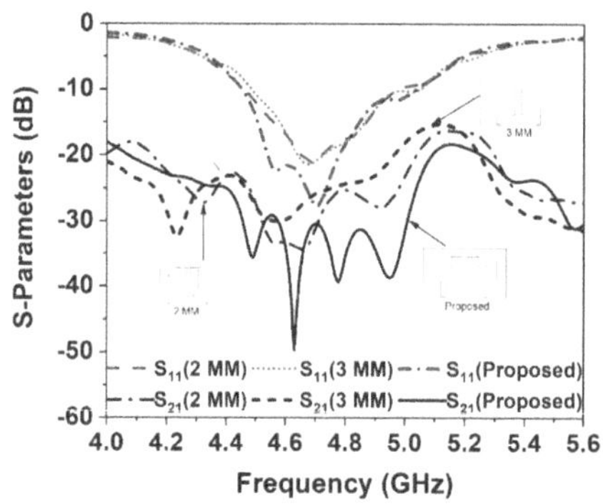

Fig. 9 Parametric investigation of MIMO antenna with T-structured shield at different thicknesses.

3.3 Electric Field Analysis

The EM energy in the adjacent two MIMO elements is coupled through the dielectric medium and space, namely surface and space wave coupling, respectively [24, 25]. The proposed MM-based T-structured shielding technique significantly suppressed the **EM coupling energy** (space and surface wave coupling) between the antenna elements. As the proposed MM shield is incorporated in the middle of the two antenna elements, it can cancel out the EM coupling energy when the coupling wave interacts with the MM shield due to the formation of an evanescent medium [26]. Hence, the isolation between two antennas is enhanced greatly. The **electric field distribution** of the initial to proposed MIMO antennas is also investigated to understand the MM influences in isolation. Figure 10 (a–d) demonstrates the port-1 electric field (V/m) distribution for the initial to proposed MIMO antennas at 4.9 GHz. The colo r ramp on the middle of Figure 10 indicates the energy intensity among antennas.

From Figure 10(a), for the initial antenna, it can be noticed that the coupled E-field density between adjacent radiators is very high, indicating lower isolation. When antenna 1 is energized, a strong E-field density is seen in antenna 2, resulting in high mutual coupling between two antenna elements. Hence, the antenna performance deteriorates owing to the serious mutual EM coupling effect. After employing a single-layered MM array between antenna elements with modified T-shaped DGS, in stage 2 (Figure 10(b)), the coupling energy becomes lower, which enhances the isolation.

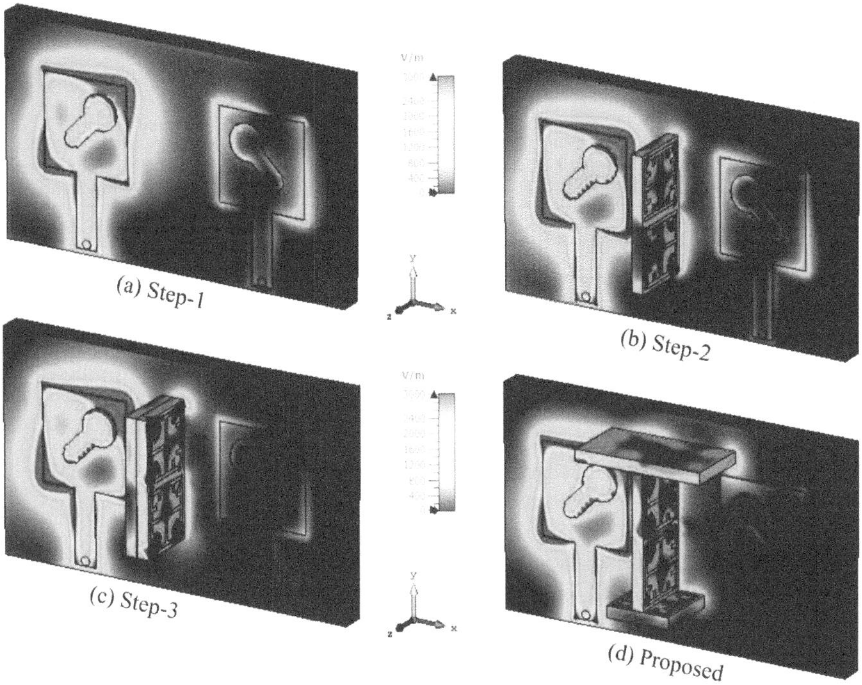

Fig. 10 E-field distribution form initial to proposed design: (a) design 1 (initial), (b) design 2, (c) design 3, (d) proposed design.

However, the isolation enhancement is lower compared to the initial antenna. In the next stage (step 3), a two-layered MM shield is mounted in the middle of the radiators, shielding the coupling field and protecting port 2. After introducing the developed unique T-structured MM shield between two MIMO elements, the electric current is dispersed to unit cells of the proposed MM wall and shields antenna 2 from the coupling field, as revealed in Figure 10(d). The T-structured shield stops the EM energy propagation from antenna 1 to antenna 2. Consequently, the coupled E-field density on antenna 2 is greatly decreased, which improves isolation significantly, whereas, a strong field is observed on the MM T-structured shield, forming a **shielding region** for antenna 2 [7]. These findings support the idea that the proposed unique MM shield achieves excellent band-stop functionality in the targeted n79 5G band [25]. Therefore, the proposed shielding technique improves isolation between antenna components in a compact MIMO system by significantly reducing mutual coupling.

3.4 Measurement Results

Once the antenna design, analysis, and optimization are completed with the desired performance, the designed geometric parameters are extracted from the CST for fabrication purposes. The cutting-edge manufacturing technology facilitates the **fabrication** of prototypes with a decreased fabrication cost. The designed T-structured MM-integrated MIMO antenna is fabricated to verify its simulation performance. Figure 11(a, b) exhibits the photos of the printed MIMO antenna prototype with various assembly components. The designed symmetric MM is fabricated and assembled by adhesive in the form of a T-structured MM shield, which is incorporated in the middle position of the two MIMO antennas.

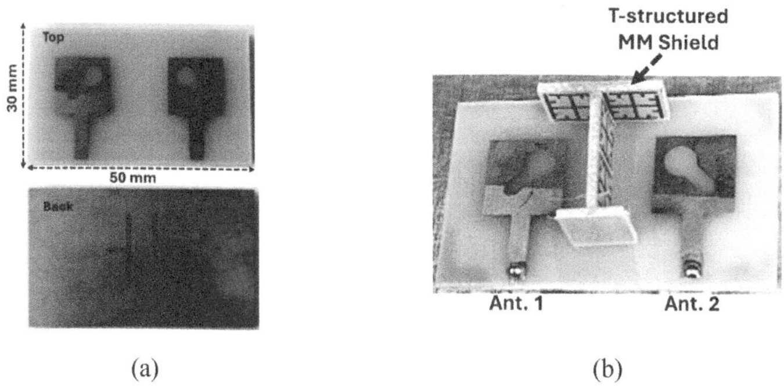

Fig. 11 Pictures of fabricated prototype: (a) Two-port MIMO configuration, (b) MIMO prototype with T-structured MM shield.

The reflection and isolation performance of the developed MIMO antenna with a T-structured MM shield is measured using the Agilent P-series VNA (Agilent N5227A), as revealed in Figure 12. The analyzer has two measuring ports and supports the frequency from 10 MHz – 67 GHz. Figure 12 represents the **experimental arrangement** for the developed MIMO prototype.

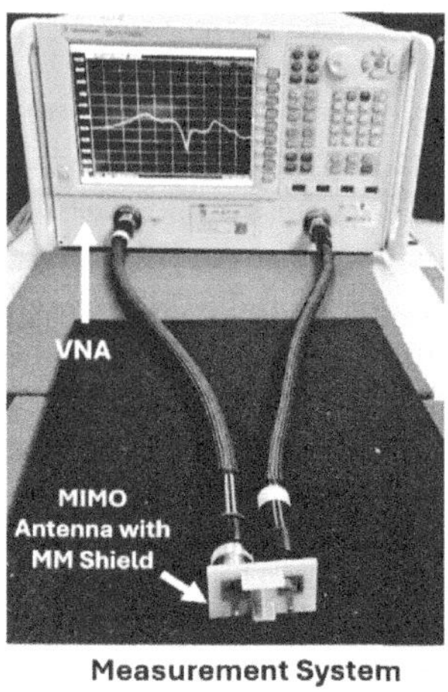

Fig. 12 Measurement setup for the developed MIMO antenna with T-structured MM shield.

Scattering Parameters: Figure 13(a, b) presents the developed innovative T-structured symmetric MM shield's outstanding contributions to the 5G MIMO system, supporting high isolation in the targeted 5G n79 band. The proposed **shielding technique** remarkably minimizes the mutual coupling of MIMO antennas, ensuring high isolation. The reflection and isolation performance of the developed antenna prototype is verified using an Agilent PNA network analyzer and the measured findings are presented in Figure 13(a, b) alongside the simulated outputs. The simulated and tested reflection (S_{11}) and isolation (S_{21}) performance of the developed MIMO antenna reveals close agreement, validating the proposed technique. The measured IBW (– 10 dB) of the developed antenna is 4.4–5.2 GHz (16.67%), which effectively covers the 5G n79 band. Figure 13(b) exhibits the isolation results, achieving a minimum measured isolation (S_{21}) of 28.8 dB over the spectrum of 4.65–5.20 GHz and 22 dB over the spectrum of 4.4–4.65 GHz. The peak measured isolation is observed at 47.1 dB with a T-shaped MM shield. The measured isolation result confirms that the developed T-structured MM shield efficiently suppressed the **mutual coupling** energy, hence enhancing isolation significantly. Therefore, the promising performance of the proposed T-structured MM shielding technique makes it potential for 5G MIMO antenna systems. The overall measured results fluctuated slightly from the simulated outcomes owing to experimental tolerance, manufacturing flaws, and various connector and soldering losses, which were not considered during the simulation.

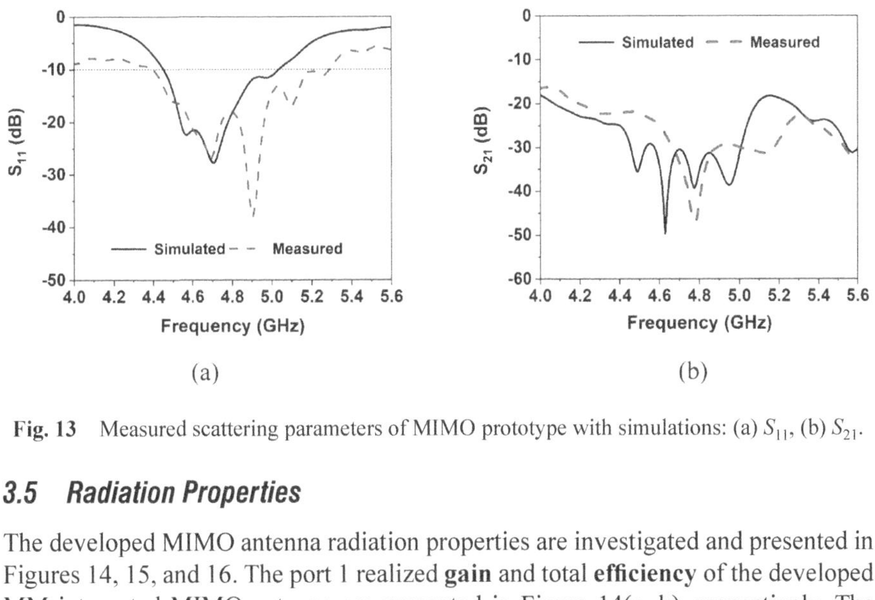

Fig. 13 Measured scattering parameters of MIMO prototype with simulations: (a) S_{11}, (b) S_{21}.

3.5 Radiation Properties

The developed MIMO antenna radiation properties are investigated and presented in Figures 14, 15, and 16. The port 1 realized **gain** and total **efficiency** of the developed MM-integrated MIMO antenna are presented in Figure 14(a, b), respectively. The proposed MM antenna exhibits a 2.74 dBi peak realized gain. It is also seen that the antenna provides almost stable gain over the desired operating bands. Figure 14(b) depicts the total efficiency of the developed MM-integrated antenna, which achieved a maximum efficiency of 59.6%.

Figure 15(a, b) exhibits the two-dimensional (2D) **radiation patterns** of the developed MM-integrated MIMO antenna for 4.7 GHz. As the developed MIMO components are symmetrical structures, only the port 1 radiation pattern is investigated and presented in Figure 15(a, b). The directional pattern is noted in both X-Z and Y-Z planes. Figure 16 presents the 3D antenna pattern for the developed MM MIMO antenna, indicating a high directivity. The 3D patterns are extracted for port 1 at 4.7 GHz and achieve a high directivity of 4.99 dBi.

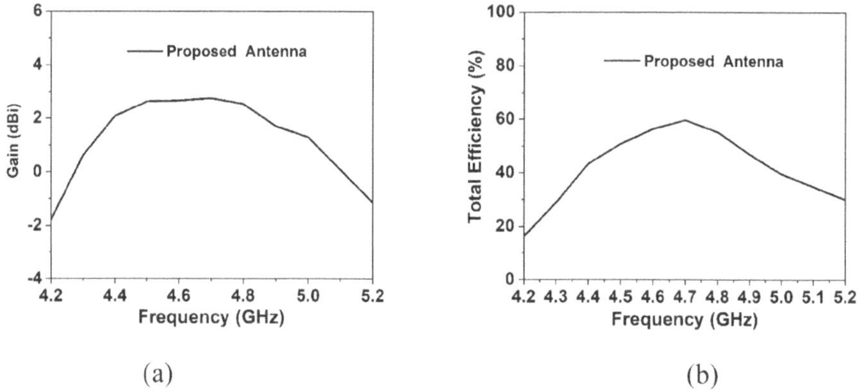

Fig. 14 (a) Realized gain, and (b) total efficiency for the developed MM-integrated MIMO antenna.

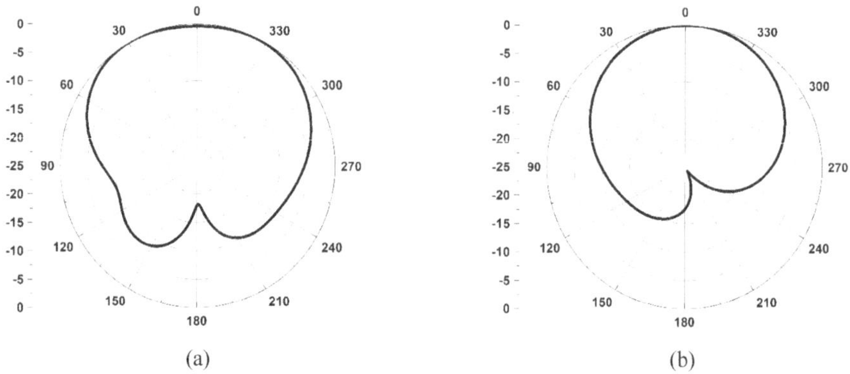

Fig. 15 2D radiation patterns of MM-integrated MIMO antenna: (a) X-Z plane, (b) Y-Z plane.

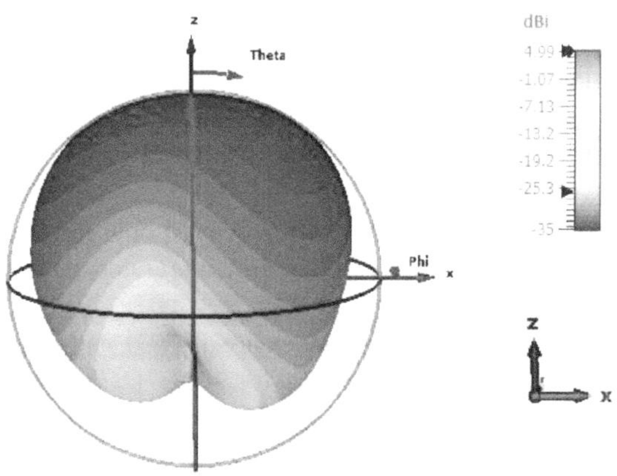

Fig. 16 **3D radiation** patterns of MM-integrated MIMO antenna.

4. MIMO Performance Parameters

The developed T-shaped MM incorporated MIMO antenna **diversity parameters**, including ECC, DG, and CCL have been estimated and assessed to confirm the antenna reliability and applicability in 5G applications.

4.1 Envelope Correlation Coefficient (ECC)

The correlation among antenna elements in a multi-antenna system can be estimated by ECC, which is a key diversity performance indicator of MIMO antenna. It presents the correlation effects between two adjacent antenna patterns in the MIMO system, indicating system capacity and reliability. The ECC should be <0.5 for adjacent two antennas [27, 28], and it can be calculated using the Equations (11) and (12) [29, 30].

$$ECC = \frac{\left|S_{ii}^{*}S_{ij} + S_{ji}^{*}S_{jj}\right|^2}{\left(1-|S_{ii}|^2-|S_{ji}|^2\right)\left(1-|S_{jj}|^2-|S_{ij}|^2\right)} \quad ...(11)$$

$$ECC = \frac{\left|\iint_0^{4\pi}[\vec{E}_i(\theta,\varphi)\times\vec{E}_j(\theta,\varphi)]\,d\Omega\right|^2}{\iint_0^{4\pi}\left|\vec{E}_i(\theta,\varphi)\right|^2 d\Omega \iint_0^{4\pi}\left|\vec{E}_j(\theta,\varphi)\right|^2 d\Omega} \quad ...(12)$$

$$\vec{E}_i(\theta,\varphi).\,\vec{E}_j(\theta,\varphi) = E_{\theta i}(\theta,\varphi)\,E_{\theta j}^{*}(\theta,\varphi) + E_{\varphi i}(\theta,\varphi)E_{\varphi j}^{*}(\theta,\varphi) \quad ...(13)$$

The reflection and transmission coefficients are presented by S_{ii} and S_{ij}, respectively, and the solid angle is Ω. The radiated pattern for antenna i and antenna j is denoted by $\vec{E}_i(\theta,\varphi)$ and $\vec{E}_j(\theta,\varphi)$, respectively, in equations (12-13). The ECC plots from S-parameters for the initial and proposed MIMO antenna systems are compared in Figure 17(a). When the proposed T-shaped MM shield is introduced, the ECC values drop significantly from 0.006 – 0.0004. The ECC of the developed MIMO antenna is less than 0.0004 throughout the operating band, which is much lower than the allowed limit (0.5). The very low ECC results ensure that the proposed MIMO antenna achieves excellent diversity performance and high channel isolation, therefore improving channel capacity and reliability [31–33].

4.2 Diversity Gain (DG)

DG is also another essential performance metric for the investigation of MIMO characteristics. It implies the reduction of transmission power with the introduction of a diversity scheme, which ensures the MIMO antenna quality and reliability in communication systems. This value should be close to 10 dB. The DG of any MIMO system can be derived using the approximation Equation (14) [30, 34].

$$DG = 10\sqrt{1-|ECC|^2} \quad ...(14)$$

Figure 17(b) reveals the DG curves from S-parameters for the initial antenna and developed T-structured MM-integrated MIMO antenna. Figure 17 (b) clearly indicates that the T-structured MM **shielding** improved the DG from 9.97 dB to more than 9.99 dB. Notably, the DG value of the developed MIMO system is better than 9.99 dB, which is very near to 10 dB.

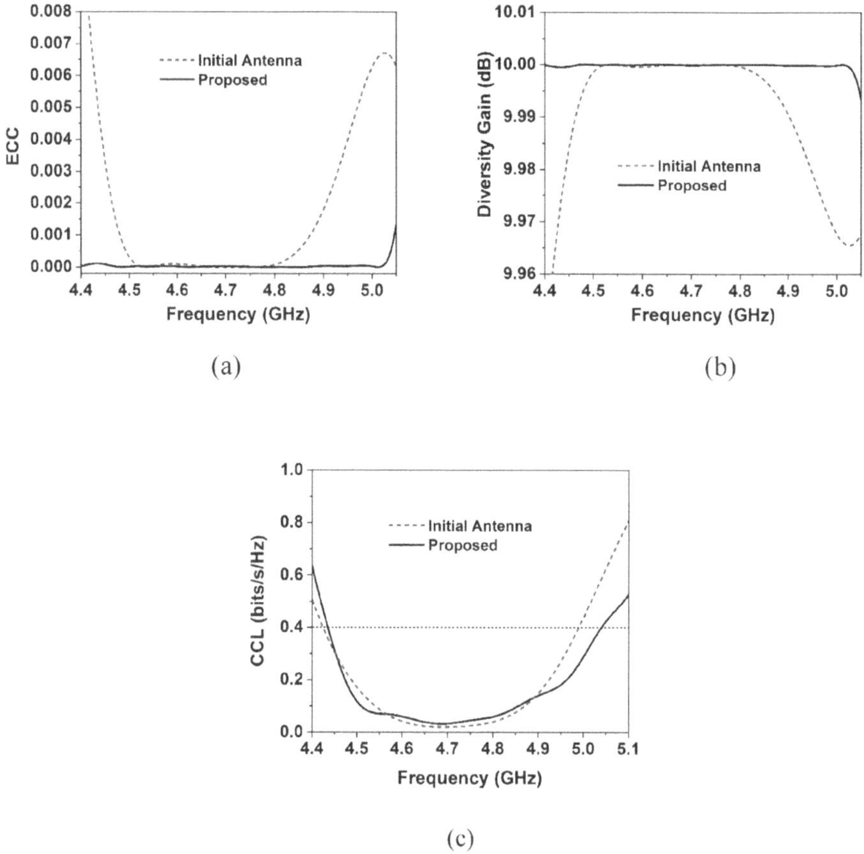

Fig. 17 Diversity performance of MIMO antenna: (a) ECC, (b) DG, and (c) CCL.

4.3 Channel Capacity Loss (CCL)

The CCL indicates the maximum possible information transmission rate or the capacity of a MIMO communication channel at which the signal can be conveyed without suffering substantial loss. The CCL should be <0.4 bits/s/Hz for a reliable MIMO antenna. The CCL is obtained using the given Equation (15) [30, 35].

$$CCL = -\log_2 det\,(\psi^R) \qquad ...(15)$$

where, (ψ^R) signifies the correlation matrix, which is defined by (16-18),

$$\psi^R = \begin{bmatrix} \rho_{11} & \rho_{12} & \rho_{13} & \rho_{14} \\ \rho_{21} & \rho_{22} & \rho_{23} & \rho_{24} \\ \rho_{31} & \rho_{32} & \rho_{33} & \rho_{34} \\ \rho_{41} & \rho_{42} & \rho_{43} & \rho_{44} \end{bmatrix} \quad ...(16)$$

$$\rho_{ii} = 1 - (|S_{ii}|^2 + |S_{ij}|^2) \quad ...(17)$$

$$\rho_{ij} = -(S_{ii}^* S_{ij} + S_{ji}^* S_{jj}), \quad ...(18)$$

where, $i, j = 1, 2, 3$, or 4 regarding port number. Figure 17(c) visualizes the extracted CCL values of the developed initial and proposed MIMO antenna. Compared to the initial antenna, the proposed antenna provides a lower CCL value. The CCL value of the proposed antenna is observed < 0.39 bits/s/Hz within the desired spectrum. Thus, the developed MIMO antenna is confirmed to have outstanding throughput.

5. Summary

This chapter described the development of a new T-structured MM-based wideband MIMO antenna for sub-6 GHz 5G band applications. The T-structured MM shield achieved a wide band-stop characteristic from 4.14–5.1 GHz and high shielding properties within the desired spectrum. The developed MIMO antenna with a T-shaped MM shield archives **high isolation** of greater than 29 dB over the desired spectrum of 4.45–5 GHz. The proposed T-structure MM shielding also improves the antenna IBW of 0.51%, covering the 5G n79 band, and reduces ECC to a very low value of < 0.0004. The compact antenna size is $0.45\lambda_{min} \times 0.74\lambda_{min} \times 0.047\lambda_{min}$ and uses inexpensive FR-4 substrate materials. The antenna design evolution, MM characteristics, and electric field distribution are explored to optimize the targeted performance and understand the MM properties with their influences on the MIMO antenna system. The MIMO antenna is performed effectively and agrees well with the experimental results. The proposed technique achieves promising isolation between antenna components of higher than 29 dB in the desired bandwidth with a small center spacing and **very low ECC** compared to state-of-the-art antennas. Therefore, the promising performance of the proposed **MM shielding** technique makes it a potential candidate for 5G MIMO antenna applications.

Reference

[1] M. Hu and Y. Li. "Wideband Back Cover Microstrip Antenna with Multiple Shorting Vias for Mobile 5G MIMO Applications," *IEEE Transactions on Antennas and Propagation*, 2023.

[2] M.A. Jensen and J.W. Wallace. "A review of antennas and propagation for MIMO wireless communications," *IEEE Transactions on Antennas and Propagation*, vol. 52, pp. 2810–2824, 2004.

[3] D. Abbasi, A. Aziz, K. Aljaloud, A.H. Alqahtani, A. Zerguine, A. Al-Garni, et al. "Higher Order MIMO Antenna Design Using Capacitively Coupled Meandered Loop Element for 5G Smartphone Applications," *IEEE Transactions on Circuits and Systems II: Express Briefs*, 2023.

[4] Y. Qin, L. Zhang, C.-X. Mao and H. Zhu. "A Compact Wideband Antenna With Suppressed Mutual Coupling for 5G MIMO Applications," *IEEE Antennas and Wireless Propagation Letters*, vol. 22, pp. 938–942, 2022.

[5] H.-H. Tran, T.T.-L. Nguyen, H.-N. Ta and D.-P. Pham. "A Metasurface-Based MIMO Antenna with Compact, Wideband, and High Isolation Characteristics for Sub-6 GHz 5G Applications," *IEEE Access,* 2023.

[6] M. Singh, W. Saleh, A. Abed and M. Fauzi. "A Review on Massive MIMO Antennas for 5G Communication Systems on Challenges and Limitations," *UKM Eng. J.,* vol. 35, pp. 95–103, 2023.

[7] F. Liu, J. Guo, L. Zhao, G.-L. Huang, Y. Li and Y. Yin. "Dual-band metasurface-based decoupling method for two closely packed dual-band antennas," *IEEE Transactions on Antennas and Propagation,* vol. 68, pp. 552–557, 2019.

[8] M.M. Bait-Suwailam. "Electromagnetic field interaction with metamaterials," *Electromagnetic Fields and Waves,* pp. 1–19, 2019.

[9] M. Ameen, O. Ahmad, and R. Chaudhary. "Wideband circularly-polarised high-gain diversity antenna loaded with metasurface reflector for small satellite applications," *Electronics Letters,* vol. 55, pp. 829–831, 2019.

[10] P.K. Panda and D. Ghosh. "Isolation and gain enhancement of patch antennas using EMNZ superstrate," *AEU-International Journal of Electronics and Communications,* vol. 86, pp. 164–170, 2018.

[11] R. Mark, H.V. Singh, K. Mandal and S. Das. "Mutual coupling reduction using near-zero ε and μ metamaterial-based superstrate for an MIMO application," *IET Microwaves, Antennas, & Propagation,* vol. 14, pp. 479–484, 2020.

[12] D. Li, Z. Szabó, X. Qing, E.-P. Li and Z.N. Chen. "A high gain antenna with an optimized metamaterial inspired superstrate," *IEEE Transactions on Antennas and Propagation,* vol. 60, pp. 6018–6023, 2012.

[13] Y. Fu, T. Shen and J. Dou. "Mutual Coupling Reduction of a Multiple-Input Multiple-Output Antenna Using an Absorber Wall and a Combline Filter for V2X Communication," *Sensors,* vol. 23, p. 6355, 2023.

[14] X. Chen, T.M. Grzegorczyk, B.-I. Wu, J. Pacheco Jr. and J.A. Kong. "Robust method to retrieve the constitutive effective parameters of metamaterials," *Physical Review E,* vol. 70, p. 016608, 2004.

[15] I. Liberal, and N. Engheta. "Near-zero refractive index photonics," *Nature Photonics,* vol. 11, pp. 149–158, 2017.

[16] O. Borazjani, M. Naser-Moghadasi, J. Rashed-Mohassel and R. Sadeghzadeh. "Design and fabrication of a new high gain multilayer negative refractive index metamaterial antenna for X-band applications," *International Journal of RF and Microwave Computer-Aided Engineering,* vol. 30, p. e22284, 2020.

[17] A. Dadgarpour, B. Zarghooni, B.S. Virdee, T.A. Denidni and A.A. Kishk. "Mutual coupling reduction in dielectric resonator antennas using metasurface shield for 60-GHz MIMO systems," *IEEE Antennas and Wireless Propagation Letters,* vol. 16, pp. 477–480, 2016.

[18] R. Mark, N. Rajak, K. Mandal and S. Das. "Metamaterial based superstrate towards the isolation and gain enhancement of MIMO antenna for WLAN application," *AEU-International Journal of Electronics and Communications,* vol. 100, pp. 144–152, 2019.

[19] I.H. Abdelaziem, A.A. Ibrahim and M.A. Abdalla. "A high gain antenna utilizing Mu-near-zero metasurface structures for 5G applications," *International Journal of Microwave and Wireless Technologies,* pp. 1–9, 2022.

[20] M.M. Masud, B. Ijaz, I. Ullah and B. Braaten. "A compact dual-band EMI metasurface shield with an actively tunable polarized lower band," *IEEE Transactions on Electromagnetic Compatibility,* vol. 54, pp. 1182–1185, 2012.

[21] L.B. Wang, K.Y. See, J.W. Zhang, B. Salam and A.C.W. Lu. "Ultrathin and flexible screen-printed metasurfaces for EMI shielding applications," *IEEE Transactions on Electromagnetic Compatibility,* vol. 53, pp. 700–705, 2011.

[22] C.A. Balanis. *Antenna Theory: Analysis and Design*: John Wiley & Sons, 2015.

[23] G. Chen, C. Guo, J. Xue, Z. Wang and M. Pang. "Miniaturized Metamaterial Ultra-Wideband Antenna for WLAN and Bluetooth Applications," *Progress In Electromagnetics Research C,* vol. 132, pp. 117–127, 2023.

[24] Z. Wang, C. Li, Q. Wu and Y. Yin. "A metasurface-based low-profile array decoupling technology to enhance isolation in MIMO antenna systems," *IEEE Access,* vol. 8, pp. 125565–125575, 2020.
[25] B. Yin, X. Feng and J. Gu. "A metasurface wall for isolation enhancement: Minimizing mutual coupling between MIMO antenna elements," *IEEE Antennas and Propagation Magazine,* vol. 62, pp. 14–22, 2019.
[26] A.K. Dwivedi, A. Sharma, A.K. Singh and V. Singh. "Metamaterial inspired dielectric resonator MIMO antenna for isolation enhancement and linear to circular polarization of waves," *Measurement,* vol. 182, p. 109681, 2021.
[27] P. Garg and P. Jain. "Isolation improvement of MIMO antenna using a novel flower shaped metamaterial absorber at 5.5 GHz WiMAX band," *IEEE Transactions on Circuits and Systems II: Express Briefs,* vol. 67, pp. 675–679, 2019.
[28] A. Iqbal, A. Smida, A.J. Alazemi, M.I. Waly, N.K. Mallat and S. Kim, "Wideband circularly polarized MIMO antenna for high data wearable biotelemetric devices," *IEEE Access,* vol. 8, pp. 17935–17944, 2020.
[29] G. Saxena, S. Kumar, S. Chintakindi, A. Al-Tamim, M.H. Abidi, W.A.M. Saif, et al. "Metasurface Instrumented High Gain and Low RCS X-Band Circularly Polarized MIMO Antenna for IoT Over Satellite Application," *IEEE Transactions on Instrumentation and Measurement,* 2023.
[30] J. Kulkarni, A. Desai and C.Y.D. Sim. "Two port CPW-fed MIMO antenna with wide bandwidth and high isolation for future wireless applications," *International Journal of RF and Microwave Computer-Aided Engineering,* vol. 31, p. e22700, 2021.
[31] M.S. Sharawi. "Printed multi-band MIMO antenna systems and their performance metrics [wireless corner]," *IEEE Antennas and Propagation Magazine,* vol. 55, pp. 218–232, 2013.
[32] J.X. Yun and R.G. Vaughan. "Multiple element antenna efficiency and its impact on diversity and capacity," *IEEE Transactions on Antennas and Propagation,* vol. 60, pp. 529–539, 2011.
[33] Y. Liu, X. Yang, Y. Jia and Y.J. Guo. "A low correlation and mutual coupling MIMO antenna," *IEEE Access,* vol. 7, pp. 127384–127392, 2019.
[34] M.G.N. Alsath and M. Kanagasabai. "Compact UWB monopole antenna for automotive communications," *IEEE Transactions on Antennas and Propagation,* vol. 63, pp. 4204–4208, 2015.
[35] I. Khan, K. Zhang, L. Ali and Q. Wu. "Enhanced Quad-Port MIMO Antenna Isolation with Metamaterial Superstrate," *IEEE Antennas and Wireless Propagation Letters,* 2023.

Chapter 7

Bandwidth, Gain and Isolation Enhancement of Miniaturized MIMO Antenna Using Metamaterials

Mohammad Tariqul Islam[1] and *Md. Mhedi Hasan*[2]

1. Introduction

In recent years, the rapid advancement of **5G communications** has revolutionized the entire world with seamless huge user connectivity, higher information rate, lower latency, improved channel capacity, and reliability [1–3]. 5G new radio (NR) sub-6 GHz spectrums (n77/n78/n79 bands) are now the most popular in various nations to roll out 5G communications due to their simplicity of integration, wide reliable coverage, and outstanding adaptability [2, 3]. The **multiple-input multiple-output (MIMO)** antenna is a potential technique to improve spectrum efficiency and reliability in 5G technology by diminishing multiple fading, boosting channel capacity, and offering massive coverage and high data rates with limited power and bandwidth [1, 3, 4]. Adopting a greater number of MIMO antenna components within a MIMO antenna system improves system reliability and capacity. Thus, the MIMO antenna element is the key step for improving communication capacity in 5G [1]. Simple and miniaturized antenna configuration is highly suitable for enhancing MIMO antenna components in a limited space. The lightweight, low-cost, and compact microstrip patch antennas are effective for compact MIMO systems [5–8]. Moreover, it is suitable for microwave integrated circuits (MIC), which is the backbone of lightweight compact devices. However, the traditional microstrip antennas have a limited bandwidth and gain [9]. Hence, antenna performance enhancement is required to think of a better prototype. Improving channel bandwidth is one of the main challenges for 5G technology. Thus, a wideband antenna design is required. Due

[1,2] Department of Electrical, Electronic and Systems Engineering, Faculty of Engineering and Built Environment, Universiti Kebangsaan Malaysia (UKM), 43600 UKM Bangi, Selangor, Malaysia.

* Corresponding author: tariqul@ukm.edu.my

to the rapidly increasing demand for high data rates and frequency flexible compact 5G smart systems, designing and integrating multiple antennas in a compact MIMO antenna system is highly challenging [3, 10–12]. Moreover, the proximity of MIMO antenna components causes significant mutual coupling between MIMO antennas, which degrades the MIMO antenna performance and communication capacity [13]. The reduction of the mutual coupling effect between neighboring MIMO elements is a crucial challenge in MIMO technology. Furthermore, large arrays, extensive frequency coverage, and high-speed compact antennas are vastly desirable in 5G to serve various spectrums, which leads to demand for extendable wideband MIMO antennas [3, 10, 11]. **Metamaterial** (MM) is a promising technique for enhancing overall antenna performance, composed of various subwavelength structural metals and dielectric substrates. [14–16]. This engineered geometry can control the electromagnetic (EM) incident waves for desired applications, which are rare in nature [17–19]. MMs are widely used in current antenna technology to improve antenna performance by integrating them into antenna substrates [13, 20–22] or as superstrates [14, 23–26]. MMs find various applications in MIMO antenna technology including isolation enhancement among MIMO components [14, 23, 27], gain enhancement [25, 28, 29], beam steering [30–32], radar cross-section reduction [24, 33], bandwidth improvement [34], etc. Moreover, the antenna miniaturization is achieved using MM with enhancing operating spectrum, which is useful for future 5G MIMO applications [35, 36]. Therefore, designing an effective wideband miniaturized massive MIMO antenna for the sub-6 GHz 5G spectrum has become an emerging research trend in contemporary high-speed communications. Furthermore, the incorporation of MM in MIMO systems with a small air gap to enhance the overall antenna performance is still a huge challenge, particularly where traditional methods are limited [37].

This chapter develops a miniaturized **MM** loaded wideband **extendable MIMO antenna** with enhanced gain, isolation, and efficiency useful for sub-6 GHz 5G communications. Two new MM structures are developed and employed in the proposed antenna to attain **miniaturization** and improve the overall antenna performance. The designed MIMO structure incorporates the coupled semicircle split resonator-based parasitic MM (PMM) loaded wide rectangular slotted antennas in a generalized mirrored side-by-side mode. The developed quad-port antenna area of $52 \times 52 \times 1.6$ mm^3, covering 3.04–5.05 GHz (bandwidth of 49.7%). The addition of PMM into the antenna substrate leads to miniaturization (52.94%) with improved matching characteristics. After that, an inter connected hook-shaped MM superstrate (MMS) is designed and employed with the MIMO system at a very low air gap of $0.07\lambda_{min}$, which significantly enhances the antenna bandwidth by 20.5%, gain by 2.4 dBi and efficiency by 6%. The isolation among MIMO components is also improved with a minimum isolation of about 15 dB. Moreover, this proposed unique MIMO structure can facilitate a greater number of MIMO elements. Besides, the proposed technique also improves the diversity performance with a low envelope correlation coefficient (ECC) <0.006, low channel capacity loss (CCL) <0.4 bits/s/Hz, and a high diversity gain (DG) >9.97 dB, proving low correlation and high channel capacity. The electric field density and surface current distribution of the developed antenna are also studied to verify the proposed MM contributions. In addition, a 16-element MIMO antenna with MM is designed and elucidated to confirm its suitability for massive

MIMO applications in 5G communications. The proposed MIMO antenna is designed and fabricated using compact FR-4 epoxy substrate and verified by measurement with acceptable agreement between simulated and measured results. Thus, the proposed MIMO structure with its improved outstanding performance leads to a promising solution for 5G sub-6 GHz communications.

2. Miniaturized Quad-Element Metamaterial MIMO Antenna

The MM is a highly effective approach for improving the antenna performance, including bandwidth, gain, and isolation between MIMO components. The effectively optimized unique structural properties can manipulate the MIMO antenna characteristics significantly for desired applications.

2.1 Metamaterial Design and Result Analysis

This chapter designed a new epsilon negative (ENG) and μ-near zero MM to enhance antenna performance, which comprises a coupled semicircle mirror symmetric resonator (Figure 1(a)). The metallic strips of the unit cell exhibit inductance and the split gaps yield capacitance. Therefore, the resonator patch behaves similarly to an LC circuit, generating a resonance effect when an incident electromotive force impinges on the MM. The MM performance can be controlled by adjusting the split

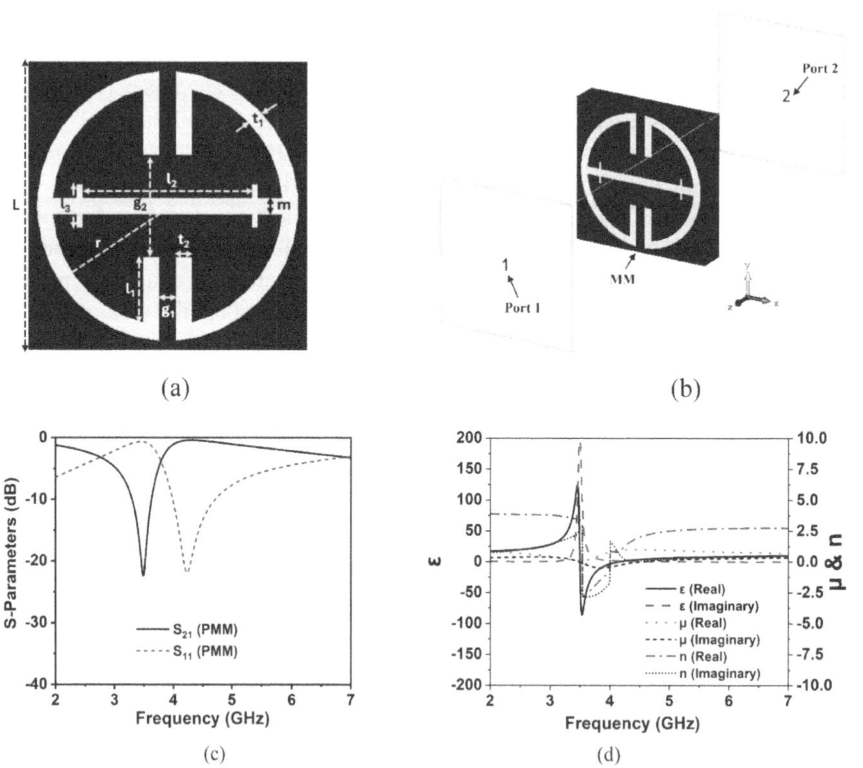

Fig. 1 (a) MM configuration, (b) simulation arrangement, (c) S-parameters, and (d) extracted parameters.

width and position as well as the length and width of the metallic rings. Numerical simulations are employed to determine the effective ring length, ring width, split gaps, split position, and connecting metal strips. The initial MM dimension is opted at the targeted bands (lower 5G bands) while maintaining the sub-wavelength principle. The electrical compact dimension of the designed MM is $0.09\lambda_{min} \times 0.09\lambda_{min} \times 0.02\lambda_{min}$, meeting the sub-wavelength principle. The suggested MM is designed with the low-priced FR-4 substrate material and simulated using the CST studio suite. A tetrahedral mesh-based frequency-domain solver, which is based on the finite element method (FEM), is utilized for simulations. FEM explores Maxwell's curl equations in 3D forms, and the full-wave solution gives S-parameters. Throughout the simulation, the excitation ports are positioned toward the z-axis, while the magnetic and electric boundaries correspond to the y-axis and x-axis, respectively. Figure 1(b) illustrates the designed MM simulation setup. After the simulation action, the scattering performance and MM characteristics of the developed MM are identified. The developed MM geometry is delineated on top of FR-4 substrate and the back side is non-copper. Two split semi-circle structures are coupled with a hook-shaped metallic strip, resonating in the targeted 5G bands, which is indicated in Figure 1(a).

Figure 1(c) exhibits the desired obtained simulated MM characteristics. The S_{11} and S_{21} properties in Figure 1(c) demonstrate a **band-stop** feature over 3.3–3.63 GHz, as well as passband characteristics below 3.3 GHz and above 3.63 GHz. Table 1 illustrates the complete design specifications of the proposed MM architecture.

Table 1 Design details of MM configuration.

Parameters	*Size (mm)*	*Parameters*	*Size (mm)*	*Parameters*	*Size (mm)*
g_1	0.5	l_2	5.18	m	0.5
g_2	2.98	l_3	1.3	r	4
l_1	1.93	L	8.5	t_1, t_2	0.5

The post-processing technique in CST is implemented to extract the developed MM properties using S-parameters, which are based on a **robust retrieval method** [38] and determined with the following equations [38].

Reflection coefficient,

$$S_{11} = \frac{R_{01}(1 - e^{i2nk_0d})}{1 - R_{01}^2 e^{i2nk_0d}} \quad ...(1)$$

Transmission coefficient,

$$S_{21} = \frac{(1 - R_{01}^2)e^{ink_0d}}{1 - R_{01}^2 e^{i2nk_0d}} \quad ...(2)$$

where, the thickness of the substrate and free space wave number are denoted by d and k_0, respectively.

$$R_{01} = \frac{z - 1}{z + 1} \quad ...(3)$$

From expressions (1) and (2):

Impedance,

$$z = \pm\sqrt{\frac{(1+S_{11})^2 - S_{21}^2}{(1-S_{11})^2 - S_{21}^2}} \qquad ...(4)$$

where, $e^{ink_0d} = X \pm i\sqrt{1-X^2}$ and $X = 1/2\, S_{21}\,(1 - S_{11}^2 + S_{21}^2)$.

Refractive index,

$$n = \frac{1}{K_0 d}\left\{\left[[\ln(e^{ink_0d})]'' + 2m\pi\right] - i\left[\ln(e^{ink_0d})\right]'\right\} \qquad ...(5)$$

where, (.)′ and (.)″ indicate the operator's real and imaginary parts, respectively, and m indicates an integer associated with the branch index of n'. The effective **permittivity**, ε, and **permeability**, μ are determined by Equations (6) and (7), respectively:

Permittivity,

$$\varepsilon = \frac{n}{Z} \qquad ...(6)$$

Permeability,

$$\mu = nz \qquad ...(7)$$

Figure 1(d) demonstrates the retrieved permittivity, permeability, and refractive index values. Negative ε and n real values are observed over 3.49–4.16 GHz and 3.54–4 GHz, respectively, while μ-near-zero value is seen around 3–4 GHz, supporting MM ENG and μ-near-zero (MNZ) features, in Figure 1(d). The high effective permittivity and refractive index generate higher electromagnetic coupling when integrated as a **parasitic MM** (PMM) in an antenna system. [39]. Additionally, the permittivity imaginary plot indicates near-to-zero values between 2–3.1 GHz and 4.5–6.5 GHz, whereas the refractive index imaginary plot indicates near-to-zero levels between 4.2–6.3 GHz. Furthermore, Figure 1(d) reveals a broad imaginary near-zero permeability level from 2–7 GHz. Table 2 demonstrates the developed MM properties. Thus, the obtained features of the developed MM can be leveraged to manipulate antenna characteristics by incorporating a parasitic element.

Table 2 Obtained properties of the developed MM structure.

Parameters	*Frequency (GHz)*	*Parameters*	*Frequency (GHz)*
Negative permittivity (real)	3.49–4.16	Near-zero permittivity (Imaginary)	2–3.1/4.5–6.5
Negative refractive index	3.54–4	Near-zero permeability (Imaginary)	2–7
Near-zero permeability (real)	3–4	Near-zero refractive index (Imaginary)	4.2–6.3

2.2 MM Loaded Quad-port MIMO Antenna

The quad-port antenna is designed with four identical PMM-loaded antennas. Figure 2(a) depicts a developed single-unit antenna with its design specifications. The antenna architecture consists of a new PMM structure with a rectangular microstrip line and a wide rectangular slot. A high-performance EM simulator software "CST studio suite" has been utilized to develop the antenna. The single-unit antenna is designed with the standard antenna equations [40] and optimized by various numerical simulations to attain the desired performance. The developed coupled semicircle mirror symmetric MM is embedded into the antenna substrate to achieve **miniaturization** with targeted performance. The small dimension of the developed single-unit antenna is $0.25\lambda_{min} \times 0.25\lambda_{min} \times 0.016\lambda_{min}$, proving its high compactness. Figure 2(b) presents the performance of a single unit antenna.

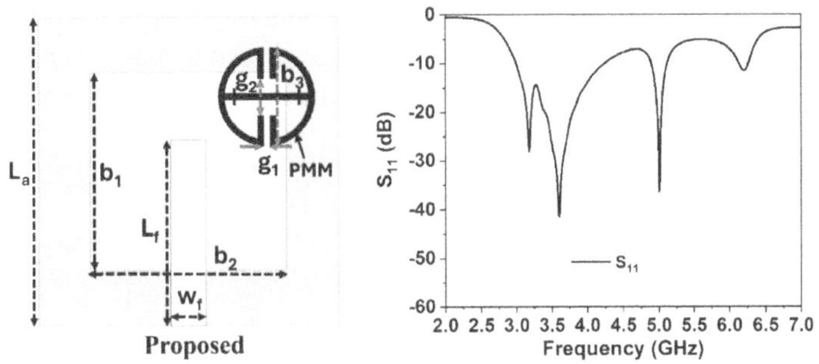

(a) (b)

Fig. 2 (a) Parasitic MM-loaded single unit antenna configuration ($L_a = 25$, $L_f = 15$, $W_f = 2.9$, $b_1 = 16$, $b_2 = 16.24$, $b_3 = 7.98$, $g_1 = 0.5$, and $g_1 = 2.98$, unit: mm) (b) S_{11} performance.

The development phases of the proposed **quad-port antenna** configuration are revealed in Figure 3(a). Four identical parasitic MM-loaded wide rectangular slotted microstrip-fed antennas are employed in the proposed MIMO configuration, which are arranged in a simple mirrored side-by-side fashion. First, antennas 1 and 2 are positioned side by side along the X-axis, and an isolated slot is introduced between two identical antennas, as illustrated in Figure 3(a). Then, a mirror copy of antennas 1 and 2 is arranged along the Y-axis integrating another isolation slot between them. The proposed MIMO configuration is designed using an FR-4 printed substrate with dimensions of $52 \times 52 \times 1.6$ mm^3, proving its compactness. The proposed MIMO configuration is planned with a unique and simple structure, not like complex and orthogonal structures, supporting an increased number of MIMO elements. Thus, the proposed MIMO technique can be extended to 8, 16, 32, or more ports due to its unique structural simplicity. Figure 3(a) represents the proposed quad-element MIMO antenna design stages with comprehensive design specifications, whereas Figure 3(b, c) demonstrates reflection coefficients and isolation findings for all design stages, respectively. Initially, four conventional patch antennas are employed in the MIMO system within a compact space, which is depicted in Figure 3(a) (stage 1). The antennas are fed by the microstrip line, and each single antenna is designed using standard equations as given in [42]. In MIMO architecture, antennas 1 and 2 are placed side by side and are mimed along the vertical direction. However, this MIMO configuration has no resonance in the desired 5G spectrum. Then, the introduction of the wide rectangular slot on the substrate top, along with the optimized back copper, leads to resonance at 6.8 GHz with –10 dB IBW of 6.57–6.99 GHz, in Figure 3(b) (stage 2). In the third stage, again etching the back copper with incorporating a plus shaped isolating slot ($s = 2$ mm) among MIMO components, shifting the resonance significantly to the lower range (3.8 GHz) with improving IBW of 17.7%, covering 5G n77/n78 spectrum. Hence, the antenna is miniaturized by 44.12%. The degree of miniaturization is determined by $\rho = \dfrac{f_1 - f_2}{f_1}$, where f_1 and f_2 represent the resonant frequencies before and after miniaturization, respectively [35]. However, the impedance matching, BW, operating range, and other antenna performances do not comply with the desired design specifications. Finally, incorporating the proposed PMMs into the MIMO system leads to miniaturization and **wideband** characteristics. Moreover, the proposed PMM contributes additional resonance and enhances IBW by

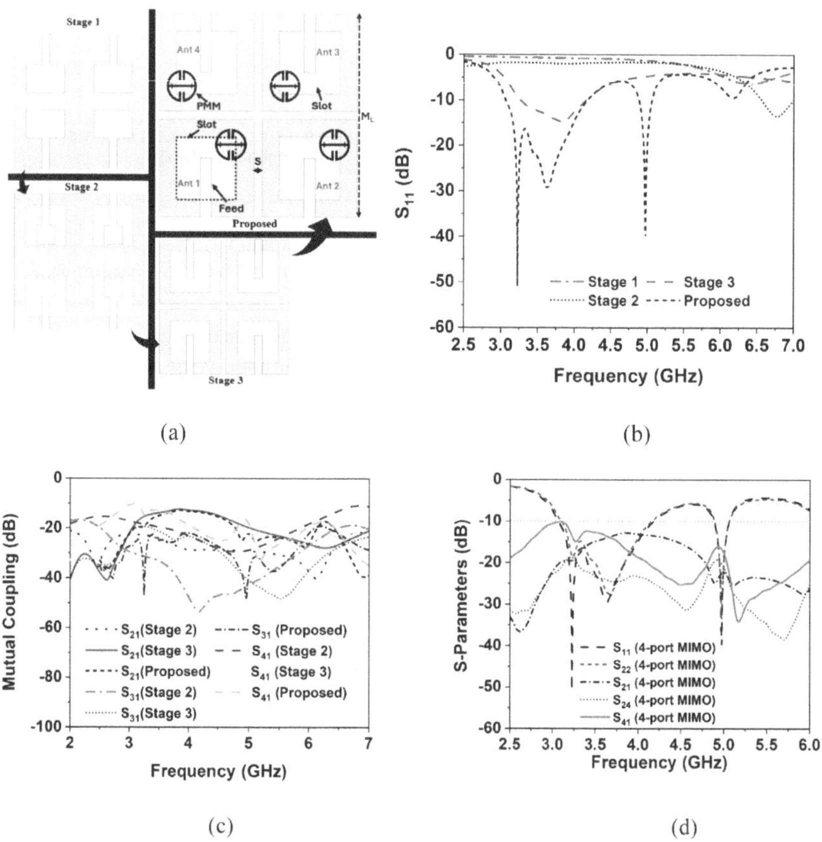

Fig. 3 (a) MIMO antenna structure and design stages, (b) simulated S_{11}, and (c) mutual coupling for various MIMO design stages, (d) S-parameters for the developed MIMO configuration.

5.3% with excellent matching characteristics. The loading of a PMM in proximity to the feedline leads to strong electromagnetic coupling, which enhances impedance BW with improved matching characteristics and provides additional resonance at 5 GHz. Moreover, the PMM contributes to generating the lowest resonance at 3.2 GHz, achieving a **miniaturization** of 52.94% compared to the conventional structure (stage 2). The proposed PMM approach in MIMO configuration greatly miniaturized the antenna by 15.8% as compared to MIMO without PMM. Furthermore, the proposed design strategy improves the isolation among MIMO antennas. The developed quad-port MIMO design specifications are $M_L = 52$ mm, and $s = 2$ mm, which are depicted in Figure 3(a).

Figure 3(d) exhibits the simulation performance of the developed PMM-loaded **MIMO configuration**. The designed antenna works from 3.1–4.16 GHz and 4.89–5.06 GHz with a maximum IBW of 29.2%, depicted in Figure 3(d). Also, the **isolation** metrics illustrated in Figure 3(d), imply a minimum isolation of about 10 dB throughout the operating spectrum. Moreover, the developed antenna reveals a maximum gain of 2 dBi and a total efficiency of 60%. The strong mutual coupling between MIMO elements in the miniature MIMO system has a substantial impact on antenna radiation performance. Although the PMM unit cell miniaturized the designed MIMO antenna, the isolation, coverage area, and radiation properties do not meet the desired 5G applications. Moreover, sub-6 GHz 5G communications demand highly isolated and high-performance MIMO systems to ensure reliable services. Therefore, an efficient MIMO performance enhancement technique is required to further boost the performance of the designed MIMO antenna, including isolation, operating frequency, gain, efficiency, and diversity performance.

3. MIMO Antenna Configuration with MM Superstrate

3.1 MM Superstrate Design and Properties Analysis

A unique mirror symmetric geometry-based **MMS** is proposed to enhance the designed MIMO antenna performance. An interconnected hook-shaped circular resonator is designed using FR-4 substrate material with an electrical size of $0.12\lambda_{min} \times 0.12\lambda_{min} \times 0.02\lambda_{min}$, which meets the sub-wavelength principle. Figure 4(a) exhibits the unit cell configuration with different design specifications and array arrangements. Two metal split circular rings are coupled with the metal strips on the top side of the FR-4 material and the bottom side is non-copper. The design structure looks like a **coupled hook-shaped architecture**, supporting the desired resonance (Figure 4(a)). The metal strips and split gaps are adjusted for the targeted response of the developed unique MM structure. Figure 4(a) shows the developed MMS configurations (6 × 6 array) with optimized designed specifications and Figure 4(b) indicates the fabricated photograph of MMS. The 6 × 6 array of designed MM unit cells is utilized in the proposed MMS with optimal spacing among the unit cells (Figure 4(a)). The designed MMS has a physical dimension of 52 × 52 mm^2, employing 36-unit cells periodically, which is deemed as antenna size for maximum performance enhancement, ease of extension to a large MIMO system, and compactness. In Figure 4(c), the MM (unit element and array) reflection and transmission characteristics indicate a band-stop characteristic from 4.13–4.8 GHz, with passband characteristics below 4.13 GHz and above 4.8 GHz. Interestingly, Figure 4(c) exhibits almost identical characteristics for the unit cell and suggested MMS structure, proving its applicability in 5G antenna technology. Table 3 depicts the MM unit cell and MMS design parameters.

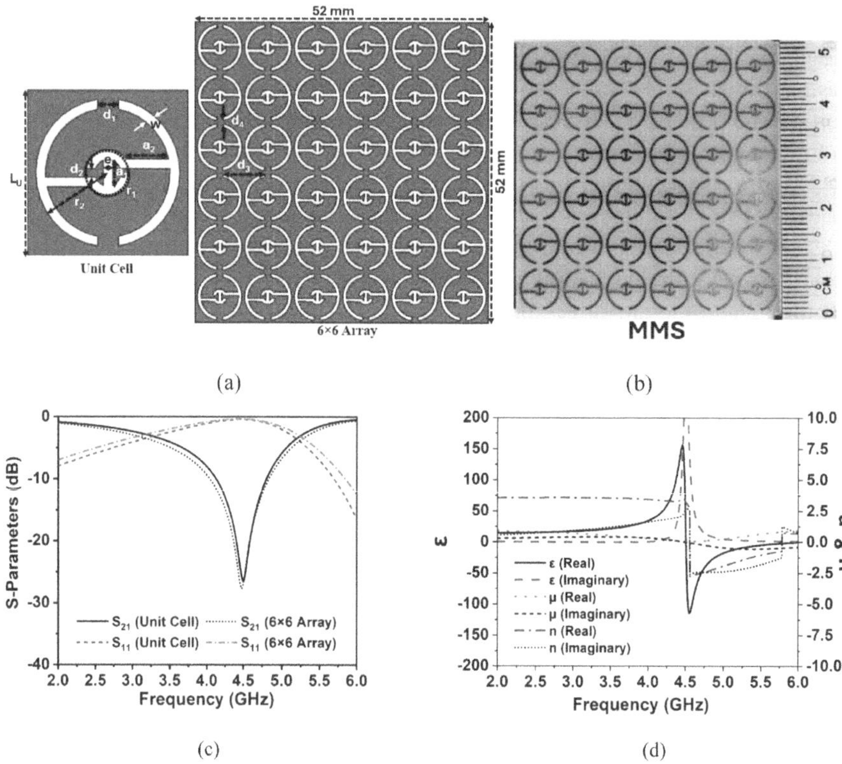

Fig. 4 (a) MMS configuration with unit cell design specifications, (b) fabricated prototype of MMS, (c) S-parameters for the MMS and unit cell configurations, and (d) **extracted parameters**.

Table 3 Design details of MMS configuration.

Parameters	*Size (mm)*	*Parameters*	*Size (mm)*	*Parameters*	*Size (mm)*
a_1	1.41	d_3	7.4	r_1	0.75
a_2	2.42	d_4	0.69	r_2	4
d_1	1.2	e	0.5	w	0.5
d_2	0.5	L_U	9		

Figure 4(d) demonstrates the permittivity, permeability, and refractive index results obtained using the CST post-processing approach. Figure 4(d) provides negative real values for ε and n at 4.5–6 GHz and 4.56–5.78 GHz, respectively, and μ-near-zero (MNZ) values at 3.3–5.3 GHz, validating the MM **epsilon negative** and **μ-near-zero** features. Moreover, the imaginary permittivity and permeability values are noted to be around zero from 3.0–4.0 GHz and 2–6 GHz, respectively. As per Maxwell's expressions,

$$\nabla \times E = i\omega\mu H \text{ (for electric } E \text{ field)} \quad ...(8)$$

and,

$$\nabla \times \mathrm{H} = -\, i\omega\varepsilon E \text{ (for magnetic } H \text{ field)} \quad ...(9)$$

where, ω is resonance frequency, the near-zero quality-based material can decrease the coupling effect in magnetic and electric fields [41]. Table 4 indicates the retrieved MMS unit cell properties. Therefore, the obtained characteristic of the proposed MMS leads to potential in a 5G MIMO system for enhancing overall antenna performance [23, 29, 41].

Table 4 Unit cell characteristics of MM superstrate.

Parameters	*Frequency (GHz)*	*Parameters*	*Frequency (GHz)*
Negative permittivity (real)	4.5–6	Near-zero permittivity (Imaginary)	3–4
Negative refractive index	4.56–5.78	Near-zero permeability (Imaginary)	2–6
Near-zero permeability (real)	3.3–5.3	–	–

Surface current, electric, and magnetic field distributions at 4.5 GHz are also explored to further comprehend the characteristics and operation of the designed MMS, as demonstrated in Figure 5(a–c). A strong current flow is exhibited in the inter-connected hook-shaped strips of each MMS resonator, as depicted in Figure 5(a). Notably, this high current path creates a strong **H-field** in the connected middle metal strips of each resonator, in Figure 5(b). The **E-field distribution** is presented in Figure 5(c), observing that the split gaps in the outer circular ring induce strong E-fields. Moreover, the gaps between consecutive resonators contribute to high electric field density. These high E-fields are induced owing to the capacitive effects on split gaps and gaps between adjacent unit cells. Hence, both gap width and ring size are optimized to get strong magnetic and electric fields for targeted resonance and applications. The distributions of magnetic and electric fields in the proposed MMS satisfy Maxwell's electromagnetic relations [19], in Figure 5.

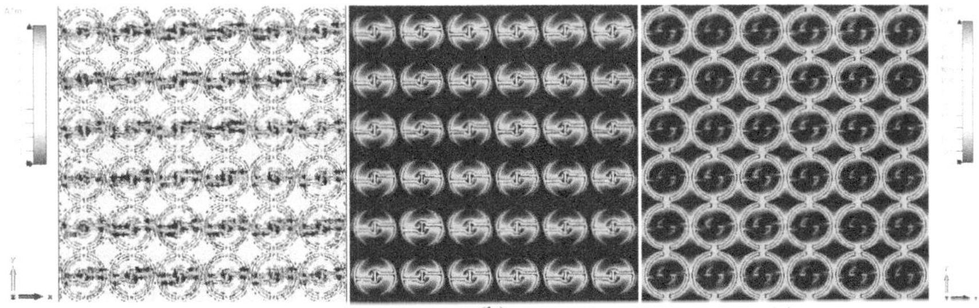

(a) (b) (c)

Fig. 5 (a) Surface current, (b) H-field, and (c) E-field distributions for the developed MMS configuration.

3.2 MIMO Antenna with MM Superstrate

The proposed MMS is mounted atop the PMM-loaded miniaturized MIMO antenna with a height (h) of $0.07\lambda_{min}$ to improve antenna bandwidth, isolation, gain, and efficiency for 5G sub-6 GHz band applications. The MMS is compactly integrated with the MIMO antenna, ensuring its compactness and matching characteristics. Figure 6(a, b) shows the 3D configuration of the developed four-element MMS-integrated MIMO antenna and side-view layout, respectively. The spacing between the antenna and MMS as well as the number of **unit cells** with their periodic intervals are crucial aspects for enhancing BW, isolation, gain, and efficiency while introducing EM coupling between the MMS and MIMO antennas. Moreover, the MMS manipulates the antenna radiation to improve MIMO performance. The proposed MMS-incorporated miniaturized MIMO antenna is optimized for maximum performance, and the structure is generalized such that it can be extended to a **massive MIMO system**. Figure 6(a) exhibits the completed design details for the desired performance.

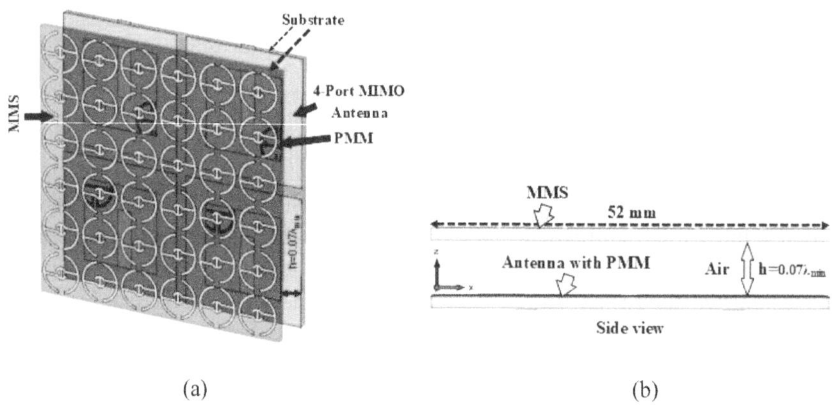

Fig. 6 MIMO antenna configuration with MMS: (a) top view, (b) back view.

4. Performance Analysis

In MIMO antenna systems, the MM structure, spacing, and arrangement of periodic unit cells, as well as the size and placement of MMS are crucial challenges to enhance BW, isolation, gain, and efficiency simultaneously while maintaining compact size. In the proposed design, the position of MMS along with the size and spacing of adjacent unit cells are optimized for achieving maximum **antenna performance**. In addition, the MMS is matched with the antenna size, which gives flexibility of extension to a large MIMO system, compactness, and ease of application. The designed MIMO antenna and MMS are fabricated to verify the real-time performance. Figure 7(a) shows the fabricated photographs of the developed prototype integrated with MMS. A polystyrene spacer is employed to hold the fabricated MMS to the antenna top. The developed MIMO antenna with its experimental arrangement is depicted in Figure 7(b).

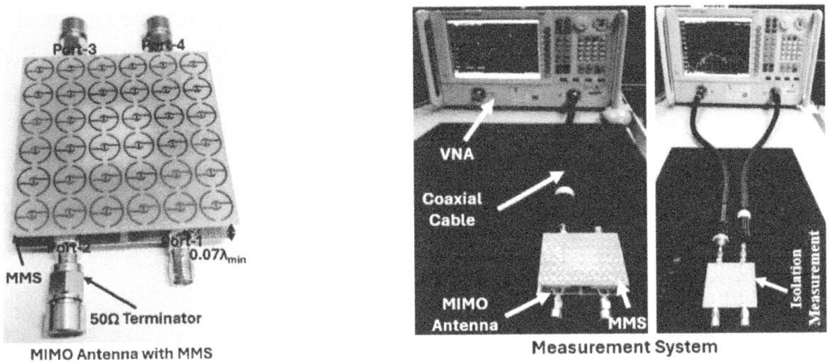

(a) (b)

Fig. 7 (a) Fabricated image of the Quad-port MM-integrated MIMO antenna, (b) **measurement** systems.

4.1 Bandwidth and Isolation Enhancement

Figures 8 and 9 exhibit the computed reflection and transmission properties for the developed MIMO antenna without (W/O) and with MMS, respectively. From Figure 8, the MIMO antenna with MMS operates at 3.04–5.05 GHz (49.7%) with an **enhanced bandwidth** of 20.5% in the desired 5G spectrum. The inclusion of unique MMS in the MIMO system enhances the operating spectrum by 0.95 GHz compared to without MMS, covering the desired complete 5G n77/n78/n79 bands. Furthermore, the proposed MMS approach significantly alleviates coupling energy between MIMO elements with a minimum isolation of about 15 dB (Figure 9). The proposed MMS technique achieves the highest **isolation** of 40.8 dB among adjacent antennas, while without MMS, the isolation is about 31.3 dB. Thus, the isolation improvement of about 9.5 dB is achieved by the MMS. The capacitance between two adjacent unit cells and split gaps of the developed MMS induces EM coupling with the antenna, which leads to antenna BW enhancement and improves isolation by reducing near-field coupling energy among MIMO components [42]. The proper combination of MMS structure, array elements, spacing among adjacent unit cells, size, and distance between MMS and antenna generates a strong coupling between MMS and antenna, greatly enhancing bandwidth and isolation. The MIMO prototypes W/O and with MMS are measured by a two-port network analyzer (Agilent N5227A) to verify the simulation performance. The measured and simulated reflection characteristics are compared in Figure 8(a, b) for the MMS-loaded and unloaded MIMO antenna. The measured IBW of MIMO with MMS is noted between 3.08–5.3 GHz (fractional BW of 52.9%), covering the desired complete 5G n77/n78/n79 bands (Figure 8). Moreover, the MMS improves the measured BW by 20%, proving the contribution of MMS. Figure 9(b) reveals the measured isolation values (mutual coupling between MIMO antennas) and provides a minimum value of >15.8 dB in the desired operating band, indicating that the proposed MMS technique effectively improves isolation by significantly diminishing the near-field coupling energy. A satisfactory correlation is identified between simulation and measurement findings, confirming the promising performance of the proposed technique in the 5G MIMO system. The simulated scattering performances deviated somewhat from the measurement due to measurement tolerance, manufacturing tolerance, and soldering losses, which were not considered in the simulation.

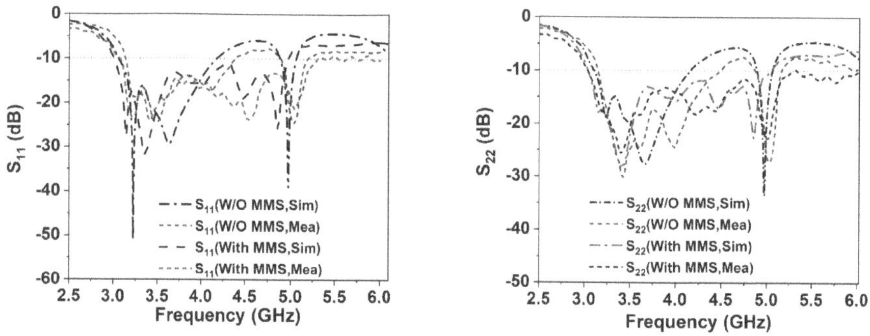

(a) (b)

Fig. 8 Simulated and measured results of the developed MIMO antenna W/O and with MMS: (a) S_{11}, (b) S_{22}.

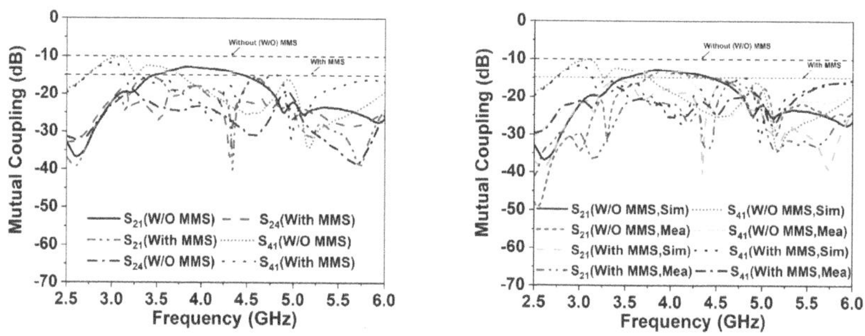

(a) (b)

Fig. 9 MC of the developed MIMO antenna W/O and with MMS: (a) simulated S_{21}, S_{24}, and S_{41}, (b) measured S_{21} and S_{41} with simulations.

Furthermore, the current density and E-field distributions of the developed antenna both W/O and with MMS, are investigated at 3.7 GHz to explore the MMS contribution in isolation, as depicted in Figure 10(a–d). In Figure 10(a), port 1 current density is demonstrated for MIMO without MMS; a noticeable coupling current is observed inside the adjacent antennas, indicating high **mutual coupling**. Therefore, the antenna radiation properties have deteriorated. Upon integrating the proposed MMS, the mutual coupling energy among adjacent antennas is significantly reduced, resulting in enhanced isolation among MIMO antennas (Figure 10(b)). When the proposed MMS is placed close to the antenna, it strives to interact with the antenna **magnetic field** and diminishes the strength of the magnetic field by creating the static field within the MMS due to its interesting near-zero properties, thus enhancing isolation between antenna elements. Moreover, the edge current between two neighboring antennas flows inversely, minimizing the coupling current [23, 29]. The E-field distribution (port 1) of the developed MIMO antenna without

MMS is revealed in Figure 10(c), the high E-field density is noticed within antennas 2, 3, and 4. However, with the presence of MMS, the coupled **electric field** density on adjacent antennas is significantly reduced, thus proving high isolation, which is indicated in Figure 10(d).

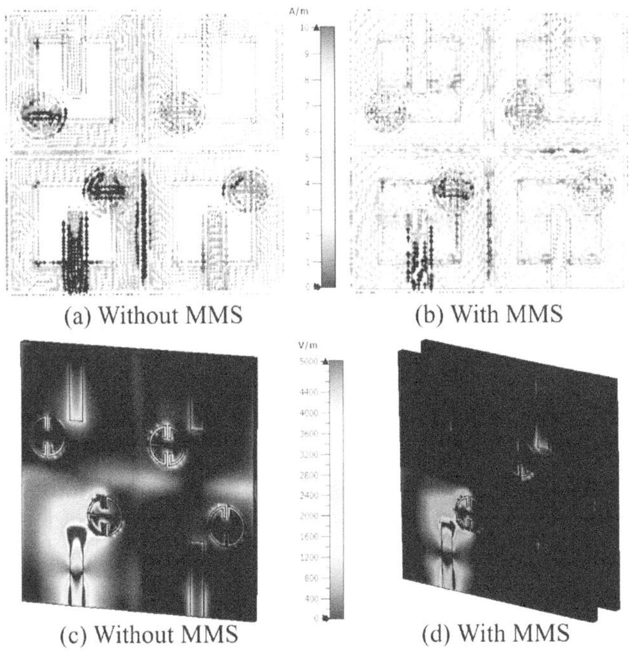

Fig. 10 (a, b) **Surface current,** and (c, d) E-field distribution.

4.2 Gain and Efficiency Enhancement

The developed antenna gain and efficiency are studied both with and without MMS at port 1, where the 50Ω terminator terminates the remaining three ports. Figure 11(a, b) depicts the simulated realized gain and total efficiency of the developed MIMO antenna without and with MMS, which are verified by measurement. It could be observed that the gain and efficiency of the developed MIMO antenna with MMS are enhanced greatly. The gain is enhanced from 2–4.4 dBi and total efficiency is increased from 60–66% after adopting the proposed MMS technique by the improvement of 2.4 dBi and 6%, respectively. The integration of MMS above the proposed MIMO antenna at a desired distance yields enhanced antenna gain and efficiency, thanks to the formation of a cavity effect. The MMS passes the EM waves in an in-phase way while combining the antenna-radiated direct and reflected waves, enhancing the antenna gain and efficiency [29, 43]. Although the proposed MMS improves antenna efficiency, the total efficiency of the developed MIMO antenna is 66%. Figure 11 also shows the measured gain and efficiency values achieved from the SATIMO near-field experimental system, indicating good agreement between measurement and simulation. The peak measured enhanced **gain** and **efficiency** are 4.8 dBi and 63%, respectively, for the developed MMS-integrated MIMO antenna, confirming its applicability in 5G bands. A slight discrepancy between simulation and measurement is seen owing to unanticipated experimental tolerance, soldering losses, and manufacturing flaws.

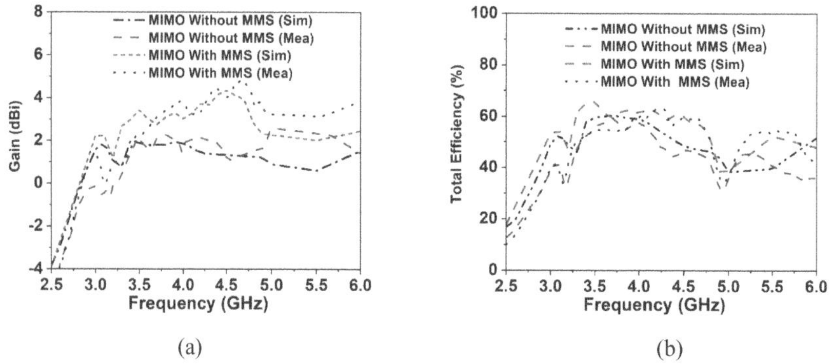

Fig. 11 Measured and simulated: (a) gain, and (b) total **efficiency** of the developed MIMO prototype W/O and with MMS.

4.3 Radiation Properties

Figure 12(a, b) reveals the two-dimensional (2-D) port-1 **radiation patterns** of the proposed quad-port antenna with and without MMS at 3.6 GHz. The MMS focuses antenna radiation in the desired direction, hence enhancing gain and directivity.

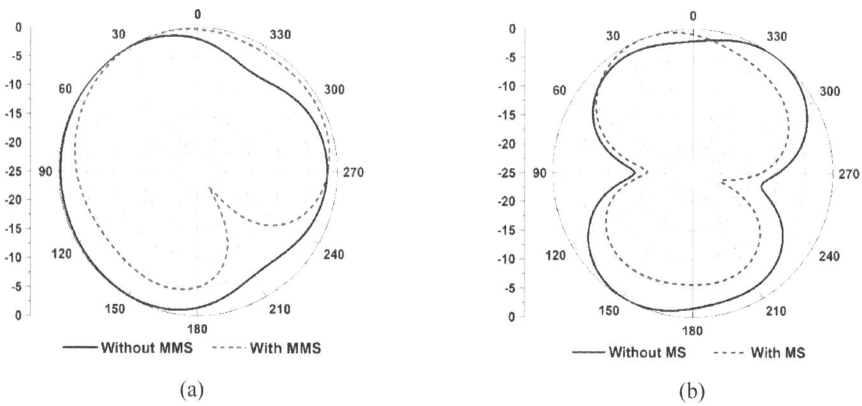

Fig. 12 2-D antenna pattern W/O and with MMS (a) X-Z plane (b) Y-Z plane.

4.4 MIMO Diversity Parameters

The correlation and port isolation characteristics of a MIMO system are crucial to confirm its reliability and applicability in 5G MIMO applications. The diversity performances, including ECC, DG, and CCL are explored to demonstrate the applicability and reliability of the developed MIMO antenna.

Envelope Correlation Coefficient (ECC): ECC states the correlation between MIMO components and this is a key diversity parameter of any MIMO antenna. The

ECC should be <0.5 for adjacent two antennas, and it is calculated using the following equations [23, 24].

$$ECC = \frac{\left| S_{ii}^{*} S_{ij} + S_{ji}^{*} S_{jj} \right|^{2}}{\left(1 - |S_{ii}|^{2} - |S_{ji}|^{2}\right)\left(1 - |S_{jj}|^{2} - |S_{ij}|^{2}\right)} \qquad ...(10)$$

$$\mathrm{ECC} = \frac{\left| \iint_{0}^{4\pi} \left[\vec{E}_i(\theta, \varphi) \times \vec{E}_j(\theta, \varphi) \right] d\Omega \right|^{2}}{\iint_{0}^{4\pi} \left| \vec{E}_i(\theta, \varphi) \right|^{2} d\Omega \iint_{0}^{4\pi} \left| \vec{E}_j(\theta, \varphi) \right|^{2} d\Omega} \qquad ...(11)$$

$$\vec{E}_i(\theta, \varphi) . \vec{E}_j(\theta, \varphi) = E_{\theta i}(\theta, \varphi)\, E_{\theta j}^{*}(\theta, \varphi) + E_{\varphi i}(\theta, \varphi) E_{\varphi j}^{*}(\theta, \varphi) \qquad ...(12)$$

The reflection and transmission coefficients are presented by S_{ii} and S_{ij}, respectively and the solid angle is Ω. The radiated pattern for antenna i and antenna j is denoted by $\vec{E}_i(\theta, \varphi)$ and $\vec{E}_j(\theta, \varphi)$, respectively. Figure 13(a) reveals the ECC values of the developed MIMO antenna W/O and with MMS, implying the effects of MMS. The developed antenna ECC is less than 0.006 over the targeted operating spectrum. The low ECC over the wide operating band of the proposed antenna indicates promising diversity performance with high reliability and channel capacity.

Diversity Gain (DG): This is another important performance indicator for exploring MIMO diversity characteristics. It implies the reduction of transmission power with the introduction of a diversity scheme, which ensures the MIMO antenna quality and reliability in communication systems. The DG of a MIMO system can be obtained using the approximation Equation (13) [13, 24].

$$DG = 10\sqrt{1 - |ECC|^{2}} \qquad ...(13)$$

Figure 13(b) exhibits the derived DG characteristics for the suggested MIMO configuration without and with MMS. The excellent high DG of over 9.97 dB is obtained by the proposed MM technique in the 5G n77/n78/n79 bands. The high DG values (almost 10 dB) indicate outstanding diversity performance.

Channel Capacity Loss (CCL): The CCL identifies the capacity of a MIMO communication link without suffering a substantial loss, and it should be <0.4 bits/s/Hz for a reliable MIMO antenna.

The CCL is obtained using the given Equation (14) [13, 24].

$$CCL = -\log_2 \, det\,(\psi^{R}) \qquad ...(14)$$

where, (ψ^{R}) signifies the correlation matrix, which is defined by,

$$\psi_R = \begin{bmatrix} \rho_{11} & \rho_{12} & \rho_{13} & \rho_{14} \\ \rho_{21} & \rho_{22} & \rho_{23} & \rho_{24} \\ \rho_{31} & \rho_{32} & \rho_{33} & \rho_{34} \\ \rho_{41} & \rho_{42} & \rho_{43} & \rho_{44} \end{bmatrix} \qquad ...(15)$$

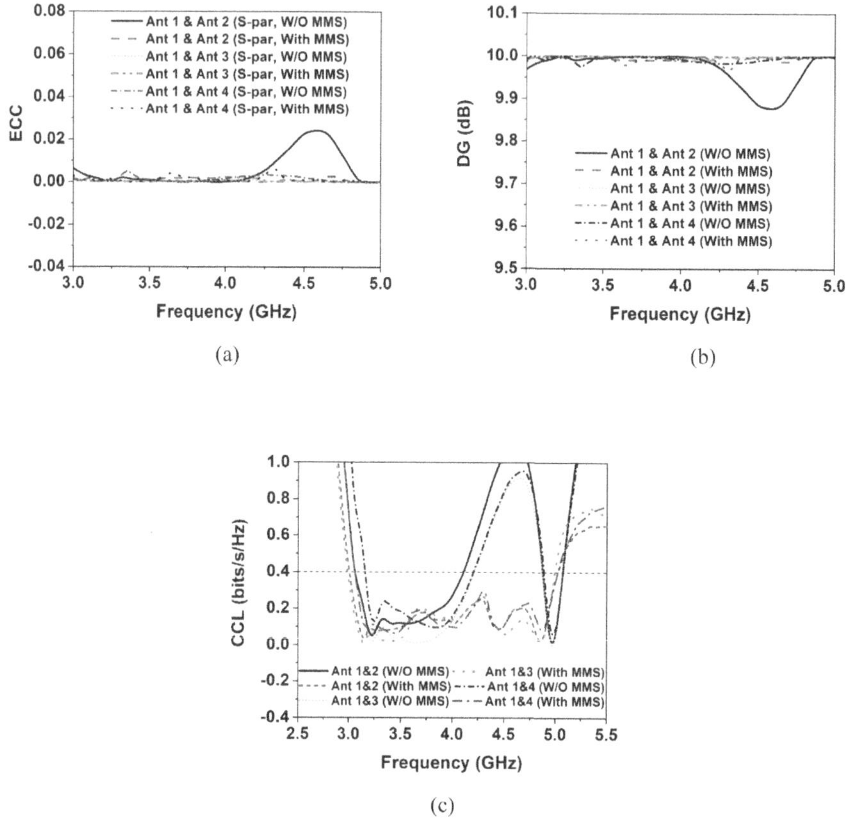

Fig. 13 **Diversity parameters** of the developed MIMO antenna: (a) ECC, (b) DG, (c) CCL between adjacent antennas.

$$\rho_{ii} = 1 - (|S_{ii}|^2 + |S_{ij}|^2) \quad ...(16)$$

$$\rho_{ij} = -(S_{ii}^* S_{ij} + S_{ji}^* S_{jj}) \quad ...(17)$$

where, $i, j = 1, 2, 3$ or 4 regarding port number. Figure 13(c) reveals the CCL curves for the developed MIMO configuration without and with MMS, implying the contribution of MMS. The CCL curves show <0.4 bits/s/Hz over the 5G bands of n77/n78/n79 with MMS, implying high data rate and throughput.

5. Enhancement of MIMO Elements

The aim of this chapter is to design a 5G sub-6 GHz frequency targeted simple **extendable MIMO structure** by avoiding orthogonal arrangement complexity. Moreover, the demand for large-array MIMO antennas has drawn more concern in emerging high-speed communications [1]. Therefore, the developed MIMO configuration is unique and simple, as well as the elements are designed by mirrored side-by-side pattern, which can easily be extended to the large array MIMO antenna. Moreover, the designed unique MM superstrate is also matched with the antenna size leading it to be readily extended to a large array of MIMO configurations.

Figure 14(a, b) reveals the 16-port MIMO configuration with MMS consisting of 25 × 6 array MM unit cells to verify the proposed technique for **massive MIMO** applications.

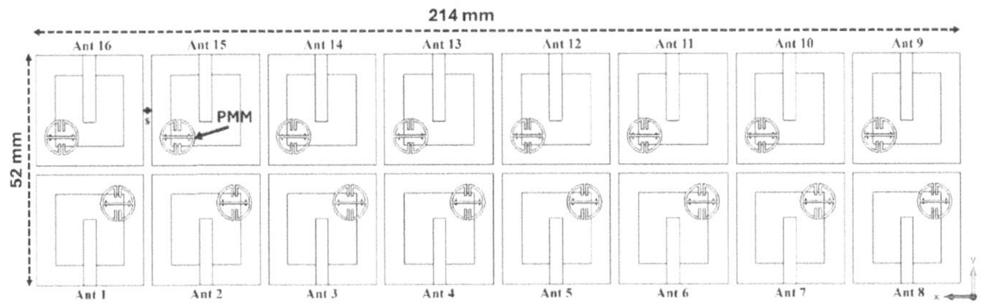

(a)

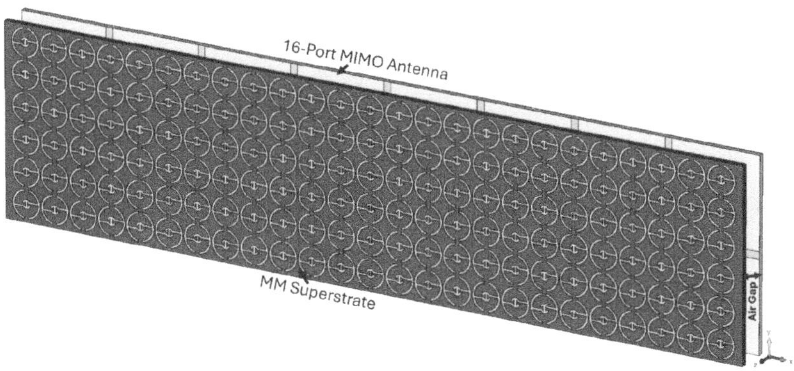

(b)

Fig. 14 (a) Proposed 16-port PMM loaded MIMO configuration, (b) **16-port MIMO configuration** with MMS.

The MMS for the 16-port MIMO configuration is designed as a quad-port antenna and matched with the antenna dimension exhibited in Figure 14(b). In this design, four-port MIMO antennas are repeated in the *X*-axis with a very compact electrical dimension of $0.52\lambda_{min} \times 2.14\lambda_{min} \times 0.016\lambda_{min}$, and a similar-sized MMS is placed above the antenna with a height of $0.07\lambda_{min}$. Notably, these design specifications, including the MMS height from the antenna, are identical to the 4-port antenna. Figure 14(b) demonstrates the proposed 16-port MIMO antenna with MMS and the values of S-parameters are presented in Figure 15(a, b) for a 16-port MIMO antenna without MMS compared with MMS. In Figure 15(a, b), the operating spectrum for S_{11} is 3.04–4.15 GHz and 4.77–4.95 GHz with a maximum IBW of 30.88%, and for S_{22} is 3–4.12 GHz and 4.78–4.94 GHz with a maximum IBW of 31.46%. After incorporating MMS, a **wide operating band** of 3.02–4.97 GHz (48.8%) and 2.99–5.05 GHz (51.24%) for S_{11} and S_{22} is achieved by the improvement of 17.92% and 19.78%, respectively, which nicely covers the targeted 5G spectrums.

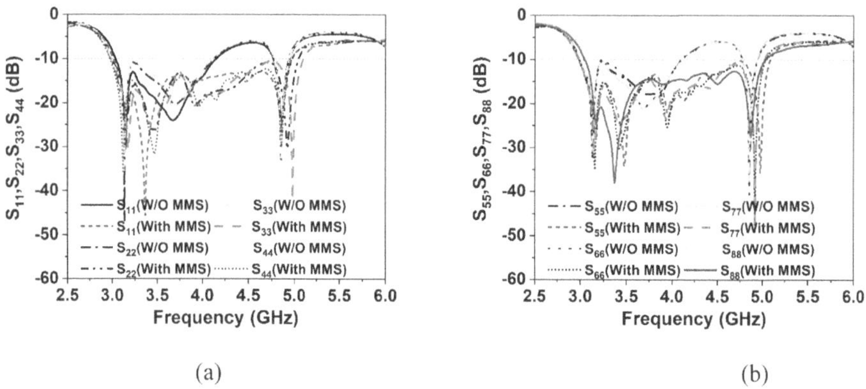

Fig. 15 Reflection characteristics of the developed 16-port antenna configuration: (a) S_{11}, S_{22}, S_{33} and S_{44}, (b) S_{55}, S_{66}, S_{77} and S_{88}.

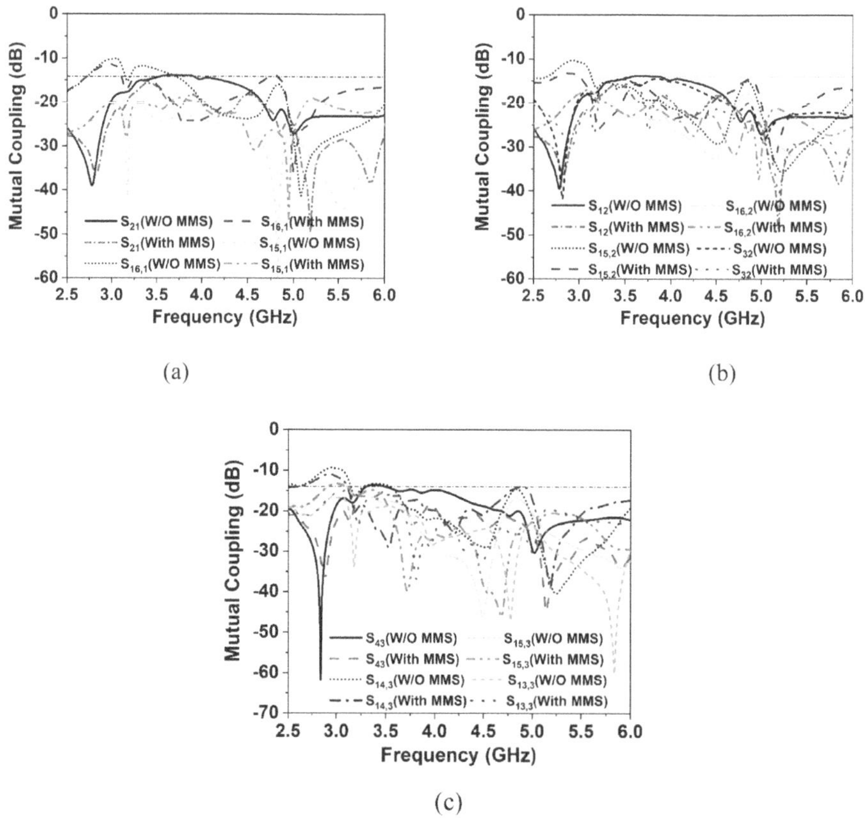

Fig. 16 MC of the developed 16-port antenna configuration: (a) port 1, (b) port 2, and (c) port 3.

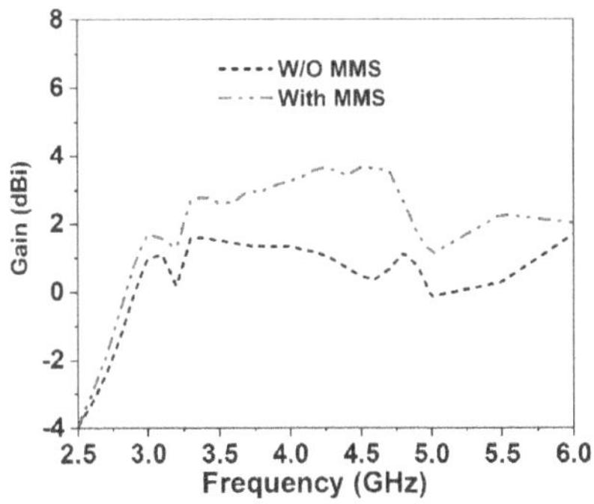

Fig. 17 Realized gain for 16-port MIMO configuration.

In addition, the proposed MMS technique substantially alleviates the **mutual coupling** energy among MIMO elements with a minimum isolation of about 14 dB, as shown in Figure 16(a, c). Figure 17 also reveals the gain curve for a 16-port MIMO system without and with MMS and shows gain enhancement by 2.1 dBi with MMS. These promising results confirm the suitability and flexibility of the proposed approach for a larger array of MIMO applications in the 5G sub-6 GHz band.

6. Summary

This chapter proposed a miniature wideband extendable MM-integrated MIMO antenna with enhanced high isolation, gain, and efficiency for 5G NR sub-6 GHz band applications. The important features of this developed antenna include high isolation within a highly **compact** dimension, a wide targeted operating spectrum, miniaturization, **enhanced gain** and efficiency, and flexibility in developing a large MIMO system. The proposed MIMO configuration achieves 52.94% miniaturization and enhanced high minimum isolation of about 15 dB within the very compact dimension of 52 × 52 mm^2. Moreover, the incorporation of MMS enhances the MIMO antenna BW by 20.5%, gain by 2.4 dBi, and efficiency by 6%. Furthermore, the proposed technique gives a very **low ECC** <0.006, high DG >9.97 dB, and low CCL <0.4 bits/s/Hz, implying low correlation, high reliability, and channel capacity. The surface current distribution and electric field density are presented and investigated to verify and understand the proposed MM working mechanism and contribution to MIMO systems. Moreover, the developed antenna performances are assessed without and with the MM structure to explore the effectiveness of the proposed technique. In addition, a 16-port MIMO antenna is designed and analyzed with MM and shows promising performance, which confirms its applicability for large array MIMO applications. The simulated MIMO performances are verified experimentally and demonstrated well consistent. Therefore, the proposed simple, **extendable**, and efficient design strategy with its reliable performance, makes it a promising prospect for 5G massive MIMO applications. This proposed technique can also be implemented for mm-wave 5G massive MIMO applications, leading the way for 6G communications.

References

[1] D. Abbasi, A. Aziz, K. Aljaloud, A.H. Alqahtani, A. Zerguine, A. Al-Garni, et al. "Higher Order MIMO Antenna Design Using Capacitively Coupled Meandered Loop Element for 5G Smartphone Applications," *IEEE Transactions on Circuits and Systems II: Express Briefs,* 2023.

[2] Y. Fang, Y. Liu, Y. Jia, J. Liang and H.H. Zhang. "Reconfigurable Structure Reutilization Low SAR MIMO Antenna for 4G/5G Full-Screen Metal Frame Smartphone Operation," *IEEE Antennas and Wireless Propagation Letters,* 2023.

[3] M. Hu and Y. Li. "Wideband Back Cover Microstrip Antenna with Multiple Shorting Vias for Mobile 5G MIMO Applications," *IEEE Transactions on Antennas and Propagation,* 2023.

[4] M.A. Jensen and J.W. Wallace. "A review of antennas and propagation for MIMO wireless communications," *IEEE Transactions on Antennas and propagation,* vol. 52, pp. 2810–2824, 2004.

[5] Z. Wang, C. Li, Q. Wu and Y. Yin. "A metasurface-based low-profile array decoupling technology to enhance isolation in MIMO antenna systems," *IEEE Access,* vol. 8, pp. 125565–125575, 2020.

[6] F. Khajeh-Khalili, M.A. Honarvar, M. Naser-Moghadasi and M. Dolatshahi. "High-gain, high-isolation, and wideband millimetre-wave closely spaced multiple-input multiple-output antenna with metamaterial wall and metamaterial superstrate for 5G applications," *IET Microwaves, Antennas, & Propagation,* vol. 15, pp. 379–388, 2021.

[7] C. Deng, X. Cao, D. Li and W. Yu. "Compact Dual-band MIMO Antenna with Shared Decoupling Structure for 5G Mobile Terminals," *IEEE Antennas and Wireless Propagation Letters,* 2023.

[8] M.Z.B. Chowdhury, M.T. Islam, H. Rmili, I. Hossain, M.Z. Mahmud and M. Samsuzzaman. "A low-profile rectangular slot antenna for sub-6 GHz 5G wireless applications," *International Journal of Communication Systems,* vol. 35, p. e5321, 2022.

[9] D. Tripathi, D. Srivastava and R.K. Verma, "Bandwidth enhancement of slotted rectangular wideband microstrip antenna for the application of WLAN/WiMAX," *Wireless Personal Communications,* vol. 119, pp. 1193–1207, 2021.

[10] Y. Qin, L. Zhang, C.-X. Mao and H. Zhu. "A Compact Wideband Antenna With Suppressed Mutual Coupling for 5G MIMO Applications," *IEEE Antennas and Wireless Propagation Letters,* vol. 22, pp. 938–942, 2022.

[11] H.-H. Tran, T.T.-L. Nguyen, H.-N. Ta and D.-P. Pham. "A Metasurface-Based MIMO Antenna with Compact, Wideband, and High Isolation Characteristics for Sub-6 GHz 5G Applications," *IEEE Access,* 2023.

[12] M. Singh, W. Saleh, A. Abed and M. Fauzi. "A Review on Massive MIMO Antennas for 5G Communication Systems on Challenges and Limitations," *UKM Eng. J,* vol. 35, pp. 95–103, 2023.

[13] F. Ez-Zaki, H. Belahrach, G. Abdelilah, S. Ahmad, A. Khabba, K. Aitbelaid, et al. "Double Negative (DNG) Metamaterial based Koch fractal MIMO antenna design for sub-6 GHz V2X communication," *IEEE Access,* 2023.

[14] I. Khan, K. Zhang, L. Ali and Q. Wu. "Enhanced Quad-Port MIMO Antenna Isolation with Metamaterial Superstrate," *IEEE Antennas and Wireless Propagation Letters,* 2023.

[15] P. Kumar, T. Ali and M.M. Pai, "Electromagnetic metamaterials: A new paradigm of antenna design," *IEEE Access,* vol. 9, pp. 18722–18751, 2021.

[16] K. Iqbal and Q.U. Khan. "Review of Metasurfaces through Unit Cell Design and Numerical Extraction of Parameters and Their Applications in Antennas," *IEEE Access,* vol. 10, pp. 112368–112391, 2022.

[17] X. Tian, Q. Zeng, S.A. Kurt, R.R. Li, D.T. Nguyen, Z. Xiong, et al. "Implant-to-implant wireless networking with metamaterial textiles," *Nature Communications,* vol. 14, p. 4335, 2023.

[18] M.B. Billa, M.T. Islam, T. Alam, A. Hoque, S. Albadran, H. Alsaif, et al. "Morphological, optical, and electrical properties of CoxCa (0.90-x) Ni0. 10Fe2O4 based flexible metamaterial substrate for industrial chemical contamination sensing," *Journal of Science: Advanced Materials and Devices,* p. 100659, 2023.

[19] M.R. Islam, M.T. Islam, M. Moniruzzaman, M. Samsuzzaman and H. Arshad. "Penta band single negative meta-atom absorber designed on square enclosed star-shaped modified split ring resonator for S-, C-, X-and Ku-bands microwave applications," *Scientific Reports,* vol. 11, p. 8784, 2021.

[20] C. Wang, X.S. Yang and B.Z. Wang. "A metamaterial-based compact broadband planar monopole MIMO antenna with high isolation," *Microwave and Optical Technology Letters,* vol. 62, pp. 2965–2970, 2020.

[21] M. Ameen and R.K. Chaudhary. "Isolation Enhancement of Metamaterial-Inspired Two-port MIMO Antenna using Hybrid Techniques," *IEEE Transactions on Circuits and Systems II: Express Briefs,* 2023.

[22] P. Garg and P. Jain, "Isolation improvement of MIMO antenna using a novel flower shaped metamaterial absorber at 5.5 GHz WiMAX band," *IEEE Transactions on Circuits and Systems II: Express Briefs,* vol. 67, pp. 675–679, 2019.

[23] F. Liu, J. Guo, L. Zhao, G.-L. Huang, Y. Li and Y. Yin. "Dual-band metasurface-based decoupling method for two closely packed dual-band antennas," *IEEE Transactions on Antennas and Propagation,* vol. 68, pp. 552–557, 2019.

[24] G. Saxena, S. Kumar, S. Chintakindi, A. Al-Tamim, M.H. Abidi, W.A.M. Saif, et al. "Metasurface Instrumented High Gain and Low RCS X-Band Circularly Polarized MIMO Antenna for IoT Over Satellite Application," *IEEE Transactions on Instrumentation and Measurement,* 2023.

[25] R. Mark, N. Rajak, K. Mandal and S. Das. "Metamaterial based superstrate towards the isolation and gain enhancement of MIMO antenna for WLAN application," *AEU-International Journal of Electronics and Communications,* vol. 100, pp. 144– 152, 2019.

[26] M. Ameen, O. Ahmad and R. Chaudhary. "Wideband circularly-polarised high-gain diversity antenna loaded with metasurface reflector for small satellite applications," *Electronics Letters,* vol. 55, pp. 829–831, 2019.

[27] M.A. Sufian, N. Hussain, H. Askari, S.G. Park, K.S. Shin and N. Kim. "Isolation enhancement of a metasurface-based MIMO antenna using slots and shorting pins," *IEEE Access,* vol. 9, pp. 73533–73543, 2021.

[28] A. Singh, A. Kumar and B.K. Kanaujia. "High gain and enhanced isolation MIMO antenna with FSS and metasurface," *Optik,* vol. 286, p. 170982, 2023.

[29] P.K. Panda and D. Ghosh. "Isolation and gain enhancement of patch antennas using EMNZ superstrate," *AEU-International Journal of Electronics and Communications,* vol. 86, pp. 164–170, 2018.

[30] J.R. Reis, M. Vala, T.E. Oliveira, T.R. Fernandes and R.F.S. Caldeirinha. "Metamaterial-inspired flat beam-steering antenna for 5G base stations at 3.6 GHz," *Sensors,* vol. 21, p. 8116, 2021.

[31] R.E.A. Shehata, M. Hindy, H. Elmekati and A. Elboushi. "Design of a Beam-Steering Metamaterial Inspired LPDA Array for 5G Applications," *Progress in Electromagnetics Research M.,* vol. 117, 2023.

[32] B.A. Esmail, S. Koziel, L. Golunski, H.B.A. Majid and R.K. Barik. "Overview of Metamaterials-Integrated Antennas for Beam Manipulation Applications: The Two Decades of Progress," *IEEE Access,* vol. 10, pp. 67096–67116, 2022.

[33] S. Pandit, A. Mohan and P. Ray. "Low-RCS low-profile four-element MIMO antenna using polarization conversion metasurface," *IEEE Antennas and Wireless Propagation Letters,* vol. 19, pp. 2102–2106, 2020.

[34] M. Ameen, O. Ahmad and R.K. Chaudhary. "Bandwidth and gain enhancement of triple-band MIMO antenna incorporating metasurface-based reflector for WLAN/WiMAX applications," *IET Microwaves, Antennas, & Propagation,* vol. 14, pp. 1493–1503, 2020.

[35] G. Chen, C. Guo, J. Xue, Z. Wang and M. Pang. "Miniaturized Metamaterial Ultra-Wideband Antenna for WLAN and Bluetooth Applications," *Progress In Electromagnetics Research C.,* vol. 132, pp. 117–127, 2023.

[36] C. Milias, R.B. Andersen, P.I. Lazaridis, Z.D. Zaharis, B. Muhammad, J.T. Kristensen, et al. "Miniaturized Multiband Metamaterial Antennas With Dual-Band Isolation Enhancement," *IEEE Access,* vol. 10, pp. 64952–64964, 2022.

[37] O. Elalaouy, M. El Ghzaoui and J. Foshi. "Enhancing antenna performance: A comprehensive review of metamaterial utilization," *Materials Science and Engineering: B.,* vol. 304, p. 117382, 2024.

[38] X. Chen, T.M. Grzegorczyk, B.-I. Wu, J. Pacheco Jr, and J.A. Kong. "Robust method to retrieve the constitutive effective parameters of metamaterials," *Physical Review E,* vol. 70, p. 016608, 2004.

[39] M.S. Islam, M.T. Islam and A.F. Almutairi. "A portable non-invasive microwave based head imaging system using compact metamaterial loaded 3D unidirectional antenna for stroke detection," *Scientific Reports,* vol. 12, p. 8895, 2022.

[40] C.A. Balanis. *Antenna Theory: Analysis and Design*: John Wiley & Sons, 2016.

[41] R. Mark, H.V. Singh, K. Mandal and S. Das. "Mutual coupling reduction using near-zero ε and μ metamaterial-based superstrate for a MIMO application," *IET Microwaves, Antennas & Propagation,* vol. 14, pp. 479–484, 2020.

[42] R. Ojo, M.F. Jamlos, P.J. Soh, M.A. Jamlos, N. Bahari, Y.S. Lee, et al. "A triangular MIMO array antenna with a double negative metamaterial superstrate to enhance bandwidth and gain," *International Journal of RF and Microwave Computer-Aided Engineering,* vol. 30, p. e22320, 2020.

[43] P. Usha and C. Krishnan. "Epsilon near zero metasurface for ultrawideband antenna gain enhancement and radar cross section reduction," *AEU-International Journal of Electronics and Communications,* vol. 119, p. 153167, 2020.

Chapter 8

Metamaterial Based Beam Tilting for mm Wave Antenna Applications

Mohammad Lutful Hakim[1], *Mohammad Tariqul Islam*[2]
and *Touhidul Alam*[3]

1. Introduction

Metamaterial (MTM) is an artificially engineered material that exhibits unnatural electromagnetic (EM) properties like negative permeability, permittivity, refractive index, magnitude, and phase change of incident EM wave [1–5]. These unnatural EM properties lead MTMs to various applications such as sensing [6, 7], absorption [8, 9], SAR reduction [10], superlenses [11], energy harvesting [12], electromagnetic interference (EMI) shielding [13, 14], antenna performance enhancement [15–18], etc. MTM can be categorized as single negative (SNG) and double negative (DNG) based on the fact that either one of permeability or permittivity is negative or both are negative [19–23]. Metallic rings and splits can be used to create an artificial MTM . The geometrical structure of the MTM is a very crucial factor for negative permeability and permittivity at resonance frequency [6, 24–28]. Nowadays, single negative MTMs are extensively investigated for sensors, EMI shielding, antenna gain enhancement, multiple-input multiple-output (MIMO) antenna isolation enhancement, antenna beam tilting, etc. [9, 13, 29–33]. A highly isolated MIMO antenna system can also boost the antenna performance through a radiation pattern-deflection mechanism [34, 35]. Moreover, this type of antenna is essential for system power frugal, interference reduction, quality service, system security, and angular coverage [36]. Recent research

[1,2] Department of Electrical, Electronic and Systems Engineering, Universiti Kebangsaan Malaysia, Bangi, Selangor, Malaysia.

[3] Pusat Sains Angkasa (ANGKASA), Institut Perubahan Iklim, Universiti Kebangsaan Malaysia, 43600 UKM Bangi, Selangor, Malaysia.

* Corresponding author: tariqul@ukm.edu.my, lutful@ukm.edu.my

shows that various methods have been implemented to deflect the radiation pattern of the antenna, such as photoconductive switches-based optical controller [37], PIN diode [38], mechanical reconfigurable structure [39], phase shifter [40], integrated lens, and MTMs integration [34, 41] etc. Frequency Selective Surfaces (FSSs) kind metasurface stand out among the various techniques for antenna performance improvement due to their unique and controllable properties, such as negative permittivity and permeability [34, 42, 43]. FSSs allow for the modification of antenna radiation patterns without relying on phase shifters, complex feeding networks, or mechanical reconfigurations. Extensive research has been conducted on using simple design structures to achieve radiation pattern control and gain enhancement. For instance, vertically stacked three-dimensional (3D) gradient index (GRIN) MTM lenses have been developed for radiation deflection [44] and [45]. However, the overall gain of the antenna is reduced. Similarly, an epsilon near-zero (ENZ) MTM-based 3D lens is used to deflect the radiation pattern at mmWave frequencies, significantly increasing antenna gain [46, 47], where antenna gain is increased significantly. A vertically placed epsilon near-zero FSS-based deflected beam slotted antenna is shown for port 1 and port 2. Additionally, a bow-tie antenna operating at sub-6 GHz 5G frequencies is developed using low permittivity FSSs, where the FSSs are positioned vertically in the propagation direction to deflect the radiation pattern and enhance the antenna gain [48]. In [49], dipole antenna radiation pattern deflection and gain enhancement techniques are developed by reconfigurable plan ar low and high-index metamaterials. However, many of these radiation deflection techniques are limited by structural complexities associated with their 3D configurations and fail to ensure stable gain across frequencies. Furthermore, most of these designs do not evaluate MIMO performance, which is a critical factor for fifth generation (5G) communications to enhance channel capacity and system reliability. Consequently, a planar antenna design capable of radiation pattern deflection, coupled with MIMO compatibility, is highly desirable for advanced 5G applications. The unusual EM properties of MTMs, such as SNG and DNG characteristics, make them a promising solution for achieving antenna gain improvements, isolation enhancements, and radiation pattern deflection by precisely controlling radiated EM waves [50]. The development of antennas for mm-wave 5G wireless communication systems presents significant challenges in current research [51]. This frequency range requires high-gain antennas to overcome the higher path loss and atmospheric absorption compared to lower frequency bands, ensuring the link budget is sustained [52]. Additionally, MIMO configurations are essential for increasing channel capacity, with isolation between MIMO elements being a critical factor. The 3rd Generation Partnership Project (3GPP) and various telecommunication industries have allocated mm-wave frequency bands for 5G applications: N258 (24.25–27.5 GHz), N257 (26.5–29.5 GHz), N260 (37.0–40.0 GHz), and N259 (39.5–43.5 GHz) [53]. Therefore, the transmission-blocking MTM mechanism is especially valuable in these frequency ranges, as it can enhance gain, isolation, and enable beam tilting. A slotted patch MTM designed for antenna gain enhancement operates in the 17.81–20.67 GHz frequency range [30]. Additionally, an interconnected circular ring resonator-based SNG MTM is introduced for 5G antenna gain enhancement in [54]. The transmission-blocking mechanism in the MTM design primarily covers the 28 GHz frequency band. However, transmission-blocking at 38 GHz is achieved using a modified peace-logo planar MTM , which enhances both gain and isolation. Additionally, the MTM is applicable in antenna beam tilting and reflection array applications, as demonstrated in various studies [55, 56] and [57], respectively. In [55], the antenna beam is tilted using a dual-layer square split-ring resonator at a resonant frequency of 10 GHz, enabling precise control over the direction of the radiated beam. Furthermore, a beam-tilting approach utilizing a uniplanar resonator is introduced for a 5G antenna in [56], operating at a resonant frequency of 29 GHz, providing efficient beam steering for high-frequency applications. The above examples indicate that while MTMs have been successfully integrated with antennas across microwave to mmWave frequencies, present designs for mmWave applications often target only a single 5G frequency band. To address this limitation, a transmission-blocking MTM that supports multiple 5G mmWave bands, including N258, N257, N260, and N259, is highly desirable.

This chapter presents the design and performance analysis of an interconnected square split ring resonator (ISSRR)-based transmission-blocking metamaterial for 5G mmWave applications. The MTM's transmission-blocking features are analyzed using an equivalent circuit model (ECM), along with studies on its MTM properties, electric and magnetic field distributions, and surface current behavior. These analyses demonstrate how the MTM significantly enhances the performance of mmWave devices at targeted frequencies. The chapter further explores the MTM's application in improving MIMO antenna performance, offering practical solutions for advanced 5G communication systems.

2. Polarization Sensitive Triple Band MTM

2.1 Design Analysis

Figure 1(a) illustrates the proposed MTM design based on ISSRRs , along with its geometrical parameters: (L = 5 mm, $R1$ = 4.5 mm, $R2$ = 2.85 mm, $R3$ = 1.9 mm, $W1$ = 0.2 mm, $W2$ = 0.2 mm, $W3$ = 0.1 mm, $G1$ = 0.10 mm, $G2$ = 0.10 mm). The unit cell is designed on a Rogers RT5880 substrate with a thickness of 0.79 mm, a dielectric constant of 2.2, and a loss tangent of 0.009. The copper patch is placed on the top surface of the substrate, while the backside remains vacant. The copper layer has a thickness of 0.035 mm and an electrical conductivity of 5.96×10^7 S/m. The design of the unit cell includes a concentric square ring resonator forming the middle ring, with the inner and outer rings consisting of square split-ring resonators. A TM wave is incident along the negative z-axis, with boundary conditions applied along the x- and y-axes. In the unit cell design, the middle ring is formed by a concentric square ring resonator, while the inner and outer rings are created using square split-ring resonators. A TM wave is applied along the negative z-axis, with boundary conditions set along the x- and y-axes. Figure 1(b) shows the simulation setup of the developed MTM. The retrieved reflection coefficient (S_{11}) reflected wave phase value and transmission coefficient (S_{21}) are presented in Figure 2, which shows $S_{21} < -10$ dB is achieved from 24.60–28.47 GHz, 36.50–39.65 GHz, and 40.77–44 GHz frequency. The MTM achieves maximum transmission-blocking resonance at frequencies of 27.09 GHz, 38.71 GHz, and 41.81 GHz. A passband of -10 dB is observed in the frequency range of 30.96–32.13 GHz. The stopband frequencies are reflected toward the positive *z*-axis, exhibiting a reflection phase shift from 90° to – 90°. The concept of the designed ISSRR unit cells towards the final design and corresponding transmission coefficient (S_{21}) is shown in Figure 3(a) and Figure 3 (b), respectively. In the first design step, a concentric square ring resonator is created, which does not exhibit any transmission blocking at the desired operating frequency. In the second step, two splits are introduced in the arms of the resonator to induce capacitance, forming a square split-ring resonator. This modification results in a transmission block frequency range from 24.7 – 28 GHz.

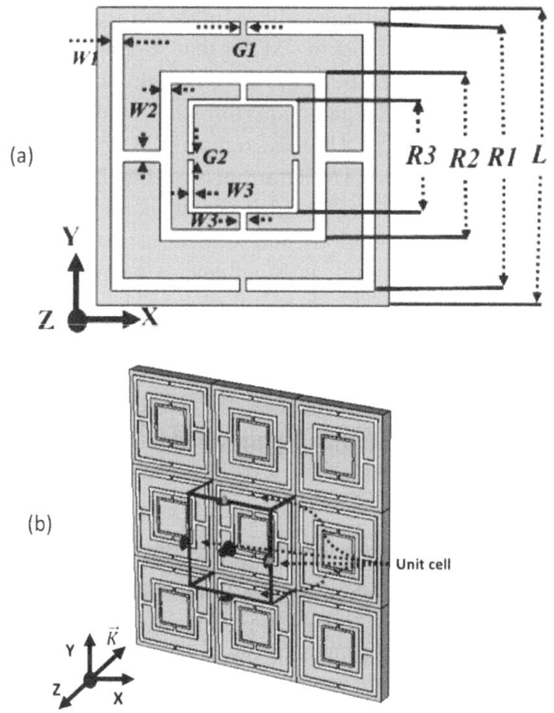

Fig. 1 (a) front view of the developed ISSRR unit cell-based MTM, and (b) simulation setup.

Subsequently, an additional inner concentric square ring resonator was introduced, resulting in a dual stopband frequency range of 24–25.2 GHz and 26.96–32.76 GHz, with a resonance observed at the 28 GHz 5G frequency band. To further enhance performance, the inner concentric square ring resonator was connected to the outer square split-ring resonator, enabling the creation of stopbands at both 28 GHz and 38 GHz. However, the stopband at 38 GHz was narrower compared to that at 28 GHz. At design step 4, the transmission coefficient ($S21$) dropped below – 10 dB for the frequency ranges 26.96–32.76 GHz and 38.94–39.6 GHz. Finally, by incorporating an additional inner square ring resonator and connecting it to the middle ring, miniaturization was achieved. This adjustment caused the resonant frequency

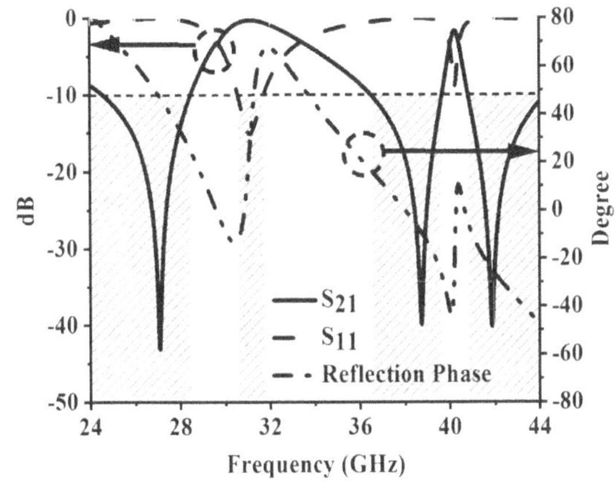

Fig. 2 Retrieved reflection coefficient (S_{11}), reflected wave phase value, and transmission coefficient (S_{21}) of the MTM

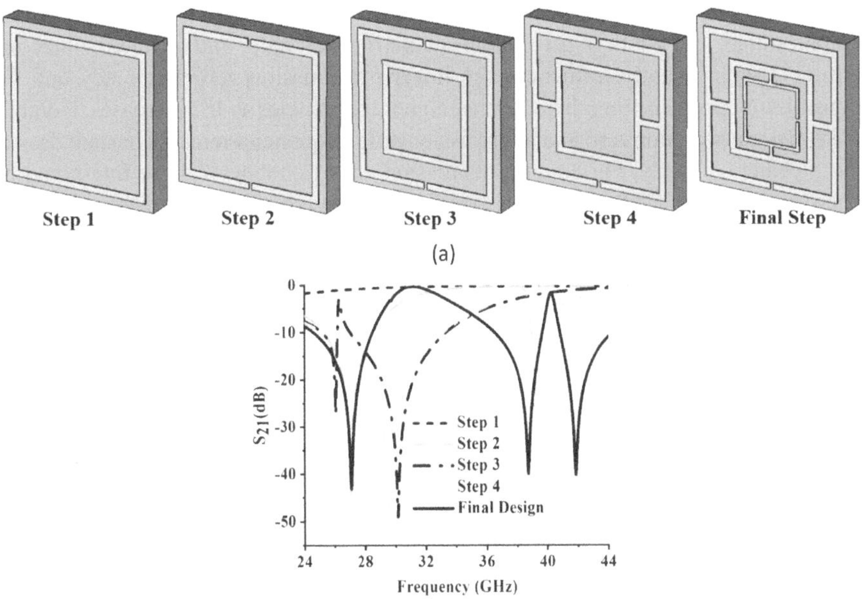

Fig. 3 (a) design evolution of the MTM, and (b) the transmission coefficient (S_{21}) of the different design steps.

of the upper band to shift slightly toward the desired lower frequency. The final design achieved stopbands in the ranges of 24.60–28.47 GHz, 36.50–39.65 GHz, and 40.77–44 GHz. The S_{21} at different design steps are listed in Table 1.

2.2 Parametric Analysis

The sensitivity of the physical parameters on S_{21} is shown by a parametric analysis and presented in Figure 4. The lower peak, middle peak transmission zero frequencies,

Table 1 Transmission coefficient (S_{21}) of the different design steps of the MTM.

Design Evolution	*Resonance Frequency*	$S_{21} < -10$ *dB*
Step 1	–	–
Step 2	27.05	25.27–28.5
Step 3	26.08 30.16	25.17–26.18 27.01–33.33
Step 4	29.53 39.21	25.27–33.15 38.97–39.65
Final Design	27.09 38.71 41.81	24.60–28.47 36.50–39.65 40.77–44

and the upper transmission pole can be adjusted by varying the width ($W1$) of the outer square ring resonator. An increase in $W1$ shifts the transmission zero frequency toward higher frequencies. Similarly, the split gap ($G1$) exhibits a comparable effect, where an increase in $G1$ also shifts the transmission zero frequency upward. However, in both cases, the upper transmission zero and lower transmission pole frequencies remain stable. The width of the middle square ring resonator and the interconnecting arm ($W2$) significantly influences the lower transmission zero frequency and the transmission pole, shifting them toward higher frequencies as $W2$ increases. Notably, the upper transmission zero and transmission pole frequencies remain constant despite these changes. Lastly, the upper transmission zero frequency can be fine-tuned by adjusting the width ($W3$) of the inner square ring resonator. As $W3$ increases, the upper transmission zero frequency shifts to higher frequencies.

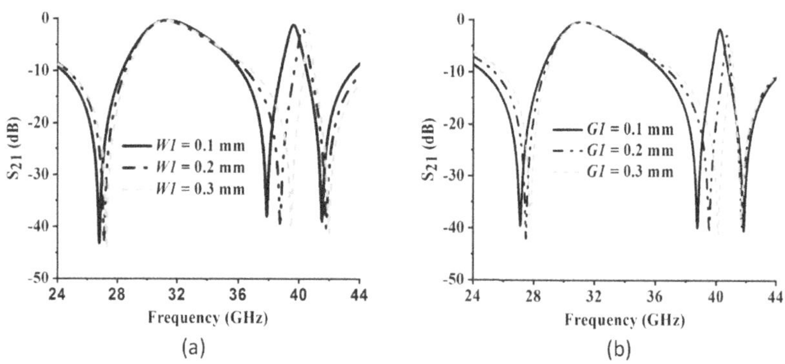

Fig. 4 Parametric variation of MTM (a) width of outer square ring resonator ($W1$), (b) split gap ($G1$), (c) width of interconnecting arm and middle square ring resonator ($W2$), and width of interconnecting arm and inner square ring resonator ($W3$).

3. Results Analysis

3.1 Equivalent Circuit Model (ECM)

The ECM model of the MTM is developed based on the [58–61], shown in Figure 5(a). The three interconnected ring resonators achieved three TZ frequencies, represented by three parallelly connected *LC* series circuits and two inductances (*L*). The inductance can be found by Equation (1) for a specific width and length [58].

$$L_n = \mu_o \frac{1}{2\pi}\left[l \ln\left(\frac{2l}{X_1} + \frac{l}{2 + 0.2235X_1 + 2K_{12}M_{12}} \right)\right] n = 1 \text{ to } 5 \qquad ...(1)$$

where, l is the length microstrip line, $X_1 = w + t$, w is the width of the microstrip line, and t is the thickness $K_{12} = 1$ when the current flows in the same direction, and $K_{12} = -1$ when the current flows in the reverse order. For these two possible cases, *I* and *II*, the mutual inductance (M_{12}) can be calculated by Equations (.2) and (3). $Y_1 = l - m$, $Y_1 = l - s$, $Y_1 = m - s$, where, m is the length associated with parallel microstrip line length, s is the overlapping length of two parallel microstrip lines, and g is the gap between the line.

$$M_{12,\text{ case I}} = \mu_o \frac{1}{4\pi}\left[l \ln\left(\frac{l}{Y_1}\right) + ml \ln\left(\frac{4mY_1}{g^2}\right) - 2m + g \right] \qquad ...(2)$$

$$M_{12,\text{ case II}} = \mu_o \frac{1}{4\pi}\left[Y_2 \ln\left(\frac{Y_2 + m}{Y_2}\right) + ml \ln\left(\frac{Y_2 + m}{Y_2}\right) + s \ln\left(4s\frac{Y_3}{g^2}\right)\right] - 2s \qquad ...(3)$$

At the peak TZ frequency, the capacitances ($C_{n,\, MTM}$) are calculated by using Equation (4), where, the peak TZ is f_{TZ} and $L_{n,\, MTM}$ is the inductance calculated by the metallic area of the unit cell.

$$C_{n,\, FSS-1} = \frac{1}{4\pi^2 L_{n,FSS-1} f_{TZ}^2}, n = 1, 2, 3 \qquad ...(4)$$

The computed ECM model is simulated by Advance Design Systems (ADS) software and is well-tuned to match the simulated results from CST. L_1, L_2, and L_3 represent the outer, middle, and inner square ring resonators. C_1, C_2, and C_3 are the capacitance at three peak transmission zero frequencies. So, three peak transmission zero frequencies at the lower, middle, and upper bands are achieved by three *LC* series circuits like L_1C_1, L_2C_2, and L_3C_3, respectively. The interconnecting arms are represented by L_4 and L_5, which significantly correspond to two transmission poles between the TZ frequency. The final values of the inductance and capacitance are $L_1 = 1.47$ nH, $L_2 = 1.37$ nH, $L_3 = 1.14$ nH, $L_4 = 0.5$ nH, $L_5 = 0.2$ nH $C_1 = 0.023$ pF, $C_2 = 0.012$ pF, and $C_3 = 0.011$ pF. The input impedance of "port 1" and "port 2" is the free space impedance $Z_o = 377\Omega$, and the substrate material is model ed by a transmission line ($Z_{TL} = Z_o/\sqrt{\varepsilon_o}$). Figure 5(b) represents the simulated transmission coefficient (S_{21}) for the CST and ECM models, where both CST and ECM simulations agree well.

3.2 MTM Property Analysis

The MTM property of the MTM is retrieved from the reflection (S_{11}) and transmission (S_{21}) coefficient [5]. The calculated MTM property is presented in Figure 6, where ε_r is the relative permittivity, μ_r is the relative permeability, K_o is the wave number,

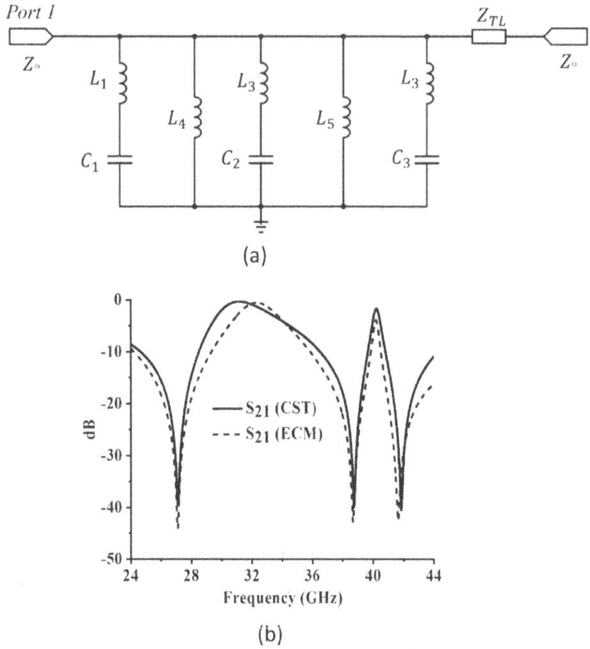

Fig. 5 (a) ECM model of the MTM, (b) retrieved transmission coefficient (S_{21}) from the ECM and CST.

d is the propagation distance, and η is the effective refractive index. The real parts of ε_r and μ_r are negative in the frequency ranges of 24–36 GHz and 27.85–44 GHz, respectively. The transmission-blocking characteristics can be understood from the retrieved alternating negative properties of permittivity and permeability at the operating frequencies of 24.60–28.47 GHz, 36.50–39.65 GHz, and 40.77–44 GHz. The near-zero relative impedance (Z) is also attained at the operating frequency of the MTM shown in Figure 6(c), which indicates the transmission-blocking at the operating frequency.

3.3 Polarization and Incident Angle Analysis

The S_{21} of the MTM for both TM and TE modes are presented in Figure 7(a). The MTM structure exhibits polarization-sensitive behavior due to its asymmetric configuration. Under TE polarization mode, the lower band and middle stopband are converted into transmission frequencies, while the upper frequency range maintains a similar stopband. The effect of the OIA at TM polarization mode is presented in Figure 7(b). In comparison, S_{21} shows a similar stopband up to 35°.

3.4 Surface Current, Electric Field, and Magnetic Field Analysis

When a plane wave incident on the MTM structure, it induces scattering fields due to the current flow in the metallic components. The current flowing through the conducting elements of the unit cell generates a magnetic field, and the variation in

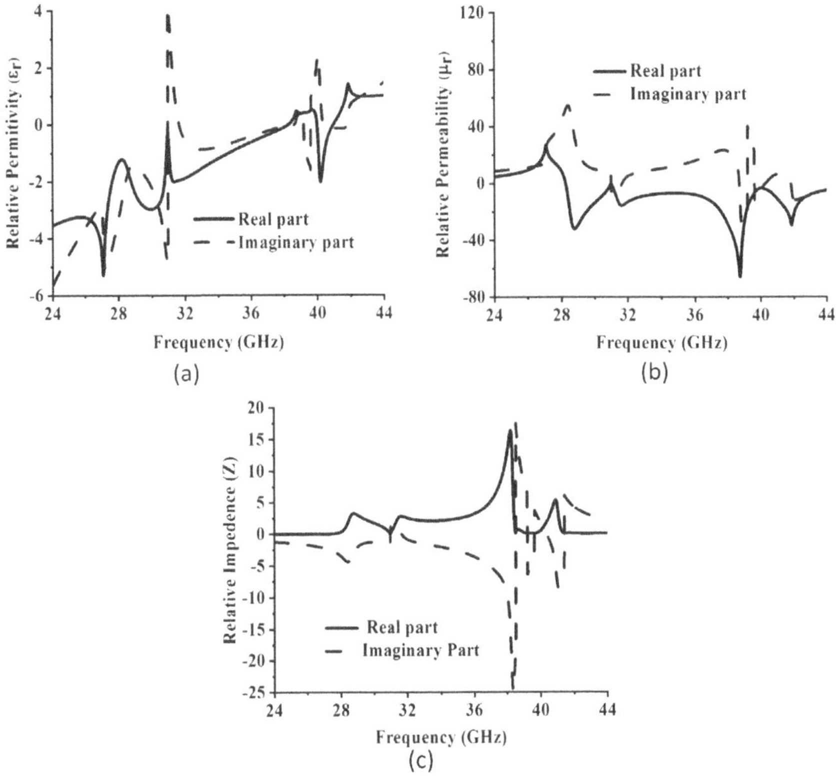

Fig. 6 Retrieved MTM property, (a) relative permittivity, (d) relative permeability, and (c) relative impedance.

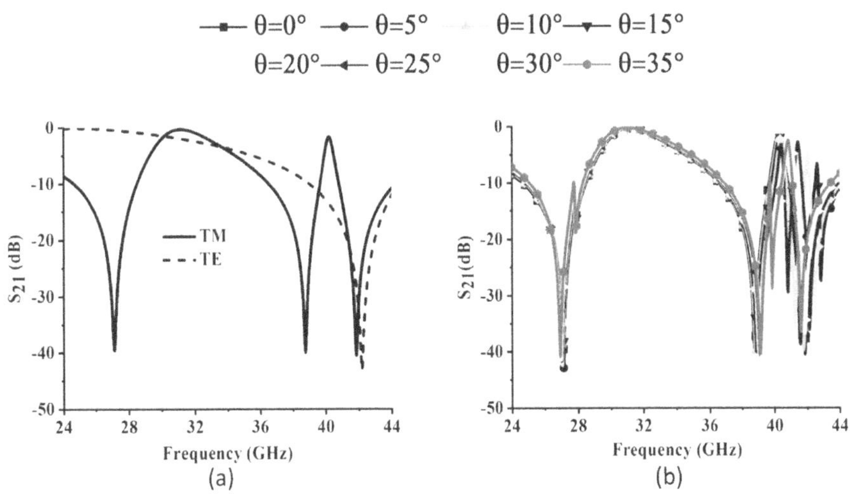

Fig. 7 (a) S_{21} of MTM at TM and TE mode, and (b) S_{21} of MTM at different OIAs at TM mode.

the magnetic field in turn produces an electromotive force. The relationship between the electric field, magnetic field, and surface current in the unit cell can be described by Maxwell's equations [62]. The surface current distribution of the MTM (unit cell) is presented in Figure 8. At a frequency of 27.09 GHz, current flows in the anti-clockwise direction on the upper side of the middle and outer rings, while it flows in the clockwise direction on the lower sides. The interconnecting arm between the middle and outer rings acts as a current source. The inner ring exhibits anti-parallel current flow relative to the associated middle ring, with the maximum current intensity occurring at the upper and lower edges where the inner and middle rings meet. This overall current distribution generates magnetic dipole resonance at 27.09 GHz.

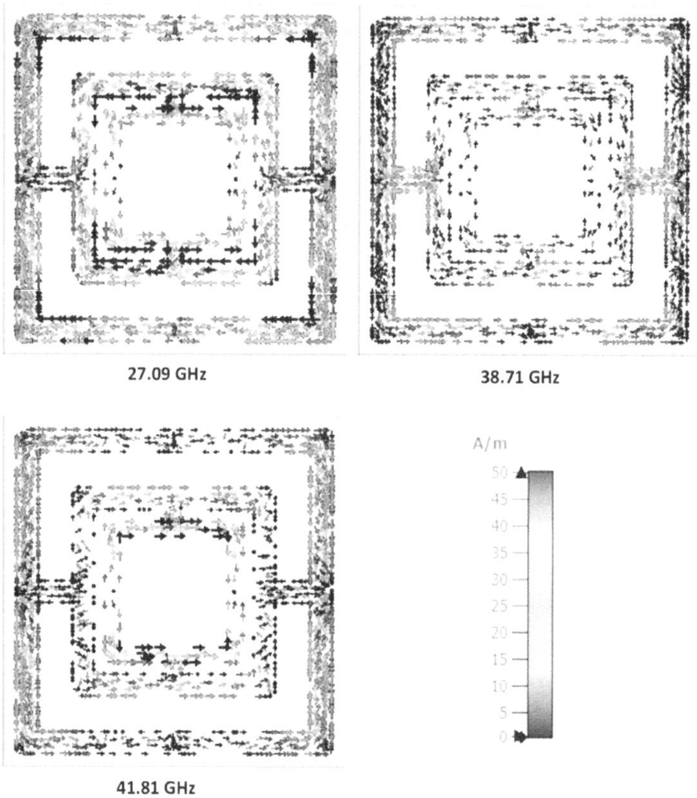

Fig. 8 Surface current distribution of the MTM unit cell at TM mode.

The middle ring exhibits anti-clockwise current flow on the upper arm and clockwise current flow on the lower arm. The interconnecting arm acts as a current source, and anti-parallel current flow occurs between the outer-middle and middle-inner rings. This overall current distribution generates magnetic dipole resonance at a frequency of 38.71 GHz. At 41.81 GHz, the interconnecting arm between the outer and middle rings continues to act as a current source, with anti-parallel current flow appearing between the middle and outer rings. The e-field distribution of the MTM at 27.09 GHz, 38.71 GHz, and 41.81 GHz frequency is presented in Figure 9. At a frequency of 27.09 GHz, the maximum electric field (e-field) intensity occurs at the splits of the outer ring and at the associated edges on the left and right sides of the inner and middle rings. In contrast, at 38.71 GHz, the maximum e-field intensity is observed on the left and right sides of the resonator's arm and at the splits of the outer ring. The highest e-field intensity also appears on the right and left sides of the resonator's splits.

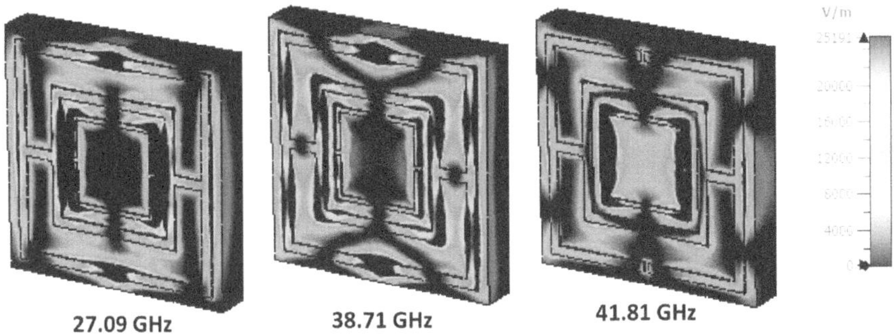

Fig. 9 The e-field intensity of the MTM unit cell at TM mode.

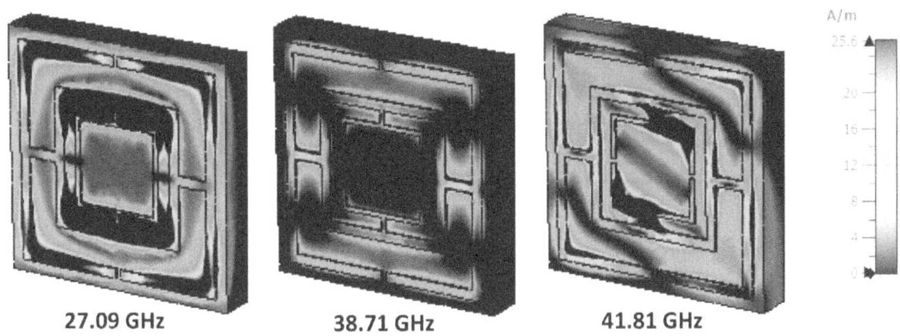

Fig. 10 The *h*-field intensity of the MTM unit cell at TM mode.

The magnetic field (H-field) of the designed MTM is closely related to the surface current distribution. The maximum H-field intensity occurs at the four corners of the outer ring and at the upper and lower regions between the middle and inner rings. The H-field is further intensified at the interconnecting arm between the outer and middle rings. The lower intensity of the H-field at 38.71 GHz compared to 27.09 GHz is attributed to the higher permittivity at 38.71 GHz, which affects the wave propagation. At 41.81 GHz, the highest H-field intensity is observed at the upper right and lower right corners of the inner and outer resonators. Figures 9 and 10 show that the h-field is annulled at the area with high e-field intensity.

4. Measurement Results

A prototype of a 6 × 5 unit cell array is fabricated, shown in Figure 11(a). Figure 11(b) demonstrates the measurement setup, The MTM is placed between two horn antennas, which are connected to a vector network analyzer (VNA). The transmission-blocking behavior is evaluated using the transmission coefficient (*S*21), and the measured *S*21 results show good agreement with the simulated data, as shown in Figure 12. The measured peak transmission zero (TZ) frequency shifts slightly toward a lower frequency compared to the simulated results, with this shift becoming more pronounced at the upper-frequency band. This discrepancy may be due to factors such as fabrication limitations, measurement errors, and the finite number of unit cells, which can lead to edge diffraction.

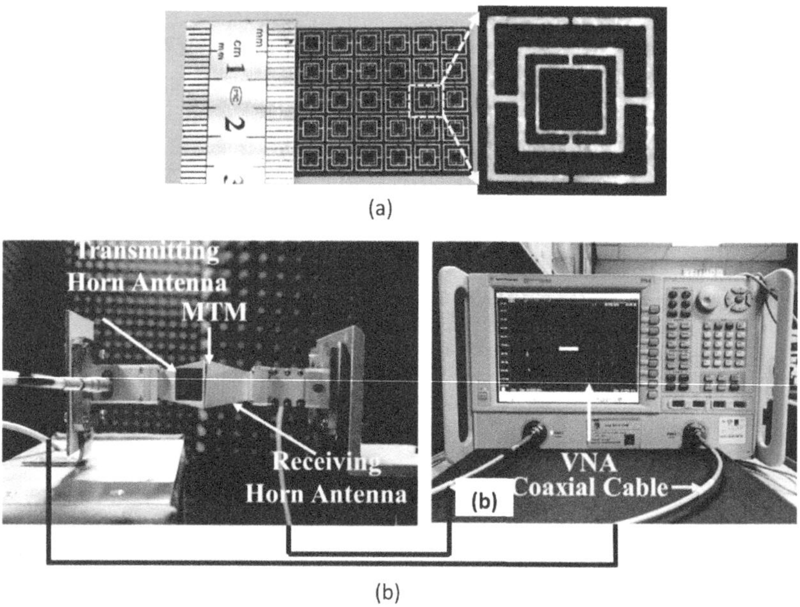

Fig. 11 (a) fabricated 6 × 5 unit cell MTM, (b) measurement setup of the MTM at the normal incident of EM wave for transmission block property.

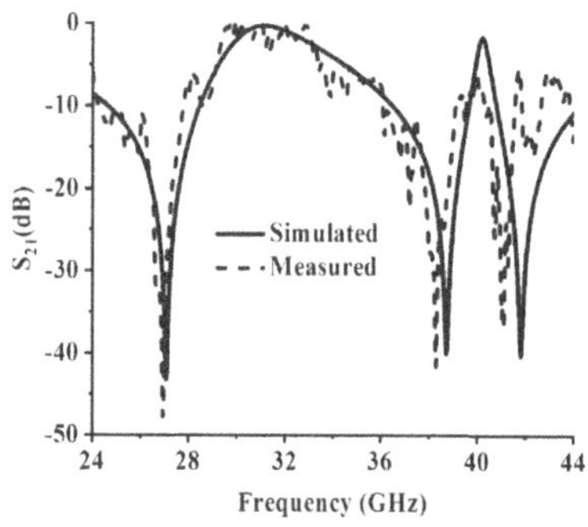

Fig. 12 Simulated and measured transmission coefficient (S_{21}) of the MTM

5. Conformal Application

The conformality of the MTM is investigated for different bending conditions, as shown in Figure 13(a), where the subtends angle $\theta\alpha$ at the center of the semicircle, and

La is the total arc length. For the 20 × 20 array size, La is replaced by $30L$. The radius (R) of the curved MTM is given by $R = 180La/\pi\ \theta\alpha$, where $\theta\alpha$ is in degree. Figure 13 (b) shows the stopband at conformal conditions, which is slightly shifted to the lower frequency. For condition 1 ($\theta\alpha$, = 140°, R = 40.92 mm), the peak TZ frequency of the lower band (27.09 GHz) shifts to 26.86 GHz, middle frequency (38.71 GHz) shifts to 38.3 GHz, and the upper band (41.81 GHz) shifts to 41.58 GHz at conformal. The shifting is slightly higher due to the high-banding angle for condition 2 ($\theta\alpha$, = 160°, R = 36.8 mm). Therefore, the stopbands for conformal MTM configurations are almost

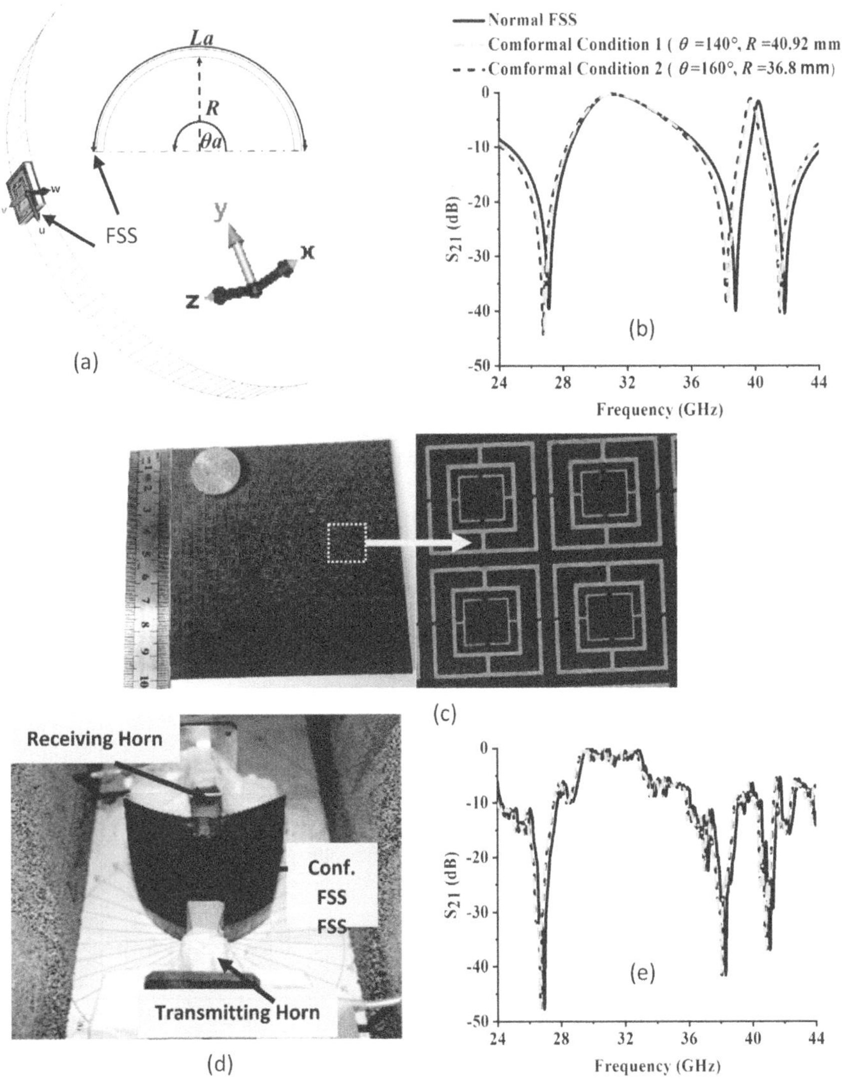

Fig. 13 (a) conformal MTM, (b) simulated S_{21} of conformal MTM at TM polarization mode, (c) fabricated 20 × 20 prototype, (d) measurement setup of conformal MTM, and (e) measured S_{21} of conformal MTM.

similar to the normal MTM configuration. Figure 13(c) shows a fabricated 20 × 20 prototype of MTM. The measurement setup of the conformal MTM is presented in Figure 13(d), where curved MTM is placed between the horn antennas. The measured results agree well with the simulation results, and frequency shifting at conformal conditions is similar to the simulations shown in Figure 13(e).

6. Performance Comparison with Others MTM

A detailed comparison of the developed MTM is presented in Table 2 with the existing polarization-sensitive MTM. The MTM unit cell design configuration, size, employed substrate materials, thickness, dielectric constant, resonant frequency, BW%, covered 5G mmWave band and conformal applications are compared. In Table 2, the research works [63, 64] are conducted in the microwave frequency range and [56, 65, 66] are in mmWave frequency range. The work [56, 65, 66] shows single operating frequencies, at mmWave frequency. Khalili et al. covers 5G mmWave n259 and n260 band [66], and Mantash et al. cover partial n257 band [56]. However, [65] did not cover any 5G mmWave band. On the other hand, the developed MTM achieved three operational frequency bands ranging from 24.60–28.47 GHz, 36.50–39.65 GHz, and 40.77–44 GHz at the mmWave range, which covers the 5G frequency band of n258 n257, n260, and n259, respectively. The narrow BW% shows high selectivity of the developed MTM. The MTM in [56, 64, 65] is not suitable for conformal applications and [66] does not perform conformal applications. Finally, the key features of the designed MTM are its multiple stopband frequencies at the 5G mmWave band and conformal applications, which makes it a potential for the 5G mmWave band.

Table 2 Comparison with the other stopband MTM.

Ref. No.	*Asymmetrical Resonator Geometry.*	*Unit Cell Size (mm^2)*	*Filter Type*	*Frequency (GHz)*	*BW (%)*	*Covers 5G Band*	*Conformal*
[63]	Conductive cylindrical fiber	9 × 9	Stop band	10.2	23.4	–	Yes
[64]	Split ring resonator	7.6 × 7.6	Stop band	10.4 14.1	38.4, 24.1	–	No
[65]	Via-based slotted patch	5.5 × 5.5	Pass band	34	13.5	–	No
[66]	Modified asymmetric resonator	6 × 5	Stop band	40	25	n259, n260,	N/A
[56]	Slotted patch	2.87 × 2.87	Stop band	29	28.2-30	Partial n257	No
Developed MTM	Interconnected square split ring resonators	5 × 5	Stop band	27.09 38.71 41.81	14.58 8.27 7.8	n257 n258 n259 n260	Yes

7. Performance Enhancement of Defected Ground Antenna

7.1 mmWave Antenna Design

The conventional patch antenna suffers from the limitation of a narrow operational bandwidth, restricting its performance across a wide range of frequencies. To address this issue, defected ground technology is employed, which effectively increases the operating bandwidth and enables support for multiple frequency bands [67]. In this context, a slotted patch defected ground (SPDG) antenna is designed for dual-band mmWave operation, with its MIMO configuration consisting of two antenna elements (Figure 14). These two antennas are placed in opposite directions to ensure coverage of both sides, optimizing the antenna's performance for MIMO applications. The antenna is fabricated on a Rogers RT5880 substrate, which has a dielectric constant of 2.2 and a loss tangent of 0.009. This choice of substrate material ensures low loss and stable performance at high frequencies, making it suitable for mmWave applications. By leveraging defected ground technology, the antenna design achieves improved bandwidth and frequency response, crucial for meeting the demands of modern communication systems, such as those used in 5G networks. The design parameters of the antenna are ($l1$ = 1.29 mm, $w1$ = 1.23 mm, $w2$ = 0.54, $l2$ = 0.69 mm, $w3$ = 1.07, $w4$ = 0.32 mm, aw = 5.03 mm, w = 0.46 mm, al = 3.48 mm, gs = 1.15 mm, wg = 3.59 mm, gl = 15 mm).

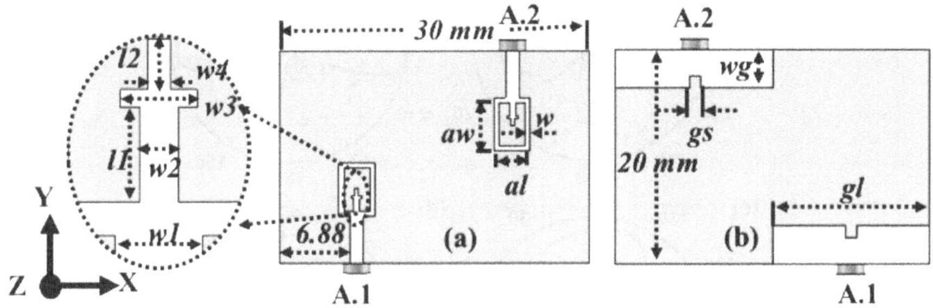

Fig. 14 Designed two-element SPDG MIMO antenna (a) front view, (b) backside view

The reflection coefficient, isolation between the MIMO elements, and gain are shown in Figure 8.15. The operational bandwidth, defined by a – 10 dB reflection coefficient, spans from 24 GHz to 29.46 GHz and from 36.2 GHz to 40 GHz. Additionally, the isolation between the MIMO elements is found to be greater than – 15 dB, which is indicative of effective decoupling and reduced mutual coupling between the antenna elements. The gain of the antenna varies from 3 – 6 dBi across the entire operational frequency range, demonstrating a reasonable and stable performance over the specified frequencies. Figure 16 shows the radiation patterns of the designed antenna in both the E-plane and H-plane are presented for frequencies of 28 GHz and 38 GHz. The patterns clearly show that the radiation from antenna one (A 1) and antenna two (A 2) covers two opposite sides in the antenna's end-fire direction. This directional radiation is beneficial for enhancing signal coverage in specific areas. The H-plane radiation pattern also reveals a significant amount of backside radiation, which is a direct result of the defected ground structure. This characteristic helps in controlling the overall radiation characteristics and improving the antenna's efficiency.

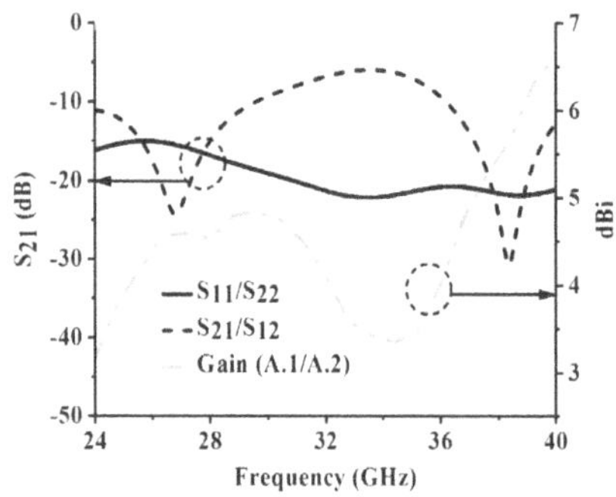

Fig. 15 Simulated reflection coefficient (S_{11}), isolation (S_{21}), and gain of designed dual-band SPDG MIMO antenna of A.1/A.2.

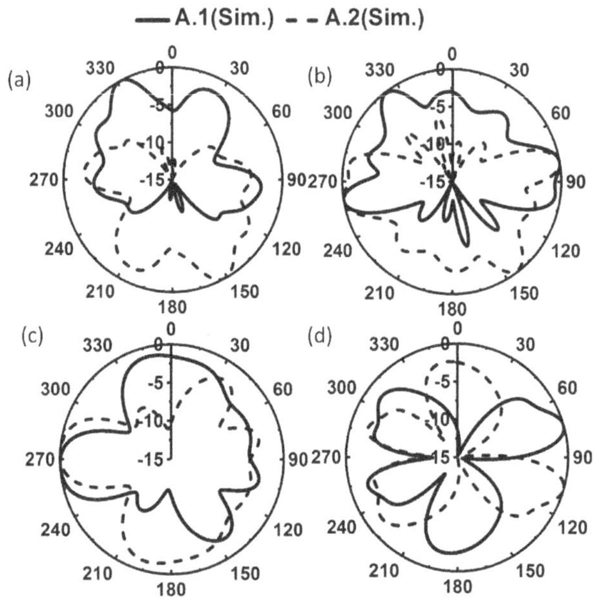

Fig. 16 E-plane radiation pattern at: (a) 28 GHz and (b) 38 GHz frequency; H-plane radiation pattern at (c) 28 GHz, and (b) 38 GHz of SPDG antenna.

7.2 Application for Gain Enhancement and Deflected Beam

The gain of the SPDG antenna can be increased by placing stopband MTM at the back side of the antenna. The MTM acts as an artificial magnetic conductor, which deflects the EM wave with – 90° to 90° phase variation, as shown in Figure 2, which makes a constructive EM wave interface toward the *z*-axis of the antenna [68]. This approach can increase the antenna gain significantly. Figure 17(a) shows a 6 × 4 unit cell MTM placed at the back side of the SPDG MIMO antenna, and Figure 17(b) illustrates the constructive wave interface by MTM towards the *z*-axis.

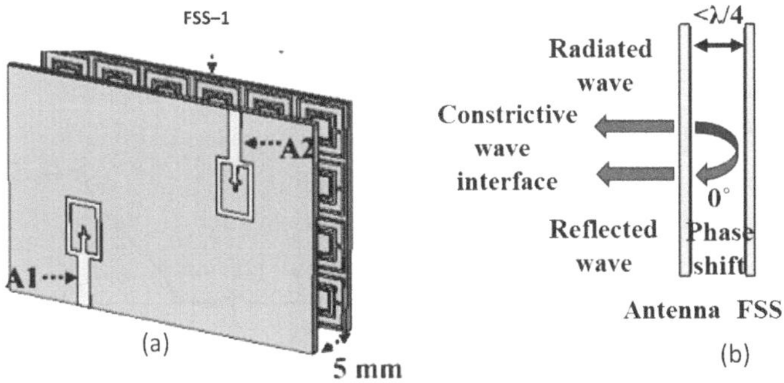

Fig. 17 (a) MTM1-backed SPDG MIMO antenna, (b) constructive wave interference mechanism by MTM1.

Traditionally, antenna radiation patterns are deflected by placing dielectric lenses in the signal path. However, this approach tends to make the antenna bulky and increases its overall complexity. In contrast, the use of MTMs is highly attractive due to their unique EM properties, which allow precise control over the propagation direction of EM waves [55]. Based on Snell's law, the EM wave traversing through different refractive index areas will be refracted at a predefined angle [55, 69]. To understand the refractive index variation, the unit cell of the MTM is simulated with boundary conditions set as perfect electric conductor (PEC) and perfect magnetic conductor (PMC) along the *y*- and *z*-axes, with the EM wave propagating along the *x*-direction, as shown in Figure 18(a). This simulation helps to comprehend the wave propagation behavior along the *x*-axis. Based on this principle, the designed MTM unit cells are arranged in series between the MIMO antenna elements to create a deflected beam antenna, as illustrated in Figure 18(b), This arrangement generates regions with different refractive indices in the end-fire direction of the antenna.

The refractive index of the unit cell is retrieved and presented in Figure 18(c), where it is shown that the refractive index of the MTM material ($n2$) is smaller than that of the substrate ($n1$) at the operational frequency bands. The deflection mechanism, which governs how the EM wave is steered, is illustrated in Figure 18(d). In this mechanism, the higher refractive index medium ($n1$) is associated with the substrate area, while the lower refractive index medium ($n2$) corresponds to the MTM area. The fabricated prototype of the deflected beam antenna is shown in Figure 19 demonstrating the practical implementation of the design.

The EM wave deflected towards a low refractive index to a high refractive index medium. The reflection coefficient of the designed normal, MTM backed and deflected beam antenna achieved a stable operational bandwidth and also simulated and measured S_{11} of the deflected beam antenna well agreed. The – 10 dB impedance bandwidth obtained from 24 GHz – 29.5 GHz and 36.7 GHz – 40 GHz is presented in Figure 20(a). The isolation (S_{21}) between the antenna elements also increases

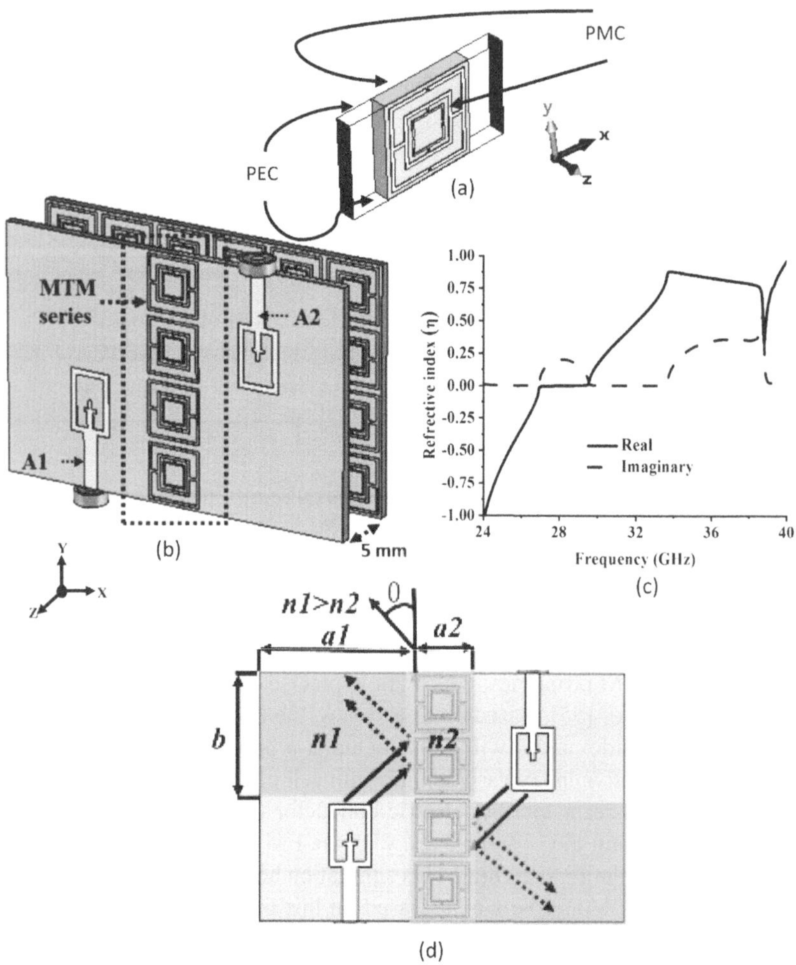

Fig. 18 (a) simulation setup for wave propagation along the *x*-axis, (b) deflected beam pattern diversity SPDG MIMO antenna, (c) effective refractive index for EM wave propagation towards the *x*-axis, and (d) deflection mechanism.

by – 5 – 10 dB by placing the MTM unit cell series between them due to the radiation pattern deflection mechanism shown in Figure 20 (b).

The resultant radiation pattern of the designed deflected beams A.1 and A.2 are presented in Figure 21. The deflection mechanism is applied along the E-plane (*xy*-axis) of the antenna, and the radiation is significantly deflected in the E-plane by 45° at both 28 and 38 GHz frequency bands shown in Figure 21(a, b). Simulated and measured antennas' radiation patterns show slight differences for different reasons, like simulations assuming ideal materials perfectly matched conditions, which does not hold in the real world. The fabrication tolerances of the antenna and calibration tolerances also create these differences. Based on the deflection mechanism explained in Figure 18 (c), the deflection angle (θ) can be calculated by Equation (19) [26, 69],

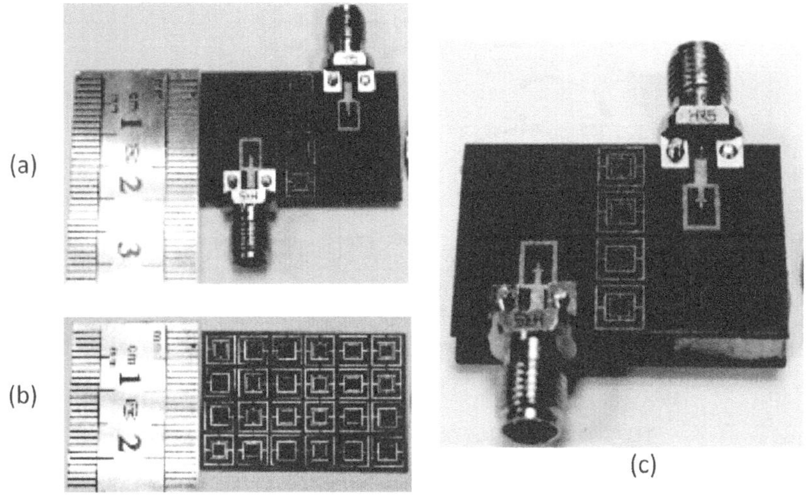

Fig. 19 Fabricated prototype: (a) MTM decoupled antenna, (b) 6 × 4 MTM, and (c) MTM-backed decoupled antenna.

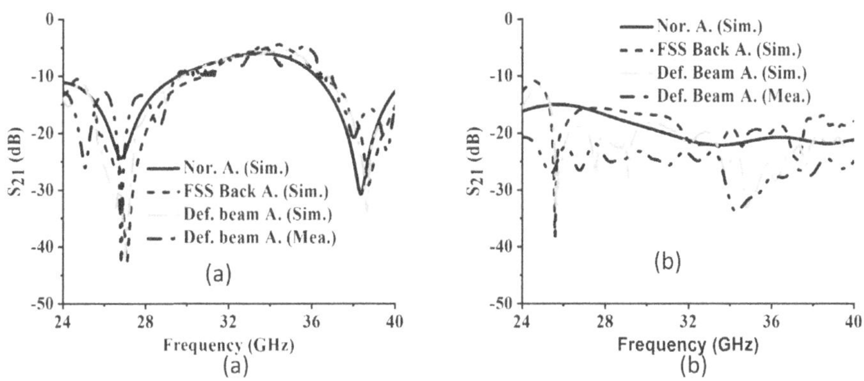

Fig. 20 Simulated and measured: (a) reflection coefficient (S_{11}) of A 1 and A2, and (b) isolation (S_{21}) between A 1 and A2 of the different configurations of MIMO SPDG antenna.

where, $w1 = \sqrt{\left(\frac{a1}{2}\right)^2 + b^2}$, $w2 = \sqrt{\left(\frac{a2}{2}\right)^2 + l^2}$, $b = 10$ mm, $a1 = 10$ mm, $a2 = 5$ mm, $n1 = 1.47$, and $n2$ is the refractive index of the substrate and MTM medium. Therefore, a deflection angle of approximately 45° can be expected. The MTM at the back of the antenna enhances the radiation intensity toward the z-axis while reducing the back-loop radiation. This can be understood from the H-plane (yz-axis) radiation pattern presented in Figure 21(c, d). The pattern diversity application of the designed antenna can be achieved by switching the antenna port. Both the MTM back side structure and deflection mechanism improve the radiation beam and significantly enhance the gain of 1–3 dBi. The gain of the different antenna configurations is illustrated in Figure 22.

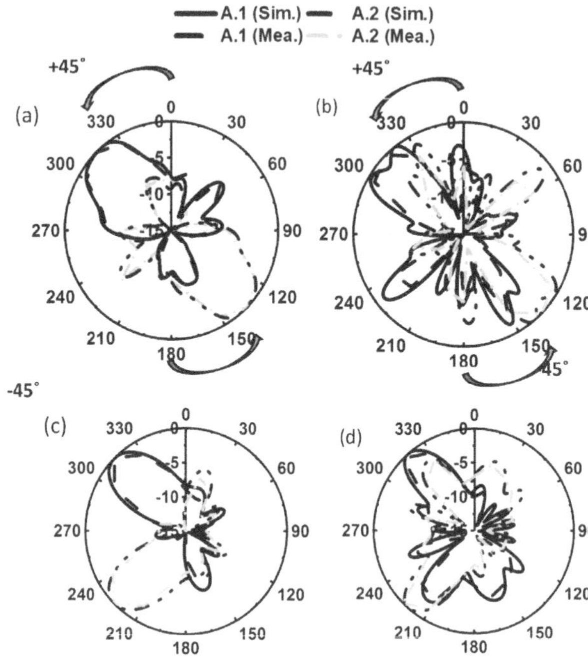

Fig. 21 Simulated and measured radiation pattern of deflected beam SPDG MIMO antenna (A 1 and A 2), E-plane at: (a) 28 GHz, and (b) 38 GHz; H-plane at: (c) 28 GHz, and (d) 38 GHz.

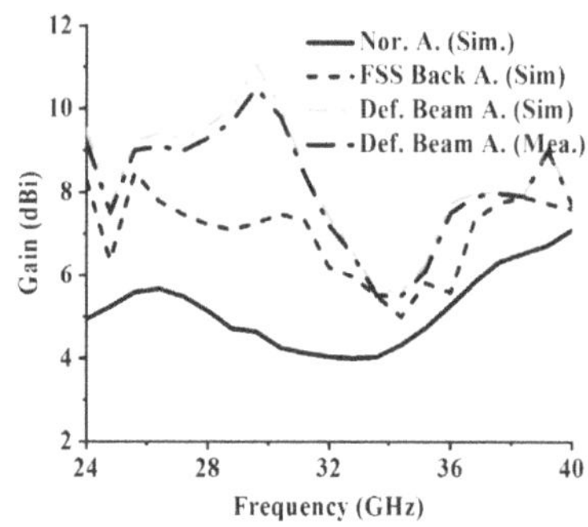

Fig. 22 Simulated gain of the normal SPDG antenna, MTM backed antenna, and simulated and measured gain of the deflected beam antenna.

The isolation enhancement and radiation pattern deflection can be understood from the electric field ($\vec{E}$) and magnetic field ($\vec{H}$) distribution. Figure 23(a) shows the $\vec{E}$ and $\vec{H}$-field of the MTM array back antenna and deflected beam antenna, where it can be seen that the MTM unit cell series significantly deflects both fields towards the substrate area due to the negative refractive index property of MTM unit cell presented in Figure 23(b).

The MIMO antenna diversity performance is analyzed by envelope correlation coefficient (ECC) and diversity of gain (DG), which is retrieved from scattering

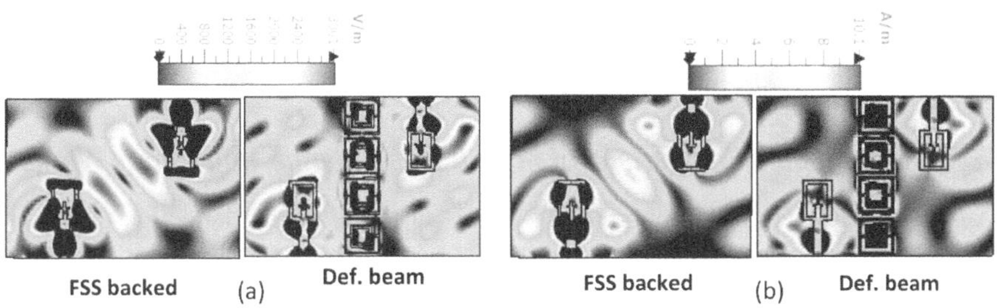

Fig. 23 (a) Electric and (b) magnetic field distributions of MTM-backed and deflected beam SPDG antenna at 28 GHz frequency.

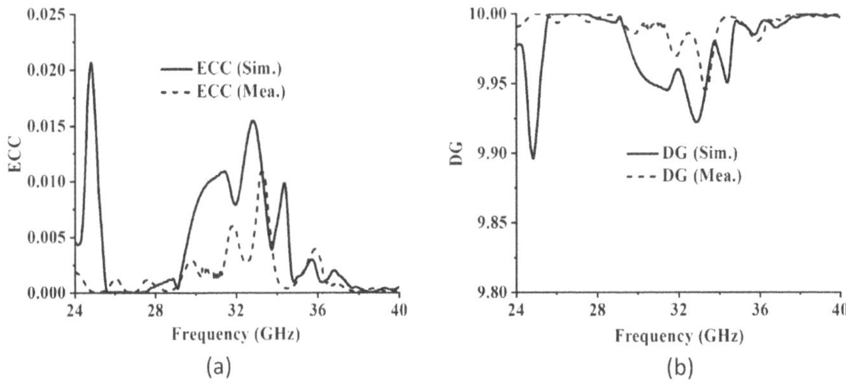

Fig. 24 (a) ECC, and (b) DG of the def. Beam SPDG MIMO antenna.

parameters presented in Figure 2, using Equations (4.8–4.9) [70]. S_{ii} and S_{jj} are the reflection coefficients, and S_{ij} and S_{ji} $(i \neq j)$ are the transmission coefficients of MIMO elements.

$$\rho_{ec}(i,j) = \frac{|S_{ii} \times S_{ij} + S_{ji} \times S_{jj}|^2}{(1-|S_{ii}|^2 - S_{ij}^2)(1-|S_{ij}|^2 - S_{jj}^2)} \quad ...(5)$$

$$DG = 10 \times \sqrt{1-|\rho_{ec}|^2} \quad ...(6)$$

The ECC and DG values of the designed deflected beam MIMO antenna are presented in Figure 24. The Envelope Correlation Coefficient (ECC) value, which indicates the correlation between the two ports, is less than 0.02, well below the acceptable value of 0.5. The Diversity Gain (DG) value exceeds 9.9 for both operating bands, demonstrating strong diversity performance of the designed antenna. The developed MTM significantly enhances the performance of the dual-band antenna, with operating frequencies ranging from 24–29.5 GHz and 36.7–40 GHz. By placing the transmission-blocking MTM at the backside of the antenna, the gain is improved. Additionally, a deflection angle of 45° is achieved by incorporating the MTM unit cells into the EM wave propagation path, creating regions with varying refractive indices. This results in a gain increase of 1–3 dBi across the operating frequencies. Furthermore, the beam deflection technique improves isolation between the antenna elements by 5–10 dB.

7.3 Performance Comparison of MTM Based Antenna with Others Works

A detailed comparison is reported in Table 3 in terms of the deflection technique, impedance bandwidth (IB), gain, and deflection angle, where the designed antenna is a competitive candidate for dual-band 5G applications. In [55], MTM utilized a bow-tie antenna to deflect the radiation beam by 17° with 2.2 dBi gain enhancement, where the antenna's operating frequency is in the microwave range (8–11 GHz). For single-band applications, the deflected beam 5G mmWave antenna is presented in [16, 56, 71]. Three layers of permeability (μ) near zero meta-lense are placed on the top of the antenna to achieve a deflection angle of 40° with a gain improvement of 2.9 dBi, but multiple layers make the antenna bulky [16]. An additional butler-matrix completes the deflection mechanism, although the deflection angle is only ±20°[71]. The MTM is employed to deflect radiation beams by −23° and +29° in [56]. However, gain improvement is limited to 0.6 dBi. In addition, the MIMO configuration has not been presented. Single element 32° beam deflection is achieved [69] at 57–64 GHz operating frequency. The isolation between the antenna elements is only < −10 dB, and other MIMO parameters were not acknowledged [72]. In MIMO configurations [16] and [71], the antenna elements were configured by 90° rotational form, which shows ECC of <0.2 and isolation of < −10 dB, respectively. Finally, the designed deflected beam antenna in this letter operated at dual frequency band at 24–29.5 GHz and 36.7–40 GHz. which achieved a larger 45° deflection angle by incorporating MTM unit cell series and overall gain improvement by MTM back layer and deflection technique is 1–4 dBi. The pattern diversity application can be obtained by passive beam switching (excitation of antenna 1 or 2). The MIMO parameters

Table 3 Comparison of MTM based deflected beam SPDG MIMO antenna with existing works.

Ref. No.	*Dimensions* $L \times W \times H$ (λ_o^3)	*Def. Technique*	*IB (−10dB) GHz*	*Gain (dBi)*	*Def. Angle*	*MIMO*	*ECC/DG*
[55]	0.98 × 1.22 × 0.029	MTMs	8–11	2.2↑	+17°	No	–
[16]	3.36 × 3.36 × 0.45	Meta-lens	24–34	2.9↑	±40°	Yes	ECC of <0.2
[71]	2.86 × 2.86 × 0.22	Butler-matrix	25.5–28.5	3.3↑	±20°	Yes	N/A
[56]	2.41 × 1.55 × 0.34	MTM	28–31	0.6↑	−23° +29°	No	–
[72]	1.7 × 1.3 × 0.58	Meta-lens	56–66	2-3↑	±25°	Yes	N/A
[69]	2.66 × 3.14 × 0.05	MTMs	57–64	2.4↑	+32°	No	–
Our work	1.6 × 2.4 × 0.02	MTMs	24–29.5 36.7–40	1-4↑	±45°	Yes	<0.001 >9.99

between these two antennas were also investigated. Therefore, compared with the listed antenna in Table 3, the designed antenna has the potential for 5G mmWave dual band pattern diversity applications.

8. Summary

This chapter presents a multiband transmission-blocking ISSRR resonator-based MTM for 5G mm-wave band applications. A transmission coefficient (S_{21}) of −10 dB is achieved across the frequency ranges of 24.60–28.47 GHz, 36.50–39.65 GHz, and 40.77–44 GHz, with resonant frequencies observed at 27.09 GHz, 38.71 GHz, and 41.81 GHz. This transmission-blocking capability is beneficial for enhancing communication quality and minimizing interference within 5G networks by focusing energy and minimizing electromagnetic interference (EMI). The designed MTM is validated by fabricating a 6 × 5 unit cell array, and measurement results agree well with the simulated one. The applications of the proposed MTM were investigated for antenna performance enhancement. To explore MTM's applications in antenna performance, a defected ground dual-band antenna was developed, utilizing the MTM array on its backside. This arrangement enhances gain, improving it by 1–3 dBi over the operational frequency ranges. Moreover, the MTM's ability to control electromagnetic wave propagation yields a 45° beam deflection angle, achieved by configuring different refractive index regions along the EM wave's path. This innovative deflection not only improves directional signal control but also increases isolation between the MIMO antenna elements by 5–10 dB across targeted frequencies, thereby boosting overall MIMO antenna efficiency and suitability for dense urban 5G environments. The MTM's versatility and performance advantages make it excellent for high-performance 5G mm-wave applications, addressing critical challenges in gain enhancement, beam control, and interference management essential for robust wireless communication systems.

References

[1] D.R. Smith, W.J. Padilla, D. Vier, S.C. Nemat-Nasser and S. Schultz. "Composite medium with simultaneously negative permeability and permittivity," *Physical Review Letters,* vol. 84, no. 18, p. 4184, 2000.

[2] Y. Soerbakti, R.F. Syahputra, M.D.H. Gamal, D. Irawan, E.H. Putra and R.S. Darwis. "Improvement of low-profile microstrip antenna performance by hexagonal-shaped SRR structure with DNG metamaterial characteristic as UWB application," *Alexandria Engineering Journal,* vol. 61, no. 6, pp. 4241–4252, 2022.

[3] M. Hakim, T. Alam, M. Baharuddin and M. Islam. "Frequency integration of dual-band hexagonal metamaterial resonator antenna for Wi-Fi and 5G wireless communication," in *Journal of Physics: Conference Series*, 2022, vol. 2250, no. 1: IOP Publishing, p. 012001.

[4] M.L. Hakim, T. Alam, A.F. Almutairi, M.F. Mansor and M.T. Islam. "Polarization insensitivity characterization of dual-band perfect metamaterial absorber for K band sensing applications," *Scientific Reports,* vol. 11, no. 1, p. 17829, 2021.

[5] M.L. Hakim, T. Alam, M.T. Islam, H. Alsaif and M.S. Soliman. "Polarization-independent fractal square splits ring resonator (FSSRR) multiband metamaterial absorber/artificial magnetic conductor/sensor for Ku/K/Ka/5G (mm-Wave) band applications," *Measurement,* vol. 210, p. 112545, 2023.

[6] M.L. Hakim, T. Alam, M.T. Islam, M.H. Baharuddin, A. Alzamil and M.S. Islam. "Quad-band polarization-insensitive square split-ring resonator (SSRR) with an inner Jerusalem cross metamaterial absorber for Ku-and K-Band sensing applications," *Sensors,* vol. 22, no. 12, p. 4489, 2022.
[7] M.R. Islam et al. "Star enclosed circle split ring resonator-based metamaterial sensor for fuel and oil adulteration detection," *Alexandria Engineering Journal,* vol. 67, pp. 547–563, 2023.
[8] B.X. Wang, C. Xu, G. Duan, W. Xu and F. Pi. "Review of Broadband Metamaterial Absorbers: From Principles, Design Strategies, and Tunable Properties to Functional Applications," *Advanced Functional Materials,* p. 2213818, 2023.
[9] M.L. Hakim et al. "Triple-band square split-ring resonator metamaterial absorber design with high effective medium ratio for 5G sub-6 GHz applications," *Nanomaterials,* vol. 13, no. 2, p. 222, 2023.
[10] M. Wang et al. "Investigation of SAR reduction using flexible antenna with metamaterial structure in wireless body area network," *IEEE Transactions on Antennas and Propagation,* vol. 66, no. 6, pp. 3076–3086, 2018.
[11] L. Chisari, E. Ricciardi, and G. Memoli. "Numerical and Experimental Analysis of a Metamaterial-based Acoustic Superlens," in *Audio Engineering Society Convention 152*, 2022: Audio Engineering Society.
[12] S. Fang, S. Zhou, D. Yurchenko, T. Yang and W.-H. Liao. "Multistability phenomenon in signal processing, energy harvesting, composite structures, and metamaterials: A review," *Mechanical Systems and Signal Processing,* vol. 166, p. 108419, 2022.
[13] M. Bilal, R. Saleem, Q.H. Abbasi, B. Kasi and M.F. Shafique. "Miniaturized and flexible FSS-based EM shields for conformal applications," *IEEE Transactions on Electromagnetic Compatibility,* vol. 62, no. 5, pp. 1703–1710, 2020.
[14] M.L. Hakim, T. Alam and M.T. Islam. "Polarization Insensitive and Oblique Incident Angle Stable Miniaturized Conformal FSS for 28/38 GHz mm-Wave band 5G EMI Shielding Applications," *IEEE Antennas and Wireless Propagation Letters,* 2023.
[15] C. Milias et al. "Metamaterial-inspired antennas: A review of the state of the art and future design challenges," *IEEE Access,* vol. 9, pp. 89846–89865, 2021.
[16] M. Borhani-Kakhki, A. Dadgarpour, M.A. Antoniades, A.R. Sebak and T.A. Denidni. "Magnetoelectric Dipole Antennas Loaded With Meta-Lens for 5G MIMO Pattern Diversity Applications," *IEEE Transactions on Antennas and Propagation,* vol. 70, no. 8, pp. 7112–7117, 2022.
[17] M.L. Bouknia et al. "Effect analysis of the general complex reciprocal gyro-bianisotropic metamaterial medium on the input impedance of a printed dipole antenna," *Alexandria Engineering Journal,* vol. 61, no. 5, pp. 3691–3696, 2022.
[18] N. Misran, S.H. Yusop, M.T. Islam and M.Y. Ismail. "Analysis of parameterization substrate thickness and permittivity for concentric split ring square reflectarray element," *Jurnal Kejuruteraan (Journal of Engineering),* vol. 23, pp. 11–16, 2012.
[19] T.J. Cui, D.R. Smith and R. Liu. *Metamaterials*. Springer, 2010.
[20] A. Musa et al. "Broadband plasmonic metamaterial optical absorber for the visible to near-infrared region," *Nanomaterials,* vol. 13, no. 4, p. 626, 2023.
[21] A. Musa et al. "Polarization independent metamaterial absorber with anti-reflection coating nanoarchitectonics for visible and infrared window applications," *Materials,* vol. 15, no. 10, p. 3733, 2022.
[22] M.K. Uddin, M.L. Hakim, T. Alam, M.T. Islam, H. Alsaif and M.S. Islam. "Polarization insensitive eye-shape enclosed split ring resonator-based multi-band metamaterial absorber/sensor for S, C, X, Ku, and K-band microwave applications," *Journal of Sandwich Structures & Materials,* p. 10996362241297689, 2024.
[23] M.K. Uddin et al. "Polarization insensitive U-coupled splits ring resonator-based hexa-band metamaterial absorber for concentration of solid material sensing applications," *Optics & Laser Technology,* vol. 181, p. 111975, 2025.

[24] R. Ghatak and A. Gorai. "Metamaterials: Engineered Materials and its Applications in High Frequency Electronics," 2021.
[25] M.L. Hakim et al. "Interconnected square splits ring resonator based single negative metamaterial for 5G (N258, N257, N260 and N259) band sensor/EMI shielding/and antenna applications," *Alexandria Engineering Journal,* vol. 81, pp. 419–436, 2023.
[26] M.L. Hakim et al. "Metamaterial physical property utilized antenna radiation pattern deflection for angular coverage and isolation enhancement of mm-wave 5G MIMO antenna system," *Radiation Physics and Chemistry,* vol. 209, p. 110998, 2023.
[27] M.L. Hakim et al. "Ultrawideband polarization-independent nanoarchitectonics: A perfect metamaterial absorber for visible and infrared optical window applications," *Nanomaterials,* vol. 12, no. 16, p. 2849, 2022.
[28] M.L. Hakim, M.T. Islam and T. Alam. "Incident Angle Stable Broadband Conformal mmWave FSS for 5G (n257, n258, n260, and n261) band EMI Shielding Application," *IEEE Antennas and Wireless Propagation Letters,* 2023.
[29] S. Tabassum, S. Nayemuzzaman, M. Kala, A. Kumar Mishra and S.K. Mishra. "Metasurfaces for sensing applications: Gas, bio, and chemical," *Sensors,* vol. 22, no. 18, p. 6896, 2022.
[30] M. Asif et al. "Design of a dual band SNG metamaterial based antenna for LTE 46/WLAN and Ka-band applications," *IEEE Access,* vol. 9, pp. 71553–71562, 2021.
[31] M.L. Hakim, M.T. Islam and T. Alam. "Ultra-Miniaturized Conformal Polarization Insensitive and Incident Angle Stable FSS for n257 Band 5G EMI Shielding Applications," *IEEE Transactions on Antennas and Propagation,* 2024.
[32] M.L. Hakim, M.T. Islam, M.S.J. Singh, T. Alam, A.M. Alenezi and M.S. Soliman. "Intensified electric field based highly sensitive backbone-shaped C and X band microwave metamaterial sensor for permittivity characterization applications," *Engineering Science and Technology, an International Journal,* vol. 60, p. 101877, 2024.
[33] A. Hanif, M.L. Hakim, T. Alam, M.T. Islam, H. Alsaif and M.S. Soliman. "Polarization insensitive oblique incident angle stable ultra-thin nano ring resonator-based metamaterial absorber for visible and near-infrared region applications," *Optics & Laser Technology,* vol. 164, p. 109494, 2023.
[34] B.A. Esmail, S. Koziel, L. Golunski, H.B.A. Majid and R.K. Barik. "Overview of Metamaterials-Integrated Antennas for Beam Manipulation Applications: The Two Decades of Progress," *IEEE Access,* vol. 10, pp. 67096–67116, 2022.
[35] R. Mittra, A. Nasri and R.K. Arya. "Wide-angle scanning antennas for millimeter-wave 5g applications," *Engineering,* vol. 11, pp. 60–71, 2022.
[36] M. Kamran Shereen, M. Khattak and G. Witjaksono. "A brief review of frequency, radiation pattern, polarization, and compound reconfigurable antennas for 5G applications," *Journal of Computational Electronics,* vol. 18, no. 3, pp. 1065–1102, 2019.
[37] I. Da Costa, A. Cerqueira, D.H. Spadoti, L. Da Silva, J.A.J. Ribeiro and S.E. Barbin. "Optically controlled reconfigurable antenna array for mm-wave applications," *IEEE Antennas and Wireless Propagation Letters,* vol. 16, pp. 2142–2145, 2017.
[38] M.S. Khan, A. Iftikhar, A.D. Capobianco, R.M. Shubair and B. Ijaz. "Pattern and frequency reconfiguration of patch antenna using PIN diodes," *Microwave and Optical Technology Letters,* vol. 59, no. 9, pp. 2180–2185, 2017.
[39] A. Martin, V. Le Neillon, A. Jouade and M. Himdi. "Mechanically reconfigurable radiation pattern slot antenna array feeded by bended sectoral horn and metalized wood splitter," *Progress in Electromagnetics Research C,* vol. 72, pp. 159–165, 2017.
[40] Y. Guo, Y. Guo, C. Li, H. Zhang, X. Zhou and L. Zhang. "Integrated optical phased arrays for beam forming and steering," *Applied Sciences,* vol. 11, no. 9, p. 4017, 2021.
[41] T. Alam, A.F. Almutairi, M. Samsuzzaman, M. Cho and M.T. Islam. "Metamaterial array based meander line planar antenna for cube satellite communication," *Scientific Reports,* vol. 11, no. 1, pp. 1–12, 2021.
[42] T. Alam et al. "Metamaterial Based Ku-Band Antenna for Low Earth Orbit Nanosatellite Payload System," *Nanomaterials,* vol. 13, no. 2, p. 228, 2023.

[43] A.M. Tamim, M.R.I. Faruque, M.U. Khandaker, M.T. Islam and D.A. Bradley. "Electromagnetic radiation reduction using novel metamaterial for cellular applications," *Radiation Physics and Chemistry,* vol. 178, p. 108976, 2021.

[44] Y. Liu et al. "Beam steering by using a gradient refractive index metamaterial planar lens and a gradient phase metasurface planar lens," *Microwave and Optical Technology Letters,* vol. 60, no. 2, pp. 330–337, 2018.

[45] D. Jia, Y. He, N. Ding, J. Zhou, B. Du and W. Zhang. "Beam-steering flat lens antenna based on multilayer gradient index metamaterials," *IEEE Antennas and Wireless Propagation Letters,* vol. 17, no. 8, pp. 1510–1514, 2018.

[46] A. Dadgarpour, M.S. Sorkherizi, T.A. Denidni and A.A. Kishk. "Passive beam switching and dual-beam radiation slot antenna loaded with ENZ medium and excited through ridge gap waveguide at millimeter-waves," *IEEE Transactions on Antennas and Propagation,* vol. 65, no. 1, pp. 92–102, 2016.

[47] V. Pacheco-Peña et al. "Mechanical 144 GHz beam steering with all-metallic epsilon-near-zero lens antenna," *Applied Physics Letters,* vol. 105, no. 24, p. 243503, 2014.

[48] A. Dadgarpour, A.A. Kishk and T.A. Denidn. "Dual band high-gain antenna with beam switching capability," *IET Microwaves, Antennas, & Propagation,* vol. 11, no. 15, pp. 2155–2161, 2017.

[49] B.A. Esmail et al. "Reconfigurable metamaterial structure for 5G beam tilting antenna applications," *Waves in Random and Complex Media,* vol. 31, no. 6, pp. 2089–2102, 2021.

[50] B.A. Esmail, S. Koziel and S. Szczepanski. "Overview of planar antenna loading metamaterials for gain performance enhancement: The two decades of progress," *IEEE Access,* vol. 10, pp. 27381–27403, 2022.

[51] W. Hong et al. "The role of millimeter-wave technologies in 5G/6G wireless communications," *IEEE Journal of Microwaves,* vol. 1, no. 1, pp. 101–122, 2021.

[52] X. Wang et al. "Millimeter wave communication: A comprehensive survey," *IEEE Communications Surveys & Tutorials,* vol. 20, no. 3, pp. 1616–1653, 2018.

[53] FCC. "America's 5G Future." fcc.gov. https://www.fcc.gov/5G (accessed 16 December 2022).

[54] H.A. Mohamed, M. Edries, M.A. Abdelghany and A.A. Ibrahim. "Millimeter-Wave Antenna With Gain Improvement Utilizing Reflection FSS for 5G Networks," *IEEE Access,* vol. 10, pp. 73601–73609, 2022.

[55] Q.-W. Lin, H. Wong and C.H. Chan. "A transparent artificial periodic structure for beam-tilting antenna," *IEEE Antennas and Wireless Propagation Letters,* vol. 20, no. 4, pp. 498–502, 2021.

[56] M. Mantash, A. Kesavan and T.A. Denidni. "Beam-tilting endfire antenna using a single-layer FSS for 5G communication networks," *IEEE Antennas and Wireless Propagation Letters,* vol. 17, no. 1, pp. 29–33, 2018.

[57] S. Dey, S. Dey and S.K. Koul. "Second-Order, Single-Band and Dual-Band Bandstop Frequency Selective Surfaces at Millimeter Wave Regime," *IEEE Transactions on Antennas and Propagation,* vol. 70, no. 8, pp. 7282–7287, 2022.

[58] S. Dey and S. Dey. "Conformal miniaturized angular stable triband frequency selective surface for EMI shielding," *IEEE Transactions on Electromagnetic Compatibility,* vol. 64, no. 4, pp. 1031–1041, 2022.

[59] D. Li, T.-W. Li, E.-P. Li and Y.-J. Zhang. "A 2.5-D angularly stable frequency selective surface using via-based structure for 5G EMI shielding," *IEEE Transactions on Electromagnetic Compatibility,* vol. 60, no. 3, pp. 768–775, 2017.

[60] T.-W. Li et al. "A novel miniaturized multiband strong coupled-FSS structure insensitive to almost all angles and all polarizations," *IEEE Transactions on Antennas and Propagation,* vol. 69, no. 12, pp. 8470–8478, 2021.

[61] X. Zhou et al. "A Novel Design of a Compact Frequency-Selective Surface With High Selectivity and Angular Stability," *IEEE Microwave and Wireless Components Letters,* vol. 32, no. 7, pp. 931–934, 2022.

[62] M.M. Bait-Suwailam. "Electromagnetic field interaction with metamaterials," *Electromagnetic Fields and Waves,* pp. 1-19, 2019.

[63] M.H. Nisanci, and F. de Paulis. "Easy-to-design-and-manufacture frequency selective surfaces for conformal applications," *IEEE Antennas and Wireless Propagation Letters,* vol. 20, no. 5, pp. 753–757, 2021.

[64] S. Abbasi, J. Nourinia, C. Ghobadi, M. Karamirad and B. Mohammadi. "A sub-wavelength polarization sensitive band-stop FSS with wide angular response for X-and Ku-Bands," *AEU-International Journal of Electronics and Communications,* vol. 89, pp. 85–91, 2018.

[65] X.-C. Zhu, P.-P. Zhang, Y.-X. Zhang, J.-X. Ge and Z.-H. Gao. "A high-gain filtering antenna based on folded reflectarray antenna and polarization-sensitive frequency selective surface," *IEEE Antennas and Wireless Propagation Letters,* vol. 19, no. 8, pp. 1462–1465, 2020.

[66] F. Khajeh-Khalili, M. A. Honarvar, M. Naser-Moghadasi and M. Dolatshahi. "Gain enhancement and mutual coupling reduction of multiple-in put multiple-output antenna for millimeter-wave applications using two types of novel metamaterial structures," *International Journal of RF and Microwave Computer-Aided Engineering,* vol. 30, no. 1, p. e22006, 2020.

[67] H.F. Abutarboush, W. Li and A. Shamim. "Flexible-screen-printed antenna with enhanced bandwidth by employing defected ground structure," *IEEE Antennas and Wireless Propagation Letters,* vol. 19, no. 10, pp. 1803–1807, 2020.

[68] P. Kumari, R.K. Gangwar and R.K. Chaudhary. "An Aperture-Coupled Stepped Dielectric Resonator UWB MIMO Antenna With AMC," *IEEE Antennas and Wireless Propagation Letters,* vol. 21, no. 10, pp. 2040–2044, 2022.

[69] B. Zarghooni, A. Dadgarpour and T.A. Denidni. "Millimeter-wave antenna using two-sectioned metamaterial medium," *IEEE Antennas and Wireless Propagation Letters,* vol. 15, pp. 960–963, 2016.

[70] A. Ahmad, D.-y. Choi and S. Ullah. "A compact two- element MIMO antenna for 5G communication," *Scientific Reports,* vol. 12, no. 1, p. 3608, 2022.

[71] A.B. Guntupalli, T. Djerafi, and K. Wu. "Two-dimensional scanning antenna array driven by integrated waveguide phase shifter," *IEEE Transactions on Antennas and Propagation,* vol. 62, no. 3, pp. 1117–1124, 2013.

[72] A. Dadgarpour, M.A. Antoniades, A.-R. Sebak and T.A. Denidni. "Two-Dimensional Millimeter-Wave Beam-Switching Antenna Using an ELC Meta-Lens," *IEEE Transactions on Antennas and Propagation,* vol. 70, no. 2, pp. 988–997, 2022.

CHAPTER 9

Machine Learning-Driven Metamaterial Absorber for RCS Reduction in the Terahertz Range

Antenna Optimization Perspective

Prince Jain[1*] and *Mohammad Tariqul Islam*[2]

1. Introduction

Metamaterials (MTMs) represent a class of artificial materials engineered to exhibit electromagnetic properties not naturally found in conventional materials [1]. MTMs have received significant attention in the microwave to terahertz (THz) frequency range due to their unique electromagnetic properties, paving the way for innovative applications ranging from antennas [2], solar cells [3], superlenses [4], resonators [5], sensing [6], and absorbers [7]. In the context of THz applications, optimizing antennas for enhanced performance and reducing **radar cross-section** (RCS) has become imperative [9, 10]. Antennas play a crucial role in transmitting and receiving electromagnetic signals efficiently, while RCS reduction techniques aim to minimize the radar signature of objects. The integration of MTM structures in antenna design has shown promise in achieving improved radiation characteristics and RCS reduction across the THz spectrum [8]. Among these applications, MTM absorbers, or MMAs, have emerged as particularly promising candidates for efficient absorption of electromagnetic radiation in the THz range. The work by Landy et al. [11], which introduced the concept of MMAs, marked a pivotal moment in MTM research, sparking widespread interest in the development of MMAs with tailored

[1] Department of Mechatronics Engineering, Parul University, Vadodara, Gujarat, India.
[2] Department of Electrical, Electronic and Systems Engineering, Faculty of Engineering and Built Environment, Universiti Kebangsaan Malaysia, Malaysia.
* Corresponding author: princeece48@gmail.com

absorption properties. Since then, researchers have proposed various types of MMAs, ranging from single-band [12] to multi band absorbers [13, 14, 15], each offering distinct advantages for specific applications.

Of particular interest are multi band MMAs with high figures of merit (FOM), as they hold promise for applications such as spectroscopic imaging [16], internet of things [17], materials detection [18, 19], and hazardous gas sensing [20]. To enhance absorption performance, researchers have explored numerous design strategies, including the integration of resonators with varying geometries and sizes within a single unit cell [21, 22] and the stacking of metallic resonators [23, 24]. Recent efforts have focused on leveraging **dipolar resonance** to achieve dual-, triple-, and quad-band absorption [25, 22], albeit facing challenges related to device size and thickness, especially at higher frequencies [26, 27]. Additionally, the combination of dipolar and LC resonances in novel structures such as T-shaped [28], Gemini-shaped [29], and #-shaped [30] resonators have shown promise in generating multi band absorption responses. However, these designs often suffer from broad linewidths of absorption peaks, limiting their suitability for practical sensor and detection applications where narrow linewidths are desirable [31, 32, 33].

Traditionally, researchers have relied on expertise to fine-tune structural parameters in metasurface absorber design [34]. The goal of this process is to achieve optimal absorptive characteristics. As a result, absorbers designed with these factors in mind frequently incur significant time and resource costs, and there is some uncertainty about whether or not the structural parameters have achieved their optimal configurations. The rapidly developing field of **machine learning** (ML) offers a potentially useful solution to this problem, which is an encouraging development. Since ML is a data-driven analysis technique, it can extract insights from datasets, build predictive models, and provide predicted outcomes in a timely and effective manner. The time-consuming and ineffective optimization processes inherent in traditional design approaches can be simplified by leveraging ML. Furthermore, ML demonstrates an impressive ability to reduce design biases caused by human subjectivity. This helps to ensure that high levels of dependability, stability, and accuracy are maintained throughout the design phase. Certain algorithmic models combined with ML have found applications in predicting the performance of a variety of optical devices, including colour filters, meta-lenses, digital encoders, and holographic imaging systems on metasurface platforms [36, 37]. This is due to the continued advancement of ML technologies. In contrast, the field of predicting the absorbing bandwidth of multiband absorbers has remained largely unexplored in terms of leveraging the potential of ML, particularly in terms of achieving optimal performance through meticulous hyperparameter tuning.

In this context, the proposed MMA stands out as a promising solution, with a compact and structurally simple design that includes a metallic board separated by a dielectric layer and two modified T-shaped **resonators**. Notably, this design yields seven distinct absorption peaks, offering versatility in absorption across a wide frequency range. Furthermore, a comprehensive parametric analysis was conducted to investigate the absorption dependence on unit cell dimensions and dielectric thickness, providing valuable insights for optimizing absorption performance. The MMA's narrow linewidth, compact dimensions, minimal thickness, and simple structural design collectively position it as a viable candidate for sensing and detection applications. To streamline the design optimis ation process, a random forest regression model is employed to predict absorption values for intermediate frequencies, reducing simulation time and resource usage while ensuring accurate performance predictions. This approach not only advances the understanding of MTM absorber design but also holds promise for optimis ing **antenna** performance and reducing RCS, thereby contributing to advancements in THz technology and electromagnetic wave management.

2. Design Simulation and Results Discussions

Figure 1(a, b) shows the perspective and top view of a three-layer **hepta-band** metamaterial absorber. This absorber is a simple structure designed to efficiently absorb **electromagnetic** radiation across multibands, particularly in the THz spectrum. It consists of three distinct layers: a top metallic layer, a dielectric middle layer, and a bottom metallic layer. The top and bottom layers are made from gold, a material chosen for its high conductivity and compatibility with THz frequencies, with a conductivity value of 4.09×10^7 S/m. The top metallic layer is engineered to interact with incident electromagnetic waves and induce resonance phenomena at specific frequencies. These resonant structures play a crucial role in the absorption mechanism of the MMA, facilitating the efficient capture and dissipation of electromagnetic energy.

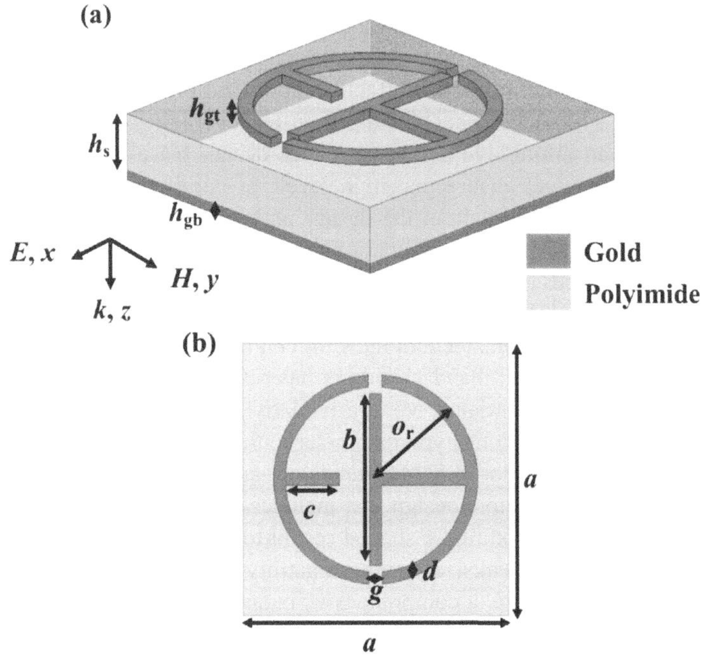

Fig. 1 (a) Perspective view, (b) Top view of the MMA.

In contrast, the bottom layer serves as a continuous metallic plane, enhancing the absorption efficiency of the device. The continuous nature of this layer ensures minimal reflection of incident waves, thereby maximis ing absorption. Sandwiched between the metallic layers is a dielectric middle layer composed of polyimide, characterised by a dielectric constant of 3 $(1 + i0.06)$ and a thickness of 9.7 μm. The dielectric layer acts as a spacer, controlling the resonant frequencies of the absorber and enabling fine-tuning of its absorption characteristics. Geometrically, the MMA is defined by several parameters, including $a = 34.4$, $b = 24$, $o = 14.8$, $c = 7.8$, $h_{gt} = 0.2$, $d = 1.8$, $g = 2$, $h_s = 9.7$, $h_{gb} = 0.5$, all specified in micrometres (μm). These dimensions are carefully chosen to ensure optimal performance of the absorber across its intended frequency bands.

At its lowest operational frequency, the absorber exhibits a thickness and compactness of 0.061 λ and 0.21 λ, respectively, where λ represents the wavelength of the incident radiation. This compact design allows the absorber to achieve high absorption efficiency, making it suitable for various applications. The absorption mechanism of the MMA is calculated as $1 - |S_{11}|^2 - |S_{21}|^2$, where S_{11} and S_{21} denote the reflection and transmission coefficients, respectively. In this **absorber** configuration, with a continuous metallic bottom layer, the transmission coefficient (S_{21}) is effectively zero, simplifying the absorption calculation to $1 - |S_{11}|^2$. This metric provides insights into the absorber's performance in capturing incident electromagnetic energy and minimis ing reflection losses.

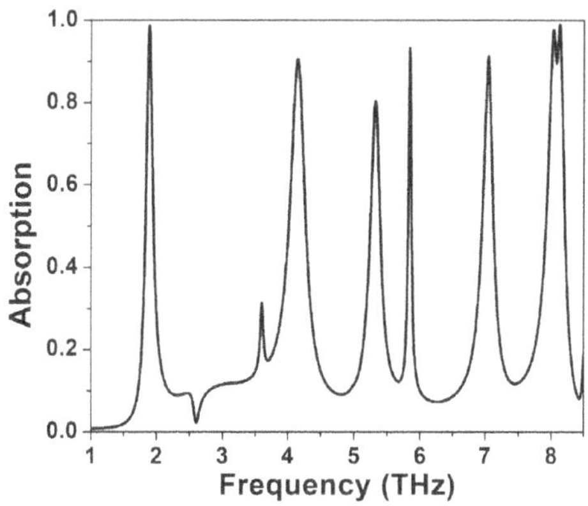

Fig. 2 Absorption response.

The proposed MMA demonstrates remarkable absorption characteristics, with multiple absorption peaks observed at distinct frequencies (Figure 2). These peaks appear at 1.89, 4.15, 5.32, 5.84, 7.04, 8.02, and 8.13THz, with corresponding absorption coefficients of 98.75%, 90.39%, 80.40%, 93.03%, 91.2%, 97.23%, and 98.25%, indicating the absorber's high efficiency in converting incident radiation into absorbed energy. This comprehensive analysis shows the hepta-band MMA's design principles and exceptional performance, highlighting its potential for THz **sensing**, **imaging**, and communication applications.

In Figures 3 and 4, a parametric analysis explores the impact of substrate thickness (h_s) and unit cell dimension (a) on absorption response. Figure 3 illustrates absorption response variations across different substrate thicknesses (h_s), revealing significant trends. Decreasing h_s from 10–9.6 μm alters absorption spectra from 1–8.5 THz. At 10 μm, distinct absorption peaks are observed. As h_s decreases to 9.9 μm, peak intensity slightly diminishes, causing a subtle overall absorption decrease. This trend continues with decreasing h_s, broadening absorption peaks, notably evident at 9.6 μm, indicating increased material-radiation interaction. These changes are most likely the result of altered dielectric properties caused by reduced substrate thickness, highlighting the material's sensitivity to environmental changes. Understanding such variations is critical for improving device performance, especially in THz applications that require precise absorption control.

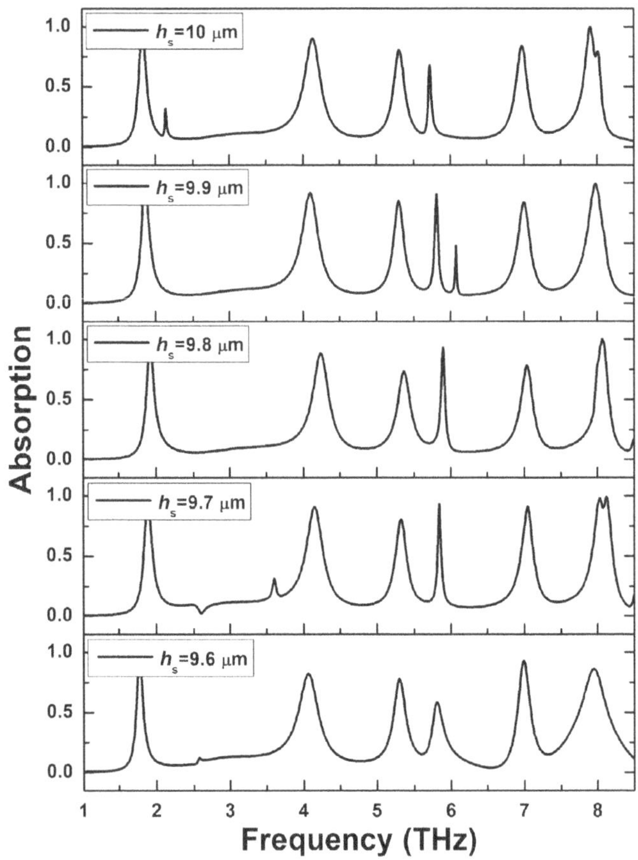

Fig. 3 Absorption response at various substrate thickness (h_s).

Figure 4 examines the absorption response with different unit cell dimensions (a), providing insights into structural effects. Absorption spectra show distinct features between 1 and 8.5 THz. Sharp peaks at 34 μm indicate a strong resonance. At 34.2 μm, there is a minor shift with a slight decrease in peak intensity due to structural geometry. Significant changes are observed at 34.4 μm, with broader peaks indicating a shift towards broadband absorption, possibly due to altered resonant modes and inter-unit cell interactions. These findings highlight the material's sensitivity to structural modifications, which is critical for tailored absorption characteristics, particularly in optimising performance for THz devices [38].

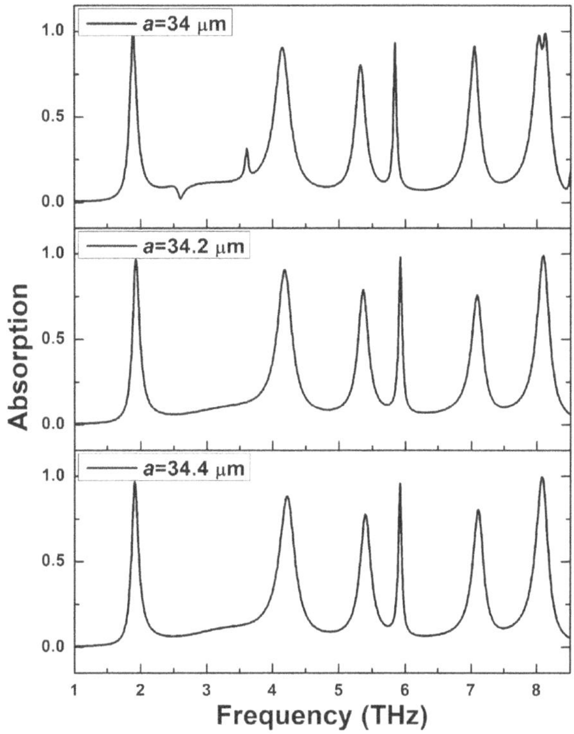

Fig. 4 Absorption response at various unit cell dimensions (a).

3. Prediction of Absorption Coefficient using Machine Learning

ML refers to various techniques for training computers to learn from data and make predictions or decisions [39]. Two fundamental types of ML are supervised learning, where the algorithm learns from labelled data, and unsupervised learning, where the algorithm learns from unlabelled data to uncover hidden patterns or structures [40, 41]. Within **supervised learning**, regression focuses on predicting continuous outcomes, such as predicting the absorption coefficient in antenna design [42], while classification is used to predict categorical outcomes, such as classifying **images** into different categories [43]. These techniques play crucial roles in various applications, from predicting stock prices to diagnosing diseases.

This section provides an overview of the role of regression models in the simulation process, emphasis ing their potential to streamline the design of effective absorbers across various proportions, including 0.2, 0.3, 0.4, 0.5, and 0.6. **Regression** analysis offers a valuable method to determine dependent parameter values based on independent parameter values, with frequency serving as the independent parameter and absorption coefficient as the dependent parameter in antenna design. Simulating experimental designs typically requires a significant amount of time and resources during the design process for complex structures. However, ML approaches for regression problems can effectively address these challenges by simplifying not only the prediction of absorption coefficients but also the identification of missing parameter values. To tackle these issues, three stages are proposed utilis ing **ML**-based regression analysis methodologies. Step 1 involves simulating the absorber's design by incrementing the frequency's step size value. Step 2 entails training the ML-based regression model using the simulated data. Finally, in Step 3, the trained regression model is applied to predict the absorption values of intermediate frequencies, thereby streamlining the design process and reducing resource requirements.

3.1 Random Forest Regression Model

In this study, regression analysis is performed using the **random forest (RF)** regression model, chosen for its ability to enhance prediction accuracy through a binary recursive partitioning mechanism [44]. This model's effectiveness lies in its capability to reduce variance more effectively by explicitly randomis ing cut-off points and attributes, coupled with ensemble averaging. The RF model's two crucial parameters are the number of attributes randomly selected at each node (K) and the minimum sample size for splitting a node (n_min), with larger values of K preferred for improved accuracy in regression problems [45]. Three attributes– substrate thickness (h_s), unit cell dimension (a), and frequency (f) — are chosen for predicting the absorption coefficient. The value of n_min is varied among the values 2, 4, 6, 8, and 10. Additionally, the RF model's performance is evaluated with different numbers of estimators (n_estimators) including 50, 100, 150, 200, and 250. Moreover, the effect of varying max_depth, another critical parameter, is considered with values of 15, 20, 25, and None. Finally, the influence of min_samples_leaf is investigated with values ranging from 1–5. In addition to these considerations, it's essential to understand how each parameter influences the RF **regression model's** predictive capabilities, particularly in the context of predicting the absorption coefficient.

- **Number of Estimators (n_estimators):** Increasing the number of estimators generally leads to a more robust model by averaging out the predictions of multiple trees, reducing the risk of overfitting. In the context of absorption coefficient prediction, higher n_estimator values may provide more stable and accurate predictions by capturing a broader range of patterns in the data.
- **Maximum Depth (max_depth):** This parameter controls the depth of each decision tree. A deeper tree can capture more intricate relationships in the data but is also prone to overfitting, especially with smaller datasets. In the context of absorption coefficient prediction, tuning max_depth allows balancing between capturing complex patterns and preventing overfitting.
- **Minimum Samples for Splitting (n_min):** Adjusting this parameter regulates the threshold for splitting nodes during the tree-building process. Setting it too low may result in overfitting, while setting it too high may lead to underfitting. In predicting the absorption coefficient, n_min influences the granularity of the decision boundaries and helps in finding the optimal balance between bias and variance.

- **Minimum Samples per Leaf (min_samples_leaf)**: Similar to n_min, **min_samples_leaf** helps control **overfitting** by setting the minimum number of samples required to form a leaf node. Higher values of min_samples_leaf can lead to simpler trees and prevent overfitting, particularly when dealing with noisy data or limited sample sizes.

By systematically exploring the effects of these parameters on the RF regression model's performance, researchers can identify the optimal configuration for predicting the absorption coefficient accurately. This comprehensive analysis enables fine-tuning the model to achieve the desired balance between prediction accuracy and generalis ation on unseen data.

Within the RF model, a collection of '*m*' unpruned regression trees is constructed, and their predictions are aggregated using arithmetic averaging to yield the final prediction. Various performance indices such as R-squared, **mean absolute error** (MAE), root mean squared error (RMSE), percentage error, and mean squared error (MSE) are utilis ed to assess the model's accuracy [46]. The predictions are evaluated under different test sizes (TS), including 0.2, 0.3, 0.4, 0.5, and 0.6, representing different proportions of training and testing data [47].

Additionally, in each test size scenario, the data generated during the simulation is divided into two distinct categories. For instance, in TS-0.2, 80% of the simulation records are randomly selected for training the RF regression model, while the remaining 20% are reserved for testing. Similar proportions are maintained for TS-0.3, TS-0.4, TS-0.5, and TS-0.6, ensuring a comprehensive assessment of the model's performance across varying proportions of training and testing data. This approach facilitates insights into the model's efficacy in predicting absorption values under different conditions, contributing to the reduction of **simulation** time and resource consumption.

Figure 5 illustrates how changes in n_min values and test sizes impact the predictive efficacy of the RF regression model. In Figure 5(a), R-squared values are represented, where darker shades indicate higher values, suggesting a better model fit. The heatmap demonstrates that smaller n_min values generally yield higher R-squared values across all test sizes, indicating improved model performance with finer splits. Moreover, larger test sizes tend to result in slightly lower R-squared values, suggesting a potential trade-off between training data size and model accuracy. Figure 5(b) displays the percentage error, with darker shades indicating lower errors. Smaller n_min values and larger test sizes typically result in lower Percentage Error, indicating improved predictive accuracy. Test size appears to have a greater impact on percentage error compared to n_min values. These findings underscore the importance of carefully selecting n_min values and test sizes to enhance the RF regression model's predictive performance across various applications. Figure 6 presents the results of using the RF regression technique to predict the **absorption** coefficient, focusing on the effect of varying the minimum number of samples required for a split (n_min) on model performance in terms of MSE (a), RMSE (b), and MAE (c) metrics across different test sizes.

Figure 6(a) depicts the **4**, with lighter shades corresponding to lower error values. The heatmap demonstrates that smaller n_min values result in lower MSE across all test sizes, implying improved model performance with finer splits. This trend occurs

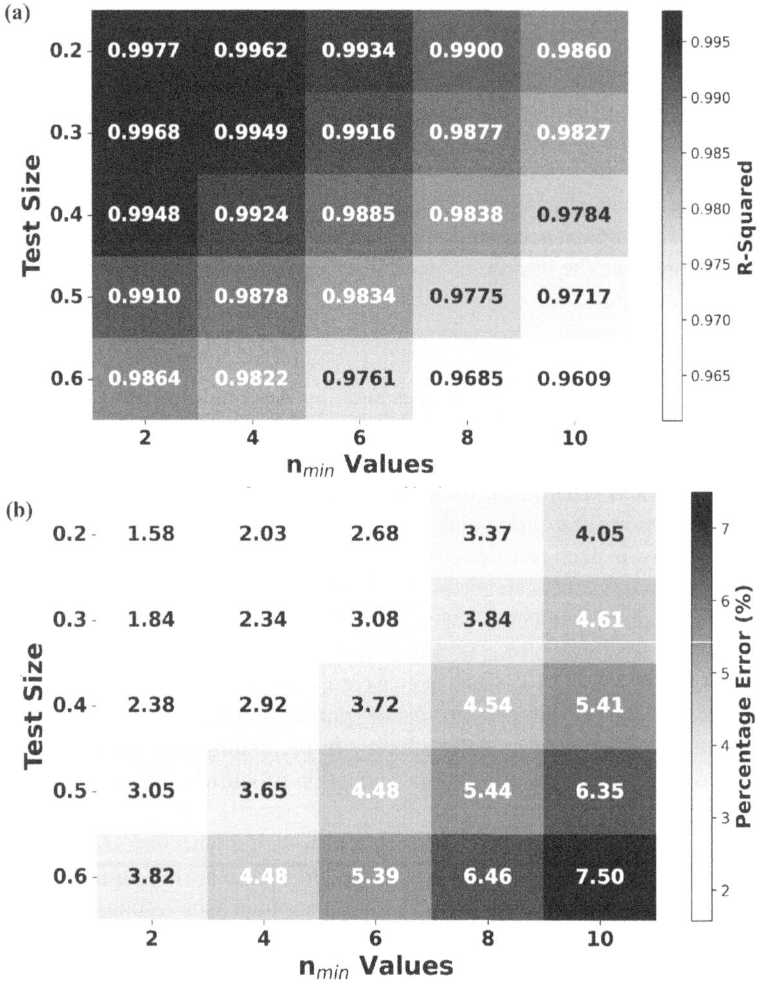

Fig. 5 Heatmaps depict the influence of varying the n_min and test size on the predictive performance of the RF regression model for (a) R-squared and (b) percentage error.

because smaller n_min values allow for more splits in decision trees, resulting in more complex models with finer details in the data, leading to lower **mean square error**. Figure 6(b) illustrates the **RMSE**, with lighter shades representing lower error values. Similar to MSE, smaller n_min values and larger test sizes generally result in lower RMSE, indicating higher predictive accuracy. Test size appears to have a greater influence on RMSE than n_min values, as larger test sizes provide more data for model evaluation, allowing for better estimation of the true error, and thus lower RMSE. Figure 6(c) displays the MAE, with lighter shades indicating lower error values. Smaller n_min values and larger test sizes typically result in lower MAE, indicating better predictive accuracy. Test size has a greater influence on MAE than n_min values, as larger test sizes enable more data points to be evaluated, resulting in a more accurate assessment of the model's absolute error performance. These

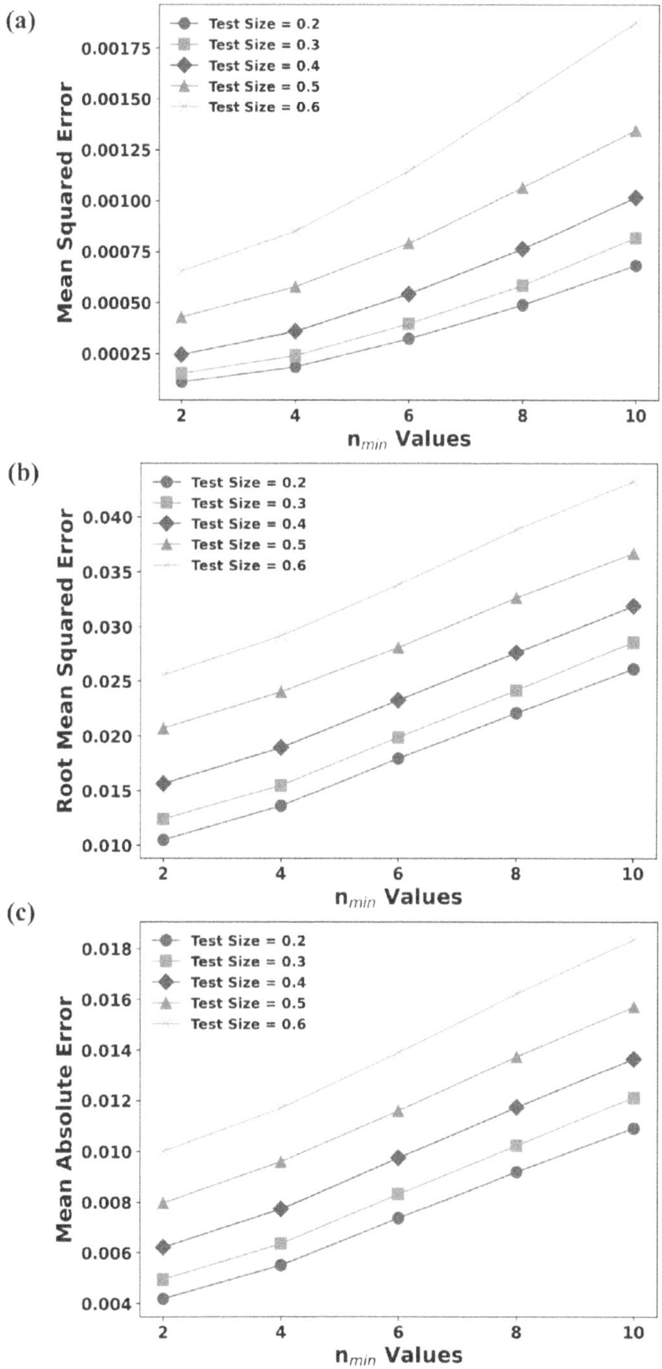

Fig. 6 Relationship between the performance metrics: (a) mean squared error (MSE), (b) root mean squared error (RMSE), and (c) mean absolute error (MAE) with varying **n_min** values for different test sizes in a predictive model.

findings underscore the importance of carefully selecting n_min values and test sizes to optimise the RF regression model's predictive performance. While smaller n_min values typically lead to better performance metrics, this decision should be balanced against factors such as computational complexity and potential overfitting. Similarly, larger test sizes can improve **error estimation** accuracy while also requiring more computational resources.

Figure 7 demonstrates the RF regression model's performance for n_min values ranging from 10 to 2. Each plot compares the actual absorption coefficients to the model's predicted values, with points closer to the diagonal indicating greater predictive accuracy. In Figure 7(a), the scatter plot for n_min = 10 shows a spread of points around the diagonal line, indicating a reasonable alignment between actual and predicted values. As n_min decreases in subsequent plots (b) through (e), there

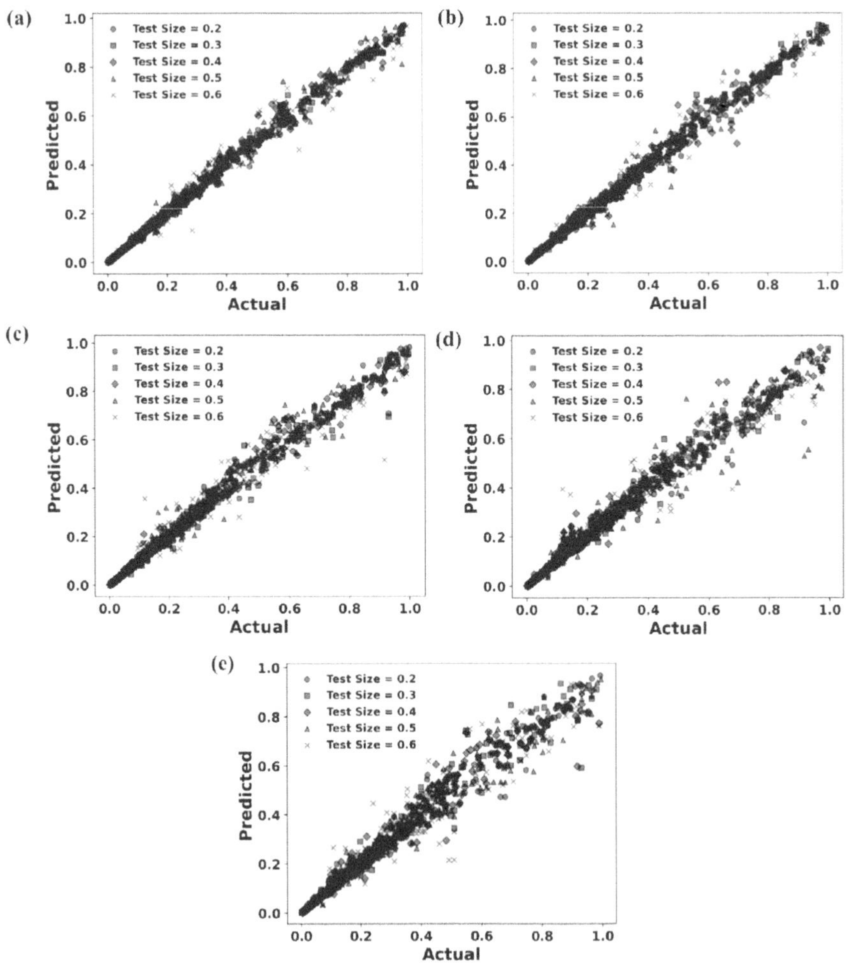

Fig. 7 Actual vs predicted values for: (a) n_min = 10, (b) n_min = 8, (c) n_min = 6, (d) n_min = 4, and (e) n_min = 2 in the RF regression model for absorption coefficient prediction.

is a noticeable trend towards tighter clustering of points around the diagonal line, implying improved predictive accuracy with finer splits. Figure 7(b), representing n_min = 8, shows a relatively consistent alignment of actual and predicted values, with slightly tighter clustering than n_min = 10. This trend continues in Figure 7(c, d, e), where the points cluster closer to the diagonal line, indicating improved predictive accuracy. The scatter plot in Figure 7(e) shows that when n_min = 2, predictive accuracy improves significantly compared to higher n_min values. This improvement can be attributed to finer splits in the decision trees of the RF regression model. Setting n_min to a lower value, such as 2, ensures that each node in the decision trees requires no more than two samples to split. As a result, decision trees can detect more nuanced patterns and relationships in the data, yielding more **accurate predictions**. Finer splits help the model adapt to the complex data distribution, improving predictive performance. Smaller n_min values allow the model to create more flexible and intricate decision boundaries, better-capturing data complexity. This flexibility lets the model capture more dataset patterns and variations to better generalize to new data. However, predictive accuracy and overfitting must be balanced. Smaller n_min values may improve model performance but may overfit training data, preventing generalis ation on unseen data. Thus, n_min should be carefully chosen to optimise model performance and **generalisation.**

Figure 8 presents the performance metrics of the RF regression model across different values of n_estimators and test sizes. In Figure 8(a), consistent shades of blue indicate stable **R-squared** values across varying n_estimators values, highlighting that changes in the number of decision trees in the ensemble do not substantially affect the model's ability to explain the variance in the target variable. The term "n_ estimators" refers to the number of decision trees in the RF ensemble. Each decision tree contributes to the final prediction, and increasing the number of trees generally leads to a more robust and stable model. In this context, the stable R-squared values across different n_estimators suggest that the model remains consistent regardless of the number of **decision trees** in the ensemble. Figure 8(b) illustrates changes in percentage error with alterations in test size. Darker shades indicate higher error rates, suggesting that adjusting the test size influences the model's predictive accuracy. Smaller test sizes have slightly higher error rates, suggesting that generalizing the model's performance to larger datasets may be difficult. The dataset's test size is the percentage used for testing and the rest for model training. Smaller test sizes mean a larger training set, which can cause overfitting in small datasets. Larger test sizes allow a more complete evaluation of the model's performance on unseen data but may reduce training data, affecting predictive accuracy. These findings emphasis e the importance of test size selection in model robustness and predictive accuracy evaluation. Researchers can evaluate the model's generalis ation and ensure reliable performance across datasets by carefully selecting a test size.

Figure 9 illustrates the performance of the RF regression model across varying values of n_estimators and test sizes. Each subplot represents the MSE, RMSE, and MAE for different test sizes ranging from 0.2–0.6. In Figure 9(a), which displays the MSE, there is little variation in error values as n_estimators change across different test sizes. The MSE remains relatively stable, indicating that altering the number of estimators in the random forest ensemble does not significantly impact the model's

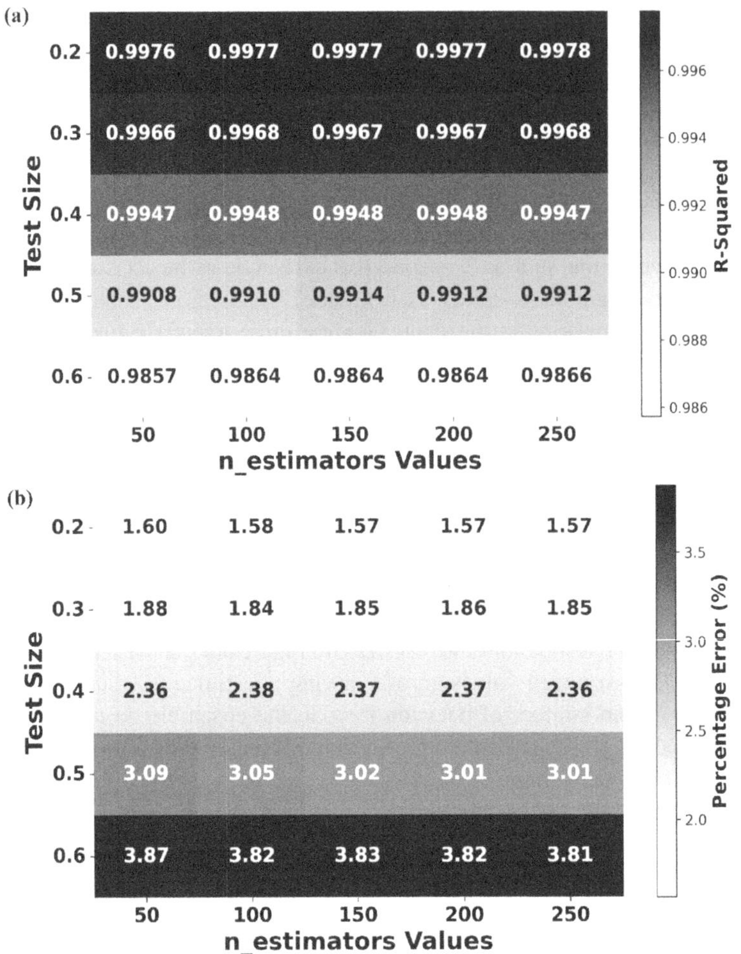

Fig. 8 (a) R-squared values, and (b) **percentage error** values for different combinations of n_estimators and test size in the RF regression model.

performance in terms of MSE. Similarly, Figure 9(b) shows the RMSE, where the error values also exhibit minimal fluctuation with changes in n_estimators. The RMSE remains consistent across different test sizes and n_estimators values, suggesting that the predictive accuracy of the model, as measured by RMSE, is not strongly influenced by the number of estimators. Figure 9(c) represents the MAE, and like the MSE and RMSE, the error values show little variation with changing n_estimators values. The MAE remains relatively constant across different test sizes and n_estimators values, indicating that adjustments to the number of estimators do not substantially affect the **model's** performance in terms of MAE. Interestingly, while n_estimators values have limited impact on the error metrics, changes in test sizes result in slight increases in error values across all three metrics. Larger test sizes lead to slightly higher error rates, suggesting a decrease in model performance. This observation underscores the importance of test size selection in evaluating the model's predictive accuracy and

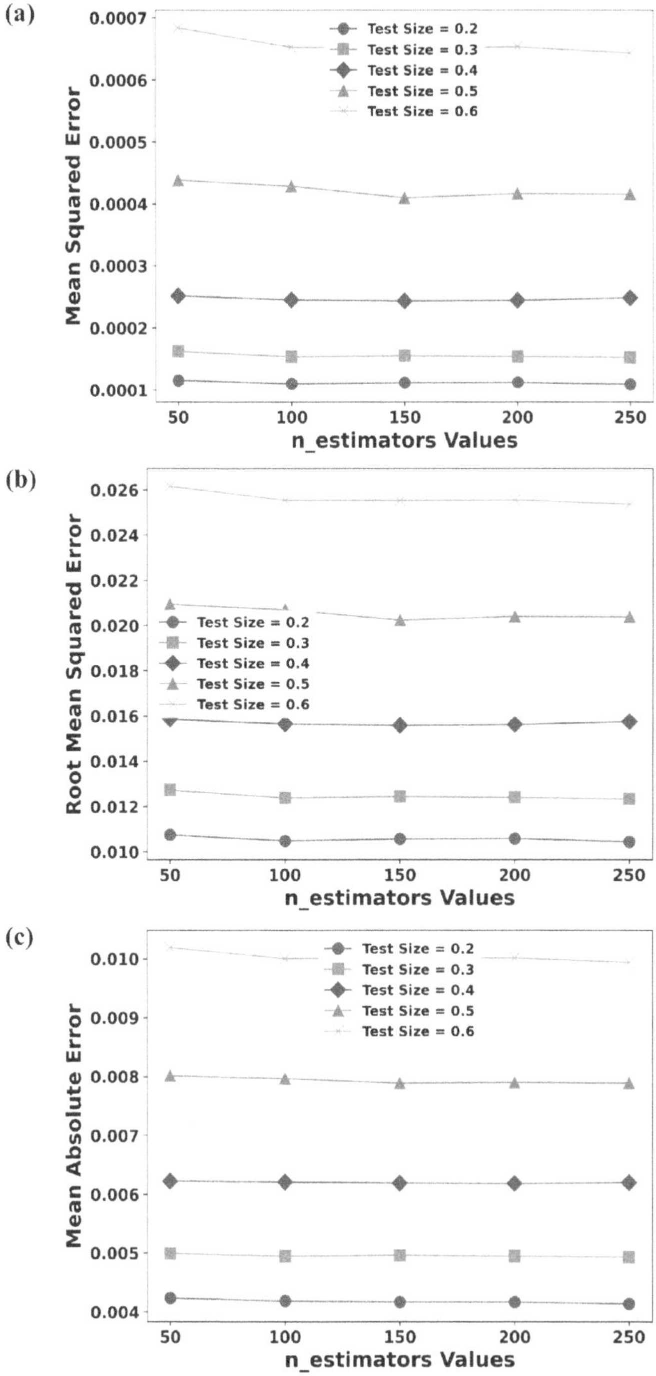

Fig. 9 Comparison of RF regression model performance with varying n_estimators values and test sizes for: (a) MSE, (b) **RMSE**, and (c) MAE.

highlights the need to balance training and testing data proportions for optimal model performance. Overall, these insights emphasize the importance of carefully choosing hyperparameters, such as n_estimators, and test sizes to ensure robust and accurate predictions while training the RF **regression model**.

Figure 10 illustrates the performance of the RF regression model across different test sizes and n_estimators values. In Figure 10(a), where **n_estimators** = 50, the scatter plot shows a spread of points around the diagonal line, indicating a reasonable alignment between actual and predicted values. Despite having a lower number of estimators, the model still demonstrates effective predictive performance across various test sizes. Figure 10(b) represents n_estimators = 100. Here, the scatter plot continues to show a consistent alignment between actual and predicted values, with points clustering closely around the diagonal line. The increase in the number of estimators leads to slightly tighter clustering of points compared to Figure 10(a), suggesting improved predictive accuracy. In Figure 10(c), with n_estimators = 150,

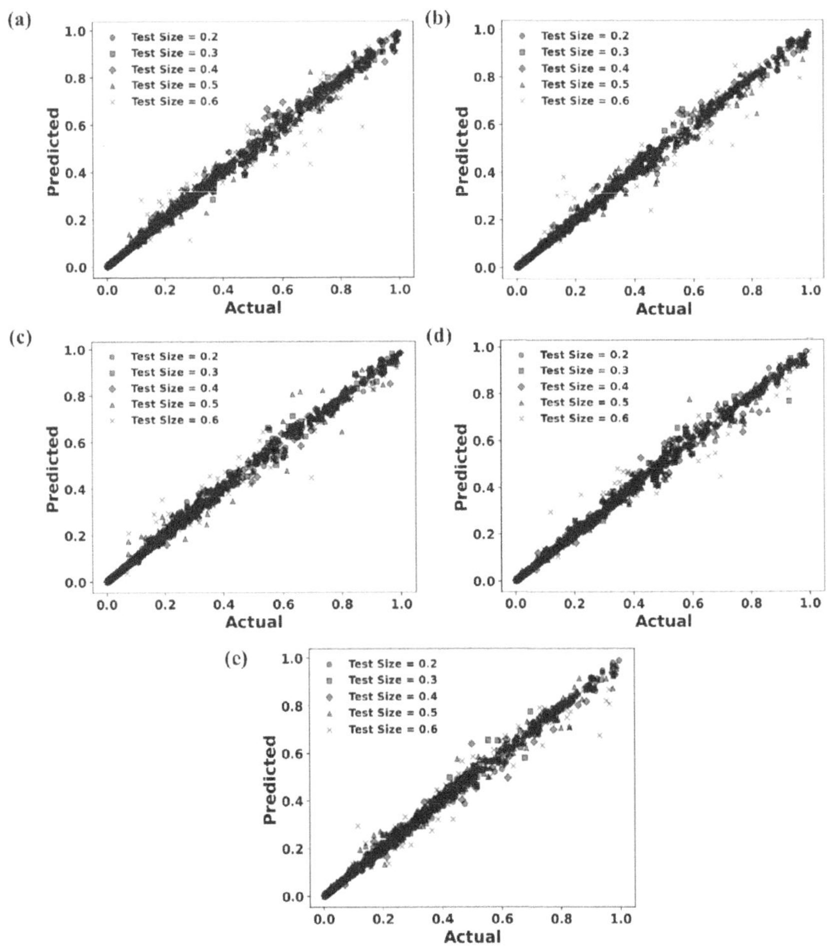

Fig. 10 Actual absorption coefficients against the predicted values with: (a) n_estimators = 50, (b) n_estimators = 100, (c) n_estimators = 150, (d) n_estimators = 200, and (e) n_estimators = 250.

the scatter plot shows further improvement in predictive **accuracy**. The points cluster even more tightly around the diagonal line compared to the previous plots, indicating enhanced alignment between actual and predicted values with the increased number of estimators. Figure 10(d) corresponds to n_estimators = 200. Here, the scatter plot reveals continued improvement in predictive accuracy, with points clustering even more closely around the diagonal line compared to Figure 10(c). The larger number of estimators further enhances the model's ability to capture the underlying patterns in the data. Finally, in Figure 10(e) with n_estimators = 250, the scatter plot demonstrates the highest level of predictive accuracy among the subplots. The points cluster very tightly around the diagonal line, indicating excellent alignment between actual and predicted values. With the largest number of estimators in this subplot, the model achieves the most precise predictions. Overall, across all subplots, variations in test sizes result in slight deviations in the spread of points but do not significantly affect the overall alignment between actual and predicted values. This observation underscores the robustness of the model across different test sizes, further highlighting its capability to generalise well to unseen data.

Figure 11(a, b) presents the model's performance metrics for various combinations of the **max_depth** hyperparameter and test size in the RF regression model. The performance is evaluated for max_depth values of 15, 20, 25, and None. Figure 11(a) shows that as the test size increases, the percentage error rises slightly across all max_depth values. Notably, for higher max_depth values such as 25 and None, the percentage error appears to decrease, indicating improved model performance, particularly evident for larger test sizes. This trend suggests that increasing the test size might lead to a broader representation of the data, resulting in a **reduction in error**, especially for models with higher max_depth values. For a test size of 0.2, the percentage error values are approximately 4.47%, 1.77%, 1.57%, and 1.58% for max_depth values of 15, 20, 25, and None, respectively. Figure 11(b) shows that higher max_depth values tend to have higher **R-squared** values, indicating a better model fit to the data.

This trend is particularly for max_depth values of 20, 25, and None. Moreover, larger test sizes generally lead to slightly higher R-squared values across different max_depth values, suggesting enhanced model performance with larger **training datasets**. This incremental improvement in R-squared values with larger test sizes may stop from the model's exposure to a more comprehensive dataset during training, allowing it to capture more intricate patterns in the data. For a test size of 0.2, the R-squared values are approximately 0.9837, 0.9975, 0.9978, and 0.9977 for max_depth values of 15, 20, 25, and None, respectively. These insights underscore the importance of max_depth selection in **optimising model** performance, with higher values contributing to better fitting of the data, as evidenced by lower percentage errors and higher R-squared values. Additionally, the influence of test size highlights the significance of adequately sizing the training dataset to ensure optimal model performance.

Figure 12(a, b, c) provides insights into the performance of a RF regression model concerning various max_depth values (15, 20, 25, and None) and test sizes, focusing on MSE, RMSE, and MAE, respectively. In Figure 12(a), the MSE decreases as the max_depth values increase, indicating improved model performance with deeper

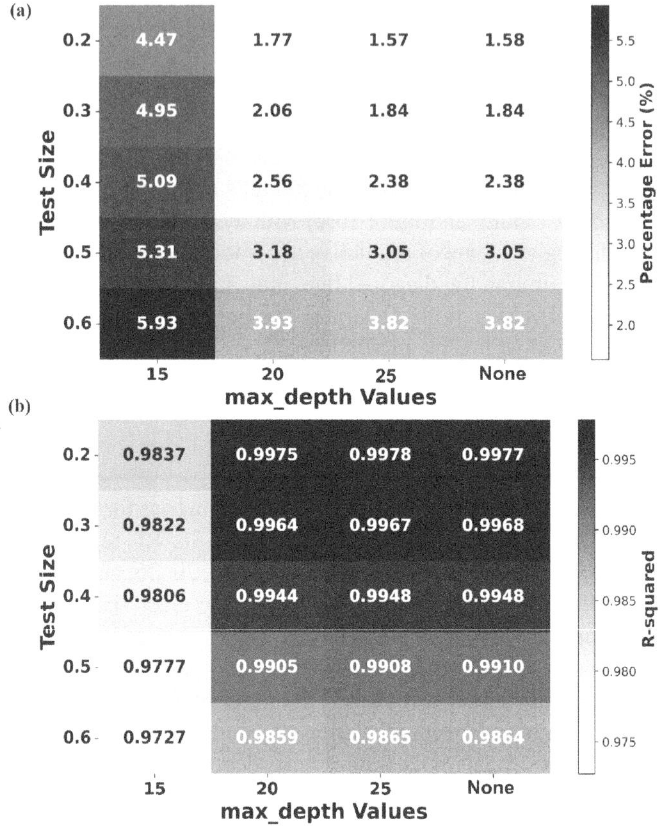

Fig. 11 (a) Percentage error, (b) R-squared values for different combinations of max_depth and **test size** in the RF regression model.

trees. This trend is consistent across different test sizes, suggesting that increasing max_depth generally leads to more **accurate predictions**. However, the rate of decrease in MSE diminishes beyond a certain max_depth threshold, emphasis ing the importance of balancing model complexity and performance.

Similarly, Figure 12(b) showcases the RMSE trends, with lower values observed for higher max_depth values. The relationship between max_depth and RMSE is similar to that of MSE, indicating that deeper trees result in better model fit. The slight fluctuations in RMSE across different test sizes highlight the impact of dataset variability on model performance. Figure 12(c) reveals the MAE patterns, demonstrating a consistent decrease in error metrics with increasing max_depth values. This decline signifies improved model accuracy and reduced discrepancies between actual and predicted values. However, similar to MSE and RMSE, the error reduction beyond a certain **max_depth** threshold underscores the need for careful model parameter selection. The relationship between max_depth and test sizes suggests that while deeper trees generally contribute to better model performance, it's crucial to consider the trade-off between model complexity and overfitting. The slight fluctuations in error metrics across test sizes indicate the importance of dataset

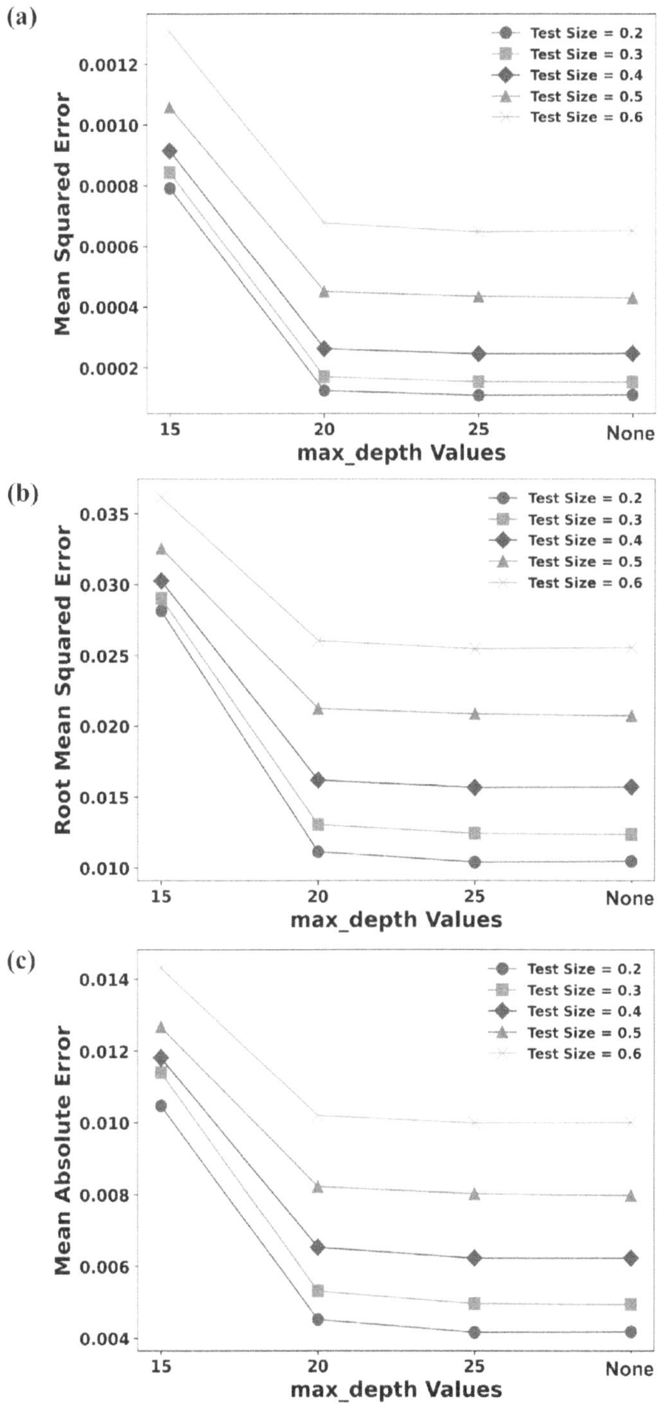

Fig. 12 Comparison of RF regression model performance with varying max_depth values and test sizes for: (a) MSE), (b) RMSE, and (c) **MAE**.

characteristics in model training and evaluation. Therefore, selecting an optimal max_depth value involves balancing the benefits of deeper trees with the risks of overfitting, ensuring robust and generalised model performance.

The scatter plots obtained from a RF regressor model trained with different max_depth values (15, 20, 25, and None) are shown in Figures 13(a–d), respectively. Figure 13(a) illustrates the relationship between actual and predicted values for max_depth = 15. A strong diagonal line alignment across all test sizes indicates a good fit between model predictions and data. However, there are slight deviations from the diagonal line, particularly noticeable for smaller test sizes, suggesting some variability in the predictions, possibly due to the limited amount of training data. In Figure 13(b), corresponding to max_depth = 20, a further improvement in the alignment between actual and predicted values compared to Figure 13(a) is noticeable. The points cluster more tightly around the diagonal line across all test sizes, indicating enhanced **predictive accuracy**. This suggests that increasing the max_depth parameter allows the model to capture more intricate patterns in the data, resulting in improved predictions. Moving on to Figure 13(c) with max_depth = 25, a similar trend of tight clustering around the diagonal line is observed, indicative of high predictive accuracy. The alignment between actual and predicted values appears to be slightly improved compared to Figure 13(b), especially for smaller test sizes. This suggests that deeper trees, given by higher max_depth values, enable the model to capture even finer details in the data, leading to more accurate predictions. Finally, in Figure 13(d) where max_depth = None, representing unconstrained tree growth, a remarkable alignment between actual and predicted values across all test sizes is observed. The points tightly cluster along the diagonal line, indicating excellent **predictive accuracy**. This suggests that allowing the trees to grow without any depth constraint enables the model to capture the most intricate patterns in the data, resulting in highly accurate predictions. Overall, these scatter plots demonstrate the impact of max_depth on the predictive performance of the RF regressor model. Deeper trees, achieved by higher max_depth values, lead to improved predictive accuracy by allowing the model to capture more complex relationships in the data.

The heatmap analysis in Figure 14 offers valuable insights into the performance of an RF model across various combinations of **min_samples_leaf** values and test sizes. Firstly, examining the R-squared heatmap (Figure 14(a)), it's evident that adjusting min_samples_leaf significantly influences the model's fit to the data. Higher min_samples_leaf values generally lead to decreased R-squared values across all test sizes, indicating a reduction in predictive accuracy as the trees become less complex. Conversely, lower min_samples_leaf values tend to yield higher R-squared values, particularly noticeable for smaller test sizes, suggesting that more intricate tree structures can capture finer patterns in the data. However, excessively low min_samples_leaf values may lead to overfitting, especially evident in larger test sizes where moderate min_samples_leaf values demonstrate better performance. Notably, extremely high min_samples_leaf values result in notably lower R-squared values across all test sizes, suggesting that overly simplified trees may fail to adequately represent the data. Similarly, the **percentage error** heatmap (Figure 14(b)) provides insights into the accuracy of predictions made by the RF model. Consistent with the R-squared heatmap, increasing min_samples_leaf values generally leads to higher

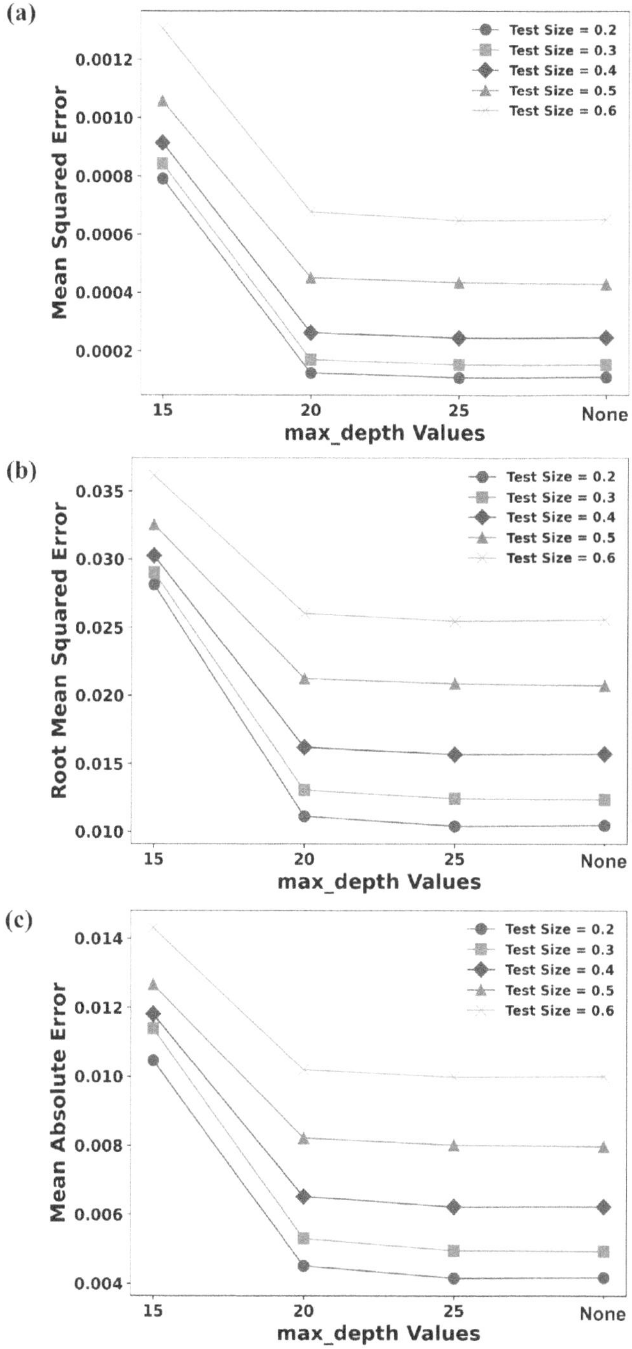

Fig. 13 Scatter plots showing the relationship between actual and predicted values obtained from a RF Regressor model trained with: (a) max_depth = 15, (b) max_depth = 20, (c) max_depth = 25, and (d) max_depth = None.

percentage errors, indicating reduced prediction accuracy. Smaller test sizes exhibit relatively lower percentage errors compared to larger test sizes for all min_samples_leaf values, likely due to the model's ability to capture the variability in smaller datasets more effectively. Moderate min_samples_leaf values (e.g., 3 or 4) often result in lower percentage errors across different test sizes, indicating improved predictive accuracy compared to extremely high or low values. Conversely, extremely high min_samples_leaf values consistently yield the highest percentage of errors across all test sizes, indicative of poor model performance resulting from oversimplified trees. The analysis shows that min_samples_leaf is crucial to balancing model complexity and predictive accuracy in **RF Regression**. Given the trade-offs between **overfitting and underfitting** across test sizes, choosing the best min_samples_leaf values (1, 2, 3, 4, 5) is essential for model performance.

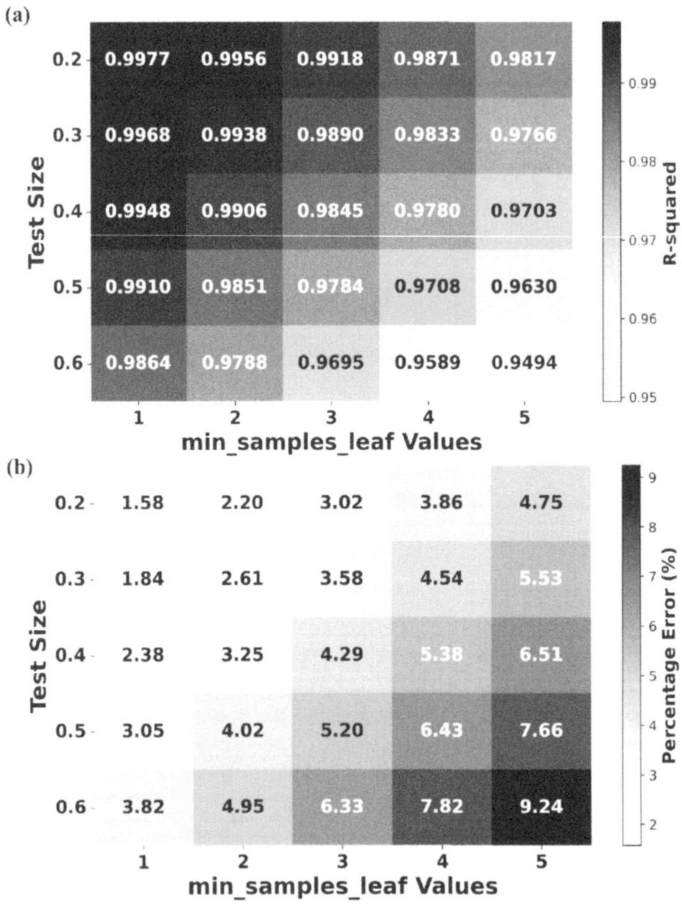

Fig. 14 (a) R-squared and (b) percentage error for different min_samples_leaf values across various test sizes in the RF regression model.

Figure 15 illustrates the performance evaluation metrics, including **MSE, MAE, and RMSE**, across varying min_samples_leaf values and test sizes within the RF

(a)

(b)

(c)

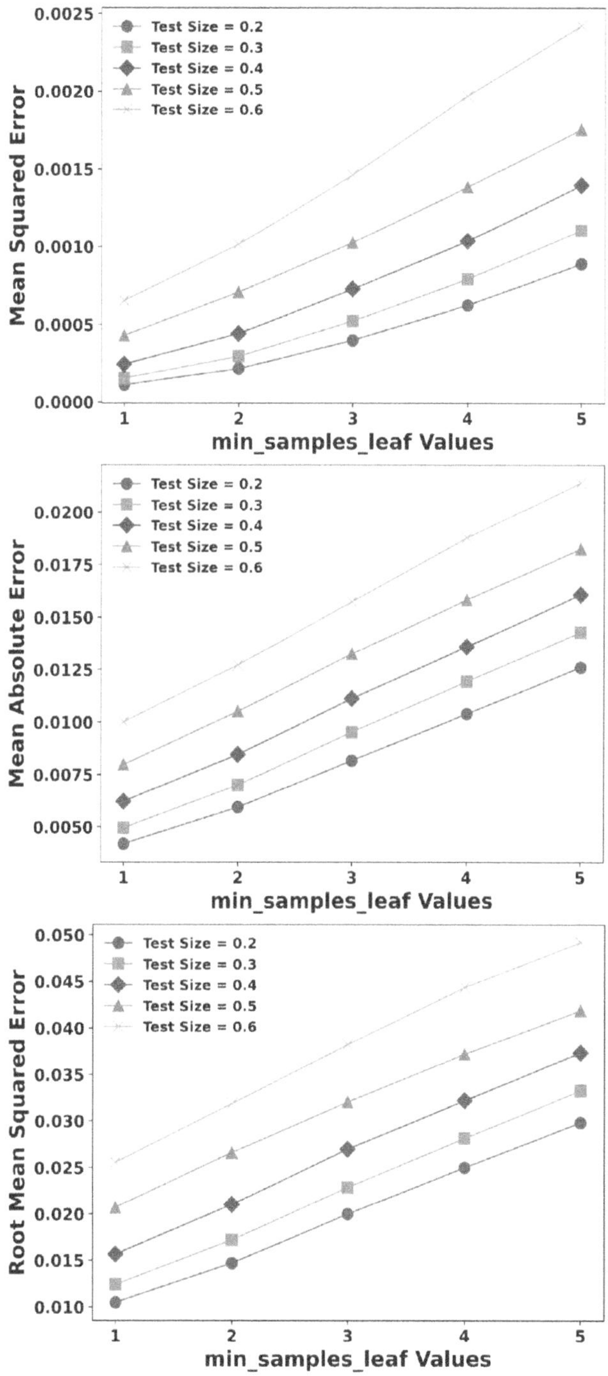

Fig. 15 (a) MSE, (b) MAE, and (c) RMSE for different min_samples_leaf values across various test sizes in the **RF regression** model.

Regression model. Analys ing the MSE plot (Figure 15 (a)), it becomes apparent that as the min_samples_leaf value increases, the MSE generally exhibits an upward trend across all test sizes. This observation suggests that higher values of min_samples_leaf contribute to increased errors in prediction, indicating a decline in the model's overall accuracy. Particularly noteworthy is the influence of test size on MSE, where smaller test sizes (0.2 and 0.3) tend to showcase lower MSE values, emphasis ing the model's superior performance in smaller datasets. Conversely, larger test sizes (0.5 and 0.6) exhibit higher MSE values, implying higher prediction errors in larger datasets, especially noticeable with higher min_samples_leaf values. MAE plot (Figure 15 (b)) shows a similar trend wherein increasing min_samples_leaf values correspond to higher MAE values across all test sizes. Lower min_samples_leaf values demonstrate comparatively lower MAE, particularly evident in smaller test sizes, indicating the model's ability to achieve more accurate predictions in smaller datasets. Conversely, larger test sizes exhibit higher MAE values, signifying increased **prediction errors** in larger datasets, especially pronounced with higher min_samples_leaf values. This suggests that the model's performance is influenced not only by the complexity of the decision trees but also by the size of the test dataset. In Figure 15(c), depicting the RMSE, a parallel trend to MSE and MAE is observed. Higher min_samples_leaf values coincide with high RMSE values across all test sizes, indicating diminished predictive accuracy with more conservative leaf node splits. Smaller test sizes show relatively lower RMSE, underscoring the model's enhanced performance in smaller datasets. Conversely, larger test sizes present higher RMSE values, reflecting increased prediction errors in larger datasets, particularly noticeable with higher min_samples_leaf values. This confirms the importance of balance between model complexity and dataset size to optimize predictive performance in random forest regression.

In Figure 16, the scatter plots illustrate the relationship between actual and predicted values for different test sizes across various min_samples_leaf values in an RF model. Each plot corresponds to a specific min_samples_leaf value, ranging from 1 to 5. Beginning with (a) min_samples_leaf = 1, tightly clustered points along a linear trend line indicate strong alignment between actual and predicted values. As the analysis progresses through (b) min_samples_leaf = 2, (c) min_samples_leaf = 3, and (d) min_samples_leaf = 4, there's a gradual widening of the point spread, suggesting increased prediction variability as the model simplifies tree structures. Finally, in Figure 16 (e) min_samples_leaf = 5, there is a further widening of the point spread, reflecting enhanced generalis ation at the cost of increased variability in predictions. These scatter plots offer valuable insights into how adjusting min_samples_leaf values influences the model's predictive performance across different test sizes, highlighting the trade-off between model complexity and generali sation. The regularis ation effect on the **RF model** widens the point spread as min_samples_leaf increases. Increased min_samples_leaf sets a minimum number of samples needed in a leaf node during tree-building. Merging nodes with fewer samples creates simpler, more generalised decision trees with simpler splits. Thus, the model becomes less sensitive to data points and noise, smoothing decision boundaries. Increased generalization reduces flexibility in capturing complex data patterns and relationships. Thus, as min_samples_leaf increases, the model's ability to fit training data decreases, resulting in a wider range of predicted values for a given set of actual values. Widening the point

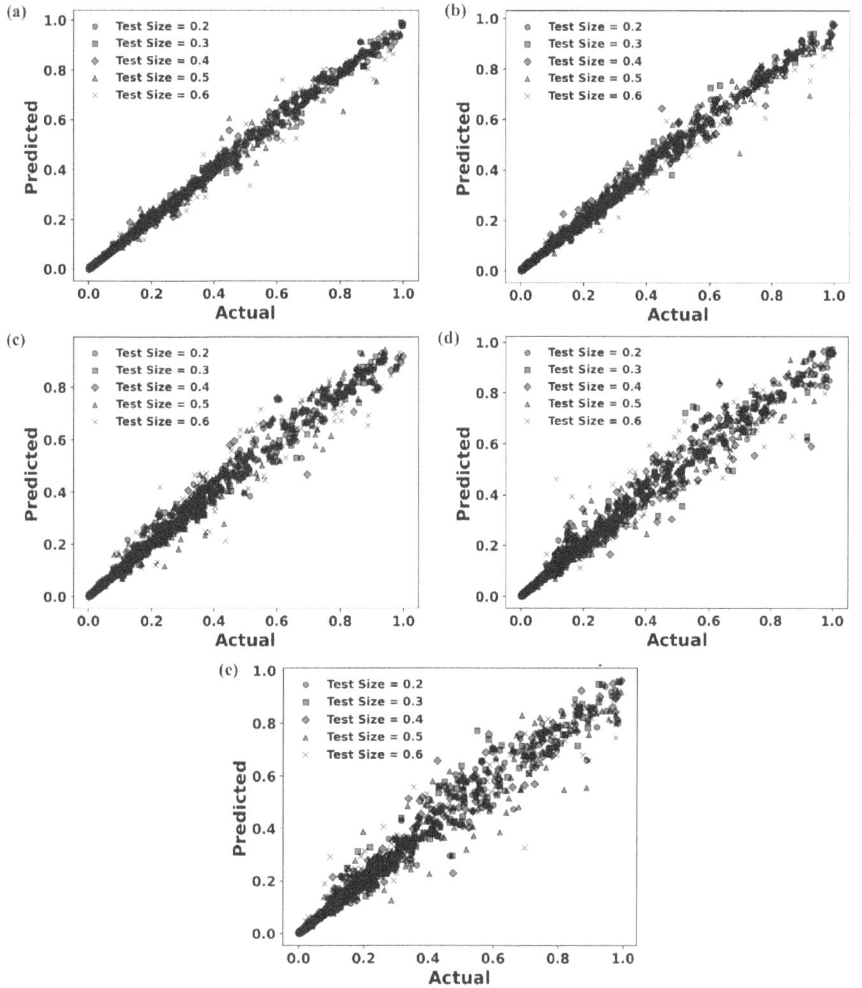

Fig. 16 Actual vs. Predicted scatter plots for different test sizes with: (a) min_samples_leaf = 1, (b) min_samples_leaf = 2, (c) min_samples_leaf = 3, (d) min_samples_leaf = 4, (e) min_samples_leaf = 5.

spread compromises model complexity and generali sation. Higher min_samples_leaf values smooth and generalise predictions but may miss subtle data patterns, increasing predicted value variability.

4. Summary

In conclusion, the study demonstrates the efficacy of employing an RF regression model for designing a hepta-band MMA optimis ed for terahertz applications, with a focus on reducing RCS. By using unit cell dimensions, substrate thickness values, and carefully tuned **hyperparameters** like n_min, n_estimators, max_depth, and min_samples_leaf, the absorber design process was enhanced while maintaining high prediction accuracy across frequencies. Furthermore, the investigation of different test sizes from 0.2–0.6 highlights the model's robustness and generalis ation performance. The RF model exhibited promising results in reducing simulation time and resource requirements by up to 60%, as validated through experimental comparisons. This methodology could serve as a valuable tool for researchers and engineers in optimis ing MTM structures for various sensing and communication applications, offering a balance between accuracy and efficiency. The ability to swiftly iterate through design iterations and predict absorption coefficients accurately contributes significantly to the advancement of THz technology. As future work, further exploration could involve refining the model architecture, exploring additional hyperparameters, and extending the analysis to different metamaterial designs and applications. Additionally, validating the model's performance across a broader range of experimental conditions and environments would enhance its reliability and applicability. Overall, the study showcases the potential of ML techniques, particularly **RF regression**, in revolutionis ing the design and optimis ation of complex MTM structures.

References

[1] R.A. Shelby, D.R. Smith and S. Schultz. "Experimental verification of a negative index of refraction," *Science* 80, vol. 292, no. 5514, pp. 77–79, 2001. doi: 10.1126/science.1058847.

[2] P. Jain, A. Thourwal, N. Sardana, S. Kumar, N. Gupta and A.K. Singh. "I-shaped metamaterial antenna for X-band applications," in *Progress in Electromagnetics Research Symposium*, May 2017, pp. 2800–2803. doi: 10.1109/PIERS.2017.8262229.

[3] Y. Wang, T. Sun, T. Paudel, Y. Zhang, Z. Ren and K. Kempa. "Metamaterial-plasmonic absorber structure for high efficiency amorphous silicon solar cells," *Nano Lett.*, 2012. doi: 10.1021/nl203763k.

[4] X. Zhang and Z. Liu. "Superlenses to overcome the diffraction limit," *Nat. Mater.*, 2008. doi: 10.1038/nmat2141.

[5] P. Jain et al. "T-Shaped Resonator for X-Band Applications," in *2017 IEEE MTT-S International Microwave and RF Conference (IMaRC)*, 2017, pp. 1–5. doi: 10.1109/IMaRC.2017.8449678.

[6] P. Jain, K. Prakash, N. Sardana, S. Kumar, N. Gupta and A.K. Singh. "Design of an ultra-thin hepta-band metamaterial absorber for sensing applications," *Opt. Quantum Electron.*, vol. 54, no. 9, p. 569, 2022. doi: 10.1007/s11082-022-03917-z.

[7] P. Jain et al. "Ultra-thin metamaterial perfect absorbers for single-/dual-/multi-band microwave applications," *IET Microwaves, Antennas, Propag.*, vol. 14, no. 5, pp. 390–396, 2020. doi: 10.1049/iet-map.2019.0623.

[8] A. Sharma, S. Ghosh and K.V. Srivastava. "A polarization-insensitive band-notched absorber for radar cross section reduction," *IEEE Antennas Wirel. Propag. Lett.*, vol. 20, no. 2, pp. 259–263, 2021/. doi: 10.1109/LAWP.2020.3047643.

[9] S. Bhattacharyya, S. Ghosh and K.V. Srivastava. "Bandwidth-enhanced metamaterial absorber using electric field-driven Lc resonator for airborne radar applications," *Microw. Opt. Technol. Lett.*, 2013. doi: 10.1002/mop.27786.

[10] S. J. Li, J. Gao, X.Y. Cao, Y. Zhao, Z. Zhang and H.X. Liu. "Loading metamaterial perfect absorber method for in-band radar cross section reduction based on the surface current distribution of array antennas," *IET Microwaves, Antennas Propag.*, vol. 9, no. 5, pp. 399–406, 2015. doi: 10.1049/iet-map.2014.0490.

[11] N.I. Landy, S. Sajuyigbe, J.J. Mock, D.R. Smith and W.J. Padilla. "Perfect Metamaterial Absorber," *Phys. Rev. Lett.*, vol. 100, no. 20, p. 207402, May 2008. doi: 10.1103/PhysRevLett.100.207402.

[12] D. Yan. "Tunable all-graphene-dielectric single-band terahertz wave absorber," Journal of Physics D, Applied Physics, 2019. doi: 10.1088/1361-6463/ab1892.

[13] Y. Zhang et al. "Five-Band Terahertz Perfect Absorber Based on Metal Layer–Coupled Dielectric Metamaterial," *Plasmonics*, 2019. doi: 10.1007/s11468-019-00956-3.

[14] X. Huang, C. Lu, C. Rong and M. Liu. "Wide-angle perfect metamaterial absorbers based on cave-rings and the complementary patterns," *Opt. Mater. Express*, vol. 8, no. 9, p. 2520, 2018. doi: 10.1364/ome.8.002520.

[15] P. Jain et al. "Dual band graphene based metamaterial absorber for terahertz applications," in *2018 IEEE 13th Nanotechnology Materials and Devices Conference, NMDC 2018*, 2019. doi: 10.1109/NMDC.2018.8605833.

[16] S.P. Rodrigues, S. Lan, L. Kang, Y. Cui and W. Cai. "Nonlinear imaging and spectroscopy of chiral metamaterials," *Adv. Mater.*, 2014. doi: 10.1002/adma.201402293.

[17] P. Jain, S. Bansal and P. Samant. "Metamaterial Absorbers for IoT Applications," in *2022 IEEE VLSI Device Circuit and System (VLSI DCS)*, 2022, pp. 300–303. doi: 10.1109/VLSIDCS53788.2022.9811494.

[18] A.H.M.I. Ferdous et al. "Optical communications with guided wave fibers with rectangular cladding for fuel adulteration detection," *J. Opt. Commun.*, 2023. doi: 10.1515/joc-2023-0058.

[19] L. Senesac and T.G. Thundat. "Nanosensors for trace explosive detection," *Materials Today*. 2008. doi: 10.1016/S1369-7021(08)70017-8.

[20] T. Ritari, H.L udvigsen, J.C. Petersen, T. Soren sen, A. Bjarklev and T.P. Hansen. "Gas sensing using air-guiding photonic bandgap fibers," in *OSA Trends in Optics and Photonics Series*, 2004. doi: 10.1364/opex.12.004080.

[21] C.M. Tran, H. Van Pham, H.T. Nguyen, T.T. Nguyen, L.D. Vu and T.H. Do. "Creating Multiband and Broadband Metamaterial Absorber by Multiporous Square Layer Structure," *Plasmonics*, vol. 14, no. 6, pp. 1587–1592, 2019. doi: 10.1007/s11468-019-00953-6.

[22] B.X. Wang, X. Zhai, G.Z. Wang, W.Q. Huang and L.L. Wang. "Design of a four-band and polarization-insensitive terahertz metamaterial absorber," *IEEE Photonics J.*, vol. 7, no. 1, 2015. doi: 10.1109/JPHOT.2014.2381633.

[23] F. Ding, Y. Cui, X. Ge, Y. Jin and S. He. "Ultra-broadband microwave metamaterial absorber," *Appl. Phys. Lett.*, 2012. doi: 10.1063/1.3692178.

[24] X.J. He et al. "Broadband and polarization-insensitive terahertz absorber based on multilayer metamaterials," *Opt. Commun.*, 2015. doi: 10.1016/j.optcom.2014.11.068.

[25] A. Bhattacharya, S. Bhattacharyya, S. Ghosh, D. Chaurasiya and K. Vaibhav Srivastava. "An ultrathin penta-band polarization-insensitive compact metamaterial absorber for airborne radar applications," *Microw. Opt. Technol. Lett.*, 2015. doi: 10.1002/mop.29365.

[26] D. Chaurasiya, S. Ghosh, S. Bhattacharyya, A. Bhattacharya and K.V. Srivastava. "Compact multi-band polarisation-insensitive metamaterial absorber," *IET Microwaves, Antennas, Propag.*, 2016. doi: 10.1049/iet-map.2015.0220.

[27] M. Yoo, H.K. Kim and S. Lim. "Angular- and Polarization-Insensitive Metamaterial Absorber Using Subwavelength Unit Cell in Multilayer Technology," *IEEE Antennas Wirel. Propag. Lett.*, 2016. doi: 10.1109/LAWP.2015.2448720.

[28] H.Y. Meng, L.L. Wang, X. Zhai, G.D. Liu and S.X. Xia. "A Simple Design of a Multi-Band Terahertz Metamaterial Absorber Based on Periodic Square Metallic Layer with T-Shaped Gap," *Plasmonics*, 2018. doi: 10.1007/s11468-017-0509-1.

[29] P. Jain et al. "Quad-band polarization sensitive terahertz metamaterial absorber using Gemini-shaped structure," *Results Opt.*, vol. 8, p. 100254, 2022. doi: https://doi.org/10.1016/j.rio.2022.100254.

[30] D. Hu, H. Wang, Z. Tang, X. Zhang and Q. Zhu. "Design of four-band terahertz perfect absorber based on a simple #-shaped metamaterial resonator," *Appl. Phys. A Mater. Sci. Process.*, 2016. doi: 10.1007/s00339-016-0357-4.

[31] A.S. Saadeldin, M.F.O. Hameed, E.M.A. Elkaramany and S.S.A. Obayya. "Highly Sensitive Terahertz Metamaterial Sensor," *IEEE Sens. J.*, vol. 19, no. 18, pp. 7993–7999, 2019. doi: 10.1109/JSEN.2019.2918214.

[32] P. Jain et al. "Machine Learning Techniques for Predicting Metamaterial Microwave Absorption Performance: A Comparison," *IEEE Access*, vol. 11, pp. 128774–128783, 2023. doi: 10.1109/ACCESS.2023.3332731.

[33] P. Jain et al. "Multiband Metamaterial absorber with absorption prediction by assisted machine learning," *Mater. Chem. Phys.*, vol. 307, p. 128180, 2023. doi: https://doi.org/10.1016/j.matchemphys.2023.128180.

[34] J. Chen, S. Hu, S. Zhu and T. Li, "Metamaterials: From fundamental physics to intelligent design," *Interdiscip. Mater.*, vol. 2, no. 1, pp. 5–29, 2023. doi: 10.1002/idm2.12049.

[35] I. Malkiel, M. Mrejen, A. Nagler, U. Arieli, L. Wolf, and H. Suchowski, "Plasmonic nanostructure design and characterization via Deep Learning," *Light Sci. Appl.*, vol. 7, no. 1, 2018, doi: 10.1038/s41377-018-0060-7.

[36] C. Liu *et al.*, "A programmable diffractive deep neural network based on a digital-coding metasurface array," *Nat. Electron.*, vol. 5, no. 2, pp. 113–122, 2022. doi: 10.1038/s41928-022-00719-9.

[37] A. Ghosh, D.J. Roth, L.H. Nicholls, W.P. Wardley, A.V. Zayats and V.A. Podolskiy. "Machine Learning-Based Diffractive Image Analysis with Subwavelength Resolution," *ACS Photonics*, vol. 8, no. 5, pp. 1448–1456, 2021. doi: 10.1021/acsphotonics.1c00205.

[38] P. Jain et al. "Machine learning assisted hepta band THz metamaterial absorber for biomedical applications," *Sci. Rep.*, 2023. doi: 10.1038/s41598-023-29024-x.

[39] M.I. Jordan and T.M. Mitchell. "Machine learning: Trends, perspectives, and prospects," *Science*, vol. 349, no. 6245. pp. 255–260, 2015. doi: 10.1126/science.aaa8415.

[40] S. Giri et al. "Emotion Detection with Facial Feature Recognition Using CNN & OpenCV," in *2022 2nd International Conference on Advance Computing and Innovative Technologies in Engineering, ICACITE 2022*, 2022. pp. 230–232. doi: 10.1109/ICACITE53722.2022.9823786.

[41] P.K. Sahoo et al. "An Improved VGG-19 Network Induced Enhanced Feature Pooling for Precise Moving Object Detection in Complex Video Scenes," *IEEE Access*, vol. 12, pp. 45847–45864, 2024. doi: 10.1109/ACCESS.2024.3381612.

[42] M.M. Khan, S. Hossain, P. Mozumdar, S. Akter and R.H. Ashique. "A review on machine learning and deep learning for various antenna design applications," *Heliyon*, vol. 8, no. 4. 2022. doi: 10.1016/j.heliyon.2022.e09317.

[43] P. Jain, J. Yedukondalu, H. Chhabra, U. Chauhan and L.D. Sharma. "EEG-based detection of cognitive load using VMD and LightGBM classifier," *Int. J. Mach. Learn. Cybern.*, 2024. doi: 10.1007/s13042-024-02142-2.

[44] A.D. Watpade et al. "Comparative analysis of machine learning models for predicting dielectric properties in MoS2 nanofiller-reinforced epoxy composites," *Ain Shams Eng. J.*, p. 102754, 2024. doi: https://doi.org/10.1016/j.asej.2024.102754.

[45] B. Shingala et al. "Random Forest Regression Analysis for Estimating Dielectric Properties in Epoxy Composites Doped with Hybrid Nano Fillers," *J. Macromol. Sci. Part B*, Vol. 63, no. 12, pp. 1297–1311, 2024. doi: 10.1080/00222348.2024.2322189.

[46] A.D. Watpade, S. Thakor, P. Sharma, D.V Shah, C.R. Vaja, and P. Jain. "Synthesis, characterization, and dielectric spectroscopy of TiO2 and ZnO nanoparticle-reinforced epoxy composites," *J. Mater. Sci. Mater. Electron.*, vol. 35, no. 7, p. 466, 2024.

[47] U. Chauhan, H. Chhabra, P. Jain, A. Dev, N. Chauhan and B. Kumar. "Chaos inspired invasive weed optimization algorithm for parameter estimation of solar PV models," *IFAC J. Syst. Control*, vol. 27, 2024. doi: 10.1016/j.ifacsc.2023.100239.

Index

For Product Safety Concerns and Information please contact our
EU representative GPSR@taylorandfrancis.com Taylor & Francis
Verlag GmbH, Kaufingerstraße 24, 80331 München, Germany